PESTICIDES IN THE ENVIRONMENT

Volume 1, Part I

PESTICIDES IN THE ENVIRONMENT

Volume 1, Part I (In Two Parts)

Edited by

ROBERT WHITE-STEVENS

CHAIRMAN, BUREAU OF CONSERVATION AND
ENVIRONMENTAL SCIENCE
COLLEGE OF AGRICULTURE AND ENVIRONMENTAL SCIENCE
RUTGERS UNIVERSITY—THE STATE UNIVERSITY
NEW BRUNSWICK, NEW JERSEY

MARCEL DEKKER, INC., New York 1971

MARCEL DEKKER, INC., 95 Madison Avenue, New York, New York 10016

LIBRARY OF CONGRESS CATALOG CARD NUMBER 77-138499
ISBN 0-8247-1782-1
PRINTED IN THE UNITED STATES OF AMERICA

PREFACE

There have been two score or more books on pesticides published within the past two decades. Most of these have been highly scientific, accurately annotated, expository textbooks dealing with specialized areas of the pesticide field. A few have been unashamedly erudite, scholarly, and to a degree esoteric.

Such specializing is the hallmark of modern science, which is frequently and erroneously interpreted to be symptomatic of a callous indifference to the whole welfare of man, and is partly responsible for the sharp decline in the public favor noted toward virtually all fields of science. The pressures and frustrations of our times are often blamed upon science today, as if it were some new political entity, a cult, or a cabal. Indeed there are those who contend that man's political and economic woes derive from the overproduction and automation developed by science; that military altercations and rivalry are propelled by the overkill armaments evolved by science; that degradation of the entire environment of the planet is the result of the chemical insults hurled at it by science. It is even proposed that science be broadly de-emphasized and humanized to fulfill the emotional needs of man rather than his pragmatic wants.

As a direct result of this concerted effort to elicit "the triumph of superstition over science," there have appeared over the past decade some dozen or more books dealing with chemicals in the environment, and ascribing to them, frequently without authentic experimental evidence, much of the pollution ills currently of concern to the general public.

Although food additives, feed additives, cosmetics, factory effluvia, detergents, and fossil fuels have each been selected as prospective targets for vilification and exploitation, it has been the pesticidal chemicals that have drawn by far the most intense criticism. Extraordinary assertions have been repeatedly made about pesticides that not only are utterly absurd in their hyperbole but are actually belied by the very evidence in the environment available for all to see.

Groups and associations have been formed in several areas of the country that have exhorted, with astonishing ferocity and tenacity, both

state and federal legislators to ban certain pesticides, especially of the organo chlorine group. As these particular pesticides have served and continue to serve mankind all over the world to an extent unexcelled by even the wonder drugs, the real motive behind this anti-pesticide movement remains both questionable and obscure.

To support the demands that these valuable and effective compounds be compulsorily removed from use, a curious array of largely irrelevant "scientific" justifications were assembled *de post facto*. Thus, as one "scientificism" was propounded before the various committee hearings and effectively disposed of by evidence that it was in fact insignificant, invalid, or an actual artifact, it was quickly dropped and replaced with another crucial "scientificism" that was vociferously asserted to be the ultimate evidence needed finally to ban the target pesticide from further use. It has even been suggested that any argument, fictitious or not, would serve provided it achieved the objective of banning the chemicals on trial.

Actually, of course, no one pesticide can be regarded as utterly indispensable, yet clearly some are more dispensable than are others. The important point remains, however, that if any useful compound, machine, tool, book, or object in our society can be effectively destroyed, banned, or disposed of upon unscientific, unjustifiable, and largely fabricated and exaggerated evidence that is promoted by a vociferous, self-interested minority, then nothing can be regarded as safe, stable, or established. Perhaps this may be "the wave of the future," but it is totally incompatible with science, with jurisprudence, and with a rational civilized society.

In the chapters of this compendium our authors have sketched in authentic, objective terms the properties, functions, utilities, and contributions of pesticidal chemicals. It is not a direct answer to the torrent of tirades against pesticides; it is merely an annotated collation of the facts as they are known to be. Hopefully this will help "to set the record straight."

Certainly any extension of factual knowledge among the public is, in the long term, beneficial to all. Only the general realization and understanding of the relative benefits and hazards of pesticidal chemicals will ultimately resolve the argument. As scientific knowledge advances, unknown hazards may perhaps become revealed but, then too, so may new benefits. At each stage of development the decision of use must be made, but it can only be a sound decision when it is based firmly upon valid experimental evidence, rather than upon public opinion which is generally uninformed.

One of the most contributory factors to the general public concern over pesticides in the environment has been the extraordinary improvement in analytical techniques.

These precise and fastidious refinements in modern chemical analytical methods are discussed in detail by Dr. Charles Van Middelem and Dr. Daniel McDougall in their respective chapters in this volume. Suffice it to say here that the present precision of chemical detection and measurement exceeds that of the 1930's by from three to six orders of magnitude (i.e., 1000 to 1 million-fold).

These powerful assay methods have revealed minute and generally insignificant quantities of some of the more stable pesticides, such as the organo-metals and chlorinated hydrocarbons, often in unexpected places, namely in Antarctic penguins, raptorial birds, and pelagic fish. Of course, these creatures also undoubtedly contain many other undetermined contaminants in miniscule quantities that could also be assayed by such refined techniques.

The environmental significance of these trace residues is, as yet, largely unknown, and much of the data reported to date can only be regarded as equivocal or at least unconfirmed. The publication of such evidence in the lay press has, however, engendered a profound consternation among the general populace, who are usually unable to discern the actual relation between dose and hazard of exposure. It has been from this area of uncertainty combined with exaggerated threat that much of the public fear over pesticides in the environment has arisen.

Twenty years ago the analyst was content to measure residues to one part per million (1 ppm), or 1 milligram (mg) in a thousand grams (1 kg) of plant or animal tissue or in 1 liter of water or other fluid (e.g., milk). As, 1 ppm of contamination was usually substantially below the permissable tolerance level established for pesticides on the basis of their toxicity and the pharmacological evidence then available, it was assumed that any residue that could not be assessed by a method with a sensitivity of 1 ppm was insignificant.

These assumptions were, however, largely based on toxicity studies, both acute and chronic, carried out on two or more species of domestic vertebrates, generally the mouse, the rat, the dog, the cat, the rabbit, the chick, etc. The allowable tolerances for a pesticide or any food additive, be it intentional or incidental, have been generally set at 1/100th, the lowest level at which an observable effect occurred on one or more of the test vertebrates. As this Minimum Effect Dose (MED_{50}) or the No Effect Dose (NED_{50}) both of which, of course, have the same values, actually constitutes the threshold level at which the compound becomes biologically active for the species studied, then one hundredth of this

MED_{50} level can be regarded as biologically inactive, and therefore safe for human exposure.

For the test species this rationale is both sound and logical, but the gap in the solidity of the argument occurs when the same experimental evidence and assumption is applied to other species of vertebrates as yet untested, particularly to man. There is actually no precise and unequivocal method to asses experimentally the tolerance of man to a new pesticide other than through the reflection of its effect upon physiologically similar vertebrates.

In spite of the virtually unblemished record provided by this 100-fold margin of safety requirement for permissible tolerances on consumer food stuffs, this gap in the wall of scientific evidence, which surrounds every registered pesticide, has been actively exploited as a basis for condemning useful pesticides. It is emphasized that even though long-term (lifetime) experiments with laboratory animals of various species reveal similar "No Effect" levels, these are considered to be unacceptable evidence that the same compound is not seriously toxic to man at much lower levels than those found to be inactive to test animals. Similarly, levels found innocuous to domestic livestock are considered to provide no assurance that even lower levels are not hazardous to related wildlife.

Of course, it is clearly recognized in all scientific research that the proof of the negative, or null hypothesis, is impossible. The null hypothesis *a priori* assumes there is no difference between two or among more than two treatments whatever they may be. The experiment is executed, the criteria of comparative measurement collected, tabulated, and reduced by analytical statistics. If by tests of significance at prescribed levels of probability (e.g., $P < 0.05$, < 0.01, or < 0.001) differences between and among the compared means are, in fact, sufficiently large to be regarded as significant, then it can be concluded that the "null hypothesis" has been shown to be wrong in this particular case and there are indeed distinguishable differences present in the evidence.

Conversely, should no such "significant" differences appear, then the null hypothesis cannot be rejected; yet, on the other hand, the data do not permit acceptance of the null hypothesis more than upon the assumptive basis originally proposed. The data have not produced proof that the treatments are the same, but merely that there is insufficient evidence to conclude that they are different.

This nice interpretation of experimental evidence is not readily understood or accepted by the general public. They assume science can "prove" the absence of a deleterious quality in a chemical as readily as it can "prove" the presence of a beneficial property. Therefore they demand proof that a new compound is noncarcinogenic (noncancer inducing), nonteratogenic (nonmonster producing), and nonmutagenic (nonmutant producing) before it is released into commercial channels. Such a demand is a complete impossibility to meet, for even if each compound were studied on 1000 species of living creatures each for 1000 years, the failure to observe these undesirable features does not constitute absolute proof of their total absence.

By the careful and considered application of biometrical statistics we can, however, compute the likelihood or probability of such a property occurring in practice when the chemical is employed at the dose level, by the established route of exposure, and in the frequency required for adequate control of the pest species. Such probabilities are derived from carefully replicated experiments wherein the test pesticide is applied over an array of dose levels, by various routes of exposure, and at differential frequencies and period durations. It is assumed that a high, nonlethal exposure rate of a new compound will necessarily be more promotive of both acute, subacute, and chronic toxicity than will other lower doses over longer times or the same doses for shorter periods.

In short, industrial hygiene instructions, methods of use, permissible tolerances, and sanitation and safety precautions, all of which must appear on the label at registration, are based on these elementary toxicological principles.

That such precepts have been and are reliable and have served their purpose well is attested to by the fact that over the past quarter century, during a time when pesticide applications to crops, to livestock, to industrial plants, to domiciles, to forests, to swamps and, in fact, to the total environment have increased over 1000-fold, there has not been one single medically annotated case of sickness, let alone cancer or death, from the correct use of any U.S. registered pesticide when it has been applied strictly in accordance with label recommendations.

There have been, and continue to be, unfortunate and, usually, totally inexcusable accidents where workers become grossly exposed due to improper and inadequate industrial hygiene; or children eat, touch, or inhale unprotected or improperly stored pesticides; or consumers are inadvertently poisoned by pesticides spilled carelessly into sacks of flour, beans, sugar, etc., which are later prepared for food. These cases are, however, no indictment of the pesticide itself or the methods employed to establish its efficacy and safety; they are purely due to the irresponsibility of the user.

It is also significant that the bulk of the 100 or so cases of fatal pesticide poisoning each year in the United States, for the past two decades, continue to be from the older metal organic and arsenical type compounds. The much maligned organo-chlorine compounds reveal an astonishingly low incidence of accidental fatalities.

In the case of DDT, which has clearly led the entire pesticide field as the major target of unmitigated condemnation, the record is actually virtually unblemished so far as human life and safety is concerned. Over the past 20 years some one billion humans have been exposed to DDT in varying degrees. This aggregates approximately 20 billion man years of exposure without a single fatality or sickness, acute or chronic, with no report of cancer or death attributable to the pesticide by reliable medical reports. To paraphrase the slogan, "no other pesticide can make that claim."

Yet it is frequently implied that we must promptly abandon DDT because there *may* be unknown low level chronic effects that *might* appear in later generations. There is, of course, no evidence for such assertions, but, as illustrated earlier, there is also no "proof" it is *not* so. Therefore, it is argued, why take the risk? However, as we know more about the metabolism and pharmacology of DDT (see the chapter by Dr. Fukuto and Dr. Sims) than of any other pesticide, it is obvious that to exchange its use for that of another pesticide necessarily enhances rather than diminishes the alleged risk. To replace that of which we know the more for that of which we know the less is clearly to reduce the probability of safety to both man and beast alike.

Why, it is frequently asked, are decisions made at administrative levels to phase out useful chemicals in our society when their alleged deleterious impact on the environment is both moot and equivocal? Such decisions have, in certain cases, been made in response to the fervor and volume of the *vox populi* rather than upon strict adherence to the experimental evidence. Scientific decisions can only be validly made when based upon statistically confident data. Solutions to natural problems are not to be derived by the democratic process. The lamentable history of Lysenkoism in the USSR from 1934 to 1964, which set Soviet biological science back for perhaps half a century and became reflected in their failing harvests, declining medicine, and widespread ignorance among students of biology, should prove to be a revelation to the Western world that scientific decisions based upon political expediency and government fiat are inevitably self-destructive.

The what, how, when, and where of pesticide use must be decided by qualified scientists in the fields of chemistry, biology, agronomy, and economics together with all the ancillary disciplines nested within their broad scope. Such decisions need to be based, as they have indeed been for over two decades, upon the integrated and collated evidence from many workers to provide for the optimum combination of efficacy, safety, reliability, economy, and environmental improvement.

The broader the spectrum of factual knowledge concerning pesticides available to the general public, the more rational will be their support of administrative decisions based upon scientific fact rather than emotional, superstitious, and prejudicial opinions promoted by irresponsible and irrelevant headlines and publicity.

This compendium, *Pesticides in the Environment*, brings together the broad horizon of concepts of many experts from various specialized disciplines related to the control of pests in the environment by the carefully metered, correct, and approved use of chemicals. Each author is distinguished for the scientific work published in his own particular area, and he has distilled the essence of the relevant information extant in the legitimate scientific literature into a format and style readily understandable to the serious and intelligent reader.

The text has been divided loosely into Volume 1, Parts I and II, which deal with the more theoretical aspects of pesticidal chemicals and Volume 2, Parts I, II, and III which deal essentially with the practical business of handling pesticides in the environment.

The style has deliberately avoided that of the classical science text, without becoming trivial or irrelevant. It is directed to the use of both students and practicing professionals in other related or even more distant fields who may have need to refer to information dealing with pesticides. Physicians, nurses, public health officials, lawyers, high school and Vo-Ag teachers, county agents, agricultural chemicals salesmen, servicemen, dealers, and formulators should all find the discussions and the tables useful for reference purposes.

If but a small fraction of the reading public, particularly the urban and suburban public, can be provided with a better and more factual understanding as to the contribution pesticidal chemicals have brought, not just to the farmer but, more importantly, to the entire nation and to many other peoples beyond our shores, then our labors will have been well rewarded and our intent achieved.

CONTRIBUTORS TO VOLUME 1, PART I

T. Roy Fukuto, *University of California, Riverside, California*

J. E. Loeffler, *Shell Development Company, Biological Sciences Research Center, Modesto, California.*

Robert L. Metcalf, *University of Illinois, Urbana, Illinois*

James J. Sims, *University of California, Riverside, California*

J. van Overbeek,* *Shell Development Company, Biological Sciences Research Center, Modesto, California*

*Present address: Texas A & M University, Institute of Life Science, College Station, Texas.

CONTENTS

CONTENTS TO VOLUME 1, PART II

Chapter 1

THE CHEMISTRY AND BIOLOGY OF PESTICIDES

Robert L. Metcalf
UNIVERSITY OF ILLINOIS
URBANA, ILLINOIS

1-1 INTRODUCTION

The history of pesticides encompasses the last 200 years, since the first use of nicotine, as a tea of tobacco for the destruction of aphids, in 1763. Especially significant milestones include the development of Paris green in 1865 for the control of the Colorado potato beetle; the use of Bordeaux mixture for downy mildew control in 1882; the marketing of potassium dinitro-*o*-cresylate, the first synthetic organic pesticide, in 1892; and the development of DDT and 2,4-D during World War II. Despite this record of progress, pesticides really came of age in the postwar period. The last 20 years has seen pesticide chemicals in the United States more than quadruple in number and more than treble in dollar sales. New pesticides such as soil fumigants, seed treatments, pre-emergent herbicides, and animal and plant systemic insecticides have revolutionized farm practices and dramatically increased crop yields. For the first time in all history the conquest of arthropod-borne diseases has become a reality with the use of residual insecticides. Epidemics of plague and typhus have been halted almost at their inception through destruction of their flea and louse vectors. Malaria has been eradicated from large areas of the globe and is under dramatic control elsewhere (97).

In bringing about this revolution in human health and in the techniques of food production, the increasing use of pesticides has also resulted in new problems. Resistant races of pests have appeared in response to intensive "natural selection." The very stability and persistence that has made DDT and dieldrin the ideal weapons for malaria eradication has caused widespread distribution throughout the environment and their concentration in animal fats and ecological magnification in food chain organisms. Failure to appreciate all of the factors involved in detoxication in the soil has led to instances of misuse of pre-emergent herbicides and to consequent crop failures.

All of these factors have contributed to the need for greater sophistication in the development and use of pesticides and to the desirability of increased pesticide selectivity. It is, therefore, in the light of this need for better scientific appreciation of modern pesticides that this chapter has been written. In it are described more than 600 distinct chemical compounds that belong to one or more classes of pesticides, namely acaricides, fungicides, herbicides, insecticides, molluscacides, nematocides, or rodenticides. Emphasis has been given to the basic chemistry of the compounds, the relationships of structure to pesticidal activity, the physiological and biochemical mechanisms of action, the general biological activity of the compounds, and their mammalian toxicology. It is hoped that this chapter will not only be useful as a condensed handbook of the properties and uses of pesticides but will also serve as a stimulus to research workers interested in investigations of mode of action and in the development of new and improved pesticidal molecules.

The following general reference works have been invaluable in the preparation of this chapter: Audus(5), Crafts(29), Hayes(55), Horsfall and Dimond(64), Johnson, Krog, and Poland(65), Martin(87, 88), Metcalf and co-workers(94, 95), and Schrader(126).

1-2 FUNGICIDES

A. Inorganic Fungicides

1. Sulfur. *Sulfur* alone and combined as *lime–sulfur* is one of the most important fungicides and is also widely used as an insecticide. Sulfur is a yellow mineral that is insoluble in water but is soluble to 2.4% in benzene and to 50% in carbon bisulfide. It ignites in the air above 261°C forming sulfur dioxide. Sulfur was the first foliar fungicide and was recommended in England by Robertson in 1824(*121*) to control powdery mildew of peach. Wettable sulfur powders or pastes of very fine particle size are widely used to control powdery mildews on fruits and vegetables and as an acaricide.

In the United States, Kenrick in 1833(*72*) proposed the use of lime and sulfur boiled together for powdery mildew of grape, and lime–sulfur solutions were used as insecticides in California in 1886 as a cattle dip and as dormant sprays for scale insects.

When one part of fresh lime, $Ca(OH)_2$, and two parts of sulfur are boiled together in 4 parts of water for about one hour, a chemical reaction occurs and a reddish liquid mixture is formed that contains calcium pentasulfide, CaS_5; calcium tetrasulfide, CaS_4; calcium thiosulfate, CaS_2O_3; and calcium sulfite, $CaSO_3$. The mixture is diluted with water to a specific gravity of 1.005 for summer spraying and to 1.035 for dormant spraying. Dry lime–sulfur preparations are also available commercially.

The mode of action of sulfur fungicides is not well understood. McCallan(*91*) concludes that it acts as a hydrogen acceptor and thus interferes with the normal redox processes of the fungal cells.

2. Copper. The use of copper fungicides began in 1761 with the use by Schulthess(*128*) of copper sulfate, $CuSO_4 \cdot 5H_2O$, to treat wheat seed, and shortly thereafter it was recommended as a wood preservative. Inorganic copper compounds that are used as fungicides include *basic copper sulfate*, $CuSO_4 \cdot 3Cu(OH)_2 \cdot H_2O$; *copper oxychloride*, $3Cu(OH)_2 \cdot CuCl_2$; *copper zinc chromate*, $15CuO \cdot 10ZnO \cdot 6CrO_325H_2O$; and *cuprous oxide*, Cu_2O; these are used as foliar fungicides for vegetables and fruits. Basic copper carbonate, $Cu(OH)_2 \cdot CuCO_3$, is used as a seed treatment.

Bordeaux mixture was described by Millardet in 1885(*101*) to control the downey mildew disease of grapes, *Plasmopara viticola*, as a variable mixture of lime and copper sulfate. The chemistry of Bordeaux mixture is complex, and the active constituents consist largely of a voluminous precipitate of tetracupric sulfate, $4CuO \cdot SO_3$, and pentacupric sulfate, $5CuO \cdot SO_3$. Other investigators have suggested that a series of compounds is formed with the general formula $Cu_4(OH)_6SO_4 \cdot XCa(OH)_2$. Bordeaux mixture is widely used as a foliar fungicide for vegetables and fruits and as a repellent for leafhoppers and other insects.

The mode of action of these copper fungicides seems to be as inhibitors of a variety of —SH containing enzymes.

B. Phenols

1. Introduction and Chemistry. The bacterial action of phenol and its derivatives is well known, and many of these compounds are also fungicidal but because of their corresponding phytotoxicity, applications are restricted to soil, wood, fabrics, and other inert substrates. The commercially useful phenol fungicides include the chlorinated phenols that are wood preservatives and slimicides for paper mills:

2,4,5-trichlorophenol m.p. 67°C, $K_a^{25°}$ 4.3 × 10⁻⁸, H_2O sol. < 0.2%
2,4,6-trichlorophenol m.p. 69°C, K_a 2.6 × 10⁻⁸, H_2O sol. <0.1%
2,3,4,6-tetrachlorophenol m.p. 69°C
pentachlorophenol m.p. 190°C, K_a 5.5 × 10⁻⁶, H_2O sol. 0.008%

These compounds are all weak acids, soluble in organic solvents, and form sodium salts which are soluble in water and are also fungicidal.

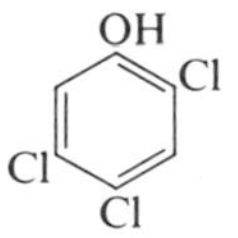

2,4,5-trichlorophenol 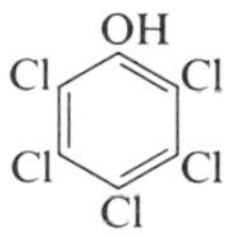   pentachlorophenol

Other more complex fungitoxic phenols are *dichlorophene*, or 2,2'-dihydroxy-5,5'-dichlorodiphenylmethane, m.p. 178°C, a fungicide and germicide for soaps and a mildewproofer; and VANCIDE BL, or 3,3',5,5'-tetrachloro-2,2'-dihydroxydiphenyl sulfide which is a fungicide and germicide used in cosmetics, soaps, polishes, and for algae control.

Phenols are pseudo acids in which the OH group is ionizable. The ionization constants (K_a) range from 1 × 10⁻⁴ for 2,4-dinitrophenol to 1 × 10⁻¹⁰ for phenol. As a result, phenols, which are generally only slightly soluble in water (pentachlorophenol to 0.008%, 4,6-dinitro-*o*-cresol to 0.0128%) and soluble in most organic solvents, readily react with sodium or potassium hydroxide to form water soluble sodium or potassium salts.

Phenols, like alcohols, are readily esterified with organic acids to form a variety of esters and form ethers with aliphatic and aromatic groups.

2. Structure–Activity Relationships. Chlorination of simple phenols significantly increases fungitoxicity, which becomes maximal with three to five chlorine atoms(50). Esterification of the phenolic group of pentachlorophenol decreases the phytotoxicity, and pentachlorophenyl acetate was found to be highly fungicidal. Other active derivatives included the *p*-chlorobenzoate and the 8-hydroxyquinolinate(159).

Fungitoxicity depends upon the presence of a free and unhindered OH-group, and Shirk(131) has shown that 2,6-dialkylphenols or 2-chloroalkylphenols, in which the OH-group is hindered by adjacent ortho-substituents, were of greatly reduced fungicidal activity.

Phenols with *p*-chloro groups were found by Marsh et al.(84,85) to be effective for the mildewproofing of fabrics, and bis-phenols such as dichlorophene were especially effective. Bridging structures that were especially effective included —CH_2—, —$CH(CH_3)$—, —$CH(C_6H_5)$—, —CH=CH—, and S, while —SO—, —SO_2—, and —C(O)—C(O)— were much less active. In tests against brown rot of peach the compound

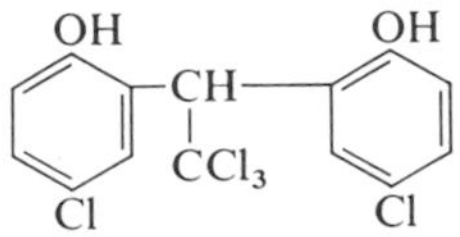

was found by Goldsworthy and Gertler(45) to be the most effective of 506 compounds tested.

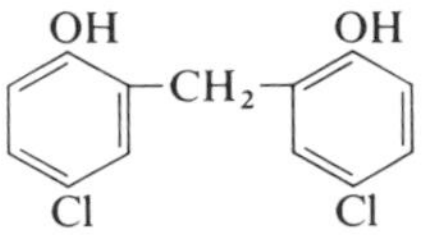

dichlorophene 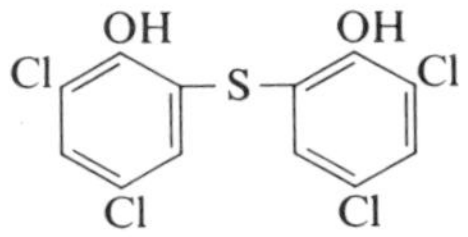   VANCIDE BL

SANTOPHEN, or 2-benzyl-4-chlorophenol, m.p. 49°C, is a broad spectrum germicide. *Ortho-phenylphenol*, m.p. 55°C, and its sodium salt are used as germicides, preservatives, and as postharvest protectants for fruits and vegetables.

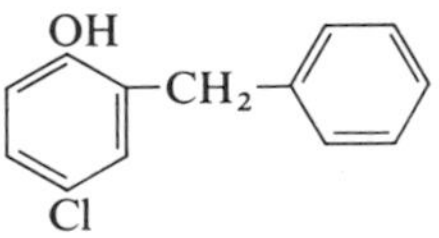

SANTOPHEN 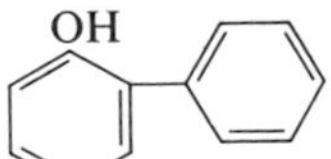   *o*-phenylphenol

The nitrophenols are also powerful fungicides. *DNOC* or *4,6-dinitro-o-cresol*, yellow crystals m.p. 86°C, and its sodium salt have been used as a spray on the orchard floor to kill overwinter forms of scab and as a dormant spray for raspberry anthracnose. However, the nitrophenols are extremely phytotoxic and are used as general herbicides (Sec. 1-4, G). Esters in which the acidity of the phenolic group is neutralized can be safely used on some plants and *dinocap* or KARATHANE, 4,6-dinitro-2-(1-methylheptyl)-phenyl crotonate, a dark viscous liquid b.p. 138–140°C/0.05 mm, is specific for the control of powdery mildew on fruit, vegetables, nursery stock, and ornamentals.

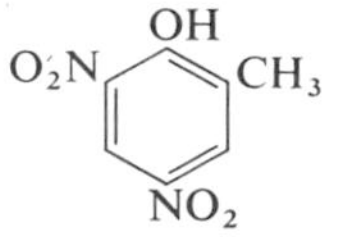

O_2N —⟨ring⟩— OH, CH_3, NO_2

DNOC

O_2N —⟨ring⟩— $OCCH=CH-CH_3$ (C=O), CHC_6H_{13}, CH_3, NO_2

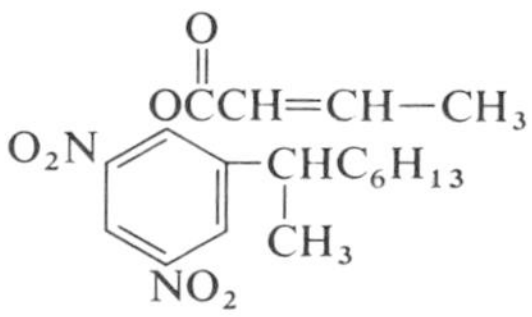

dinocap

3. Mode of Action. The fungicidal action of various phenols seems to depend upon an appropriate blend of lipophilic and hydrophobic properties in the molecule so that it will orient itself at a lipoid/water interface. As mentioned above, fungicidal activity of the phenols depends upon a free and unhindered OH-group or an ester group that is readily hydrolyzed in vitro. The phenols have been suggested as inactivators of oxidases, amylase, and catalase, and as exerting lytic action on the cell wall that causes loss of amino acids (91).

Chlorinated phenols such as pentachlorophenol and nitrated phenols such as 4,6-dinitro-o-cresol are very effective (10^{-7} to $10^{-6} M$ in vitro) as uncouplers of oxidative phosphorylation. They thus prevent the incorporation of inorganic phosphate into ATP without effecting electron transport. As a result of this action, which is believed to occur at the mitochondrial cell wall, the cells continue to respire but soon are depleted of the ATP necessary for growth.

4. Biological Activity. Phenols are generally highly toxic to most forms of life including bacteria, fungi, plants, insects, and higher animals. The variety of their fungicidal applications have been mentioned in the introduction to this section. The chlorinated phenols are used as wood preservatives and soil poisons, dichlorophene in mildew-proofing, sodium-o-phenylphenate in postharvest protection, bis-2,2-thio-(4-chlorophenol) as an antimycotic in human diseases, and dinocap for powdery mildew control on plants.

5. Toxicology. DNOC, rat oral LD_{50} 30 mg/kg, is a highly toxic compound and must be handled with great care. Pentachlorophenol, rat oral LD_{50} 180 mg/kg, is also moderately toxic and causes severe burning of the skin. The rat oral LD_{50} values for other phenols include dinocap 980, SANTOPHEN 1700, dichlorophene 2000 (dogs), sodium o-phenylphenate 2480, and VANCIDE BL 6627 mg/kg. Poisoning by the phenols may occur directly through the skin, and good hygiene and protection of the skin of workers is important.

Both DNOC and pentachlorophenol are uncouplers of oxidative phosphorylation. These toxicants may be cumulative, and fatal dosages are little larger than dosages producing no signs of illness. Symptoms include elevated temperature, sweating, dehydration, dyspnea with pain in chest or abdomen, and coma. Treatment is symptomatic(55) (see Sec. 1-5, F.)

C. Quinones

1. Introduction and Chemistry. *Chloranil*, or tetrachloro-*p*-benzoquinone, yellow crystals, m.p. 290°C, was the first successful organic fungicide. Although it was first synthesized by Erdman in 1843, it was only introduced as a fungicide by Ter Horst in 1938 and was found to be a successful seed protectant and foliar fungicide for downy mildew. Subsequent research led to the development of *dichlone*, or 2,3-dichloro-1,4-naphthoquinone, yellow crystals, m.p. 193°C(*141*) which was the most active of a number of quinones evaluated by Schoene et al.(*124, 125*),

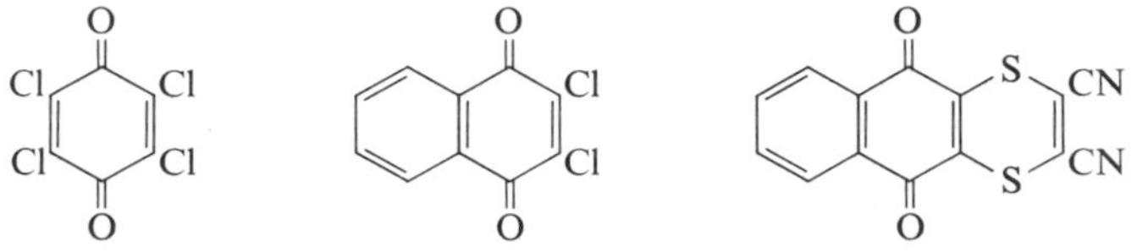

chloranil   dichlone   dithianon

and has become widely used as a seed treatment for corn and beets, a foliar fungicide for fruits and vegetables, and an algacide for water treatment. *Dithianon*, or 2,3-dicyano-1,4-dithia-anthraquinone (brown crystals, m.p. 225°C) is a new fungicide effective against apple scab and cherry leaf spot. The phytotoxicity of these compounds has been a major difficulty in their development as foliar fungicides.

Quinones are the oxidation products of phenols and are readily reduced to the corresponding bis-phenols or hydroquinones with which they form redox systems:

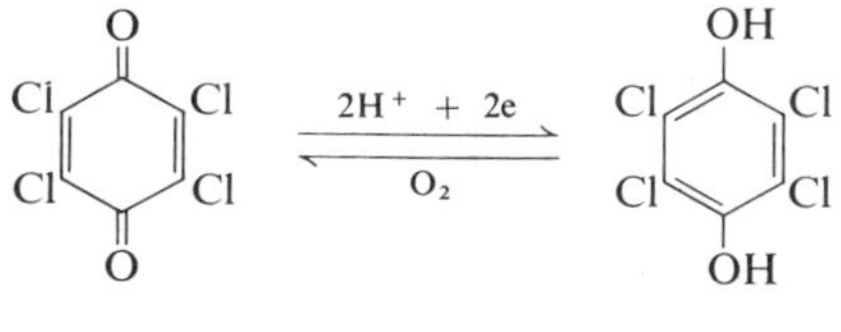

Chemically, although they are resistant to further oxidation, they are highly reactive because of the ketone group adjacent to the $C=C$ double bond, and they thus resemble the aliphatic alpha–beta unsaturated ketones. This reactivity is related to their fungicidal action.

The quinones are extremely insoluble in water, e.g., dichlone to 0.01 ppm, but are moderately soluble in organic solvents such as xylene, chloroform, and ether.

2. Structure–Activity Relationships. Fungicidal activity is found in a variety of quinones in the general order 1,4-naphthoquinone > phenanthroquinone > p-benzoquinone > anthraquinone. Activity is improved by halogen substitution in the order Cl > Br > I, which is beneficial in decreasing water solubility and phytotoxicity. Alkylation ortho to the quinone group reduces activity, possibly by steric hinderance. Some of these relationships are shown by the quantitative data of Table 1-1. Various hydroquinones and their esters, e.g., 2,3-dichloro-1,4-diacetoxynaphthalene, also show marked fungicidal activity, and there is a relationship between ease of hydrolysis and fungicidal activity. It therefore appears that the esters are hydrolyzed to the hydroquinone derivatives and subsequently oxidized by the fungus to the active quinone.

TABLE 1-1
FUNGICIDAL ACTIVITY OF QUINONES([125])

	Toxicity to *Alternaria solani*	
	LD_{50} ppm on slides	LD_{95} ppm on tomato
p-Benzoquinone	30	2000[a]
2,5-Dichloro-p-benzoquinone	11	600
2,6-Dichloro-p-benzoquinone	28	220[a]
Tetrachloro-p-benzoquinone	8	60
2,5-Dichloro-3,6-dihydroxy-p-benzoquinone	18	180[a]
1,4-Naphthoquinone	7	450[a]
2-Methyl-1,4-naphthoquinone	18	2000[a]
2-Methoxy-1,4-naphthoquinone	13	300[a]
2-Chloro-1,4-naphthoquinone	4	550[a]
2,3-Dichloro-1,4-naphthoquinone	0.9	38
2-Chloro-3-hydroxy-1,4-naphthoquinone	25	2000[a]
9,10-Phenanthraquinone	4	450
9,10-Anthraquinone	1000	
1,2-Acenaphthoquinone	10	
3,10-Pyrenequinone	400	

[a]Phytotoxic.

3. Mode of Action. The quinones are α-β- unsaturated ketones and react with sulfhydryl (—SH) groups. This reaction has been suggested as the critical biochemical lesion involving the —SH groups of enzymes such as amylase and carboxylase which are inhibited by quinones. This inhibition is reversed by cysteine or glutathione. The overall mechanism may involve binding of the enzyme to the quinone nucleus by substitution or addition at the double bond, an oxidative reaction with the —SH group, and a change in redox potential(*91*).

4. Biological Activity. Chloranil is very widely used as a seed treatment for a variety of field and vegetable crops at from 3 to 12 oz per 100 lb of seed applied dry or as a slurry. It is also used as a spray for turf, for mildews of cabbage and cantaloupe, and for celery and tobacco seedlings. It is unstable in sunlight and generally not effective as a foliar treatment. Dichlone is also used as a seed treatment at 1 to 4 oz per 100 lb of seed and as a foliar treatment for scab, blights, rots, and leaf curl of apples, cherries, plums, peaches, beans, celery, tomato, and strawberry. It has an FDA Tolerance of 3 ppm on fruits and vegetables, and 15 ppm on strawberries.

5. Toxicology. The quinone fungicides are of low to moderate acute toxicity with oral LD_{50} values to the rat of: chloranil 4000, dichlone 1300, and dithianon 1015 mg/kg. Under some conditions these compounds may cause severe skin irritation in sensitive individuals.

D. Dithiocarbamates

1. Introduction and Chemistry. The dithiocarbamate fungicides were introduced by the patent of Tisdale and Williams(*144*) which disclosed *thiram* or tetramethylthiuram disulfide, m.p. 155°C, tetramethylthiuram monosulfide, and their sodium, iron, and cadmium salts, all prepared by reacting dimethylamine and carbon bisulfide under alkaline conditions.

$$
\begin{array}{cc}
\underset{\text{thiram}}{(CH_3)_2N\overset{\displaystyle S}{\overset{\displaystyle \|}{C}}S S\overset{\displaystyle S}{\overset{\displaystyle \|}{C}}N(CH_3)_2} & \underset{\text{nabam}}{NaS\overset{\displaystyle S}{\overset{\displaystyle \|}{C}}NHCH_2CH_2NH\overset{\displaystyle S}{\overset{\displaystyle \|}{C}}SNa}
\end{array}
$$

Subsequently Dimond, Heuberger, and Horsfall(*32*) made the equally valuable discovery of the fungicidal properties of *nabam* or disodium ethylene-bis-dithiocarbamate, the reaction product of ethylenediamine and carbon bisulfide(*58*). It was further discovered that heavy metal complexes of the dithiocarbamates were unusually stable and suitable as fungicides. The zinc dimethydithiocarbamate or *ziram*, m.p. 246°C, and ferric dimethyldithiocarbamate or *ferbam*, m.p. 180°C dec., proved the most useful.

$$[(CH_3)_2 N\overset{\overset{\displaystyle S}{\|}}{C}S^-]_2 Zn^{2+} \qquad [(CH_3)_2 N\overset{\overset{\displaystyle S}{\|}}{C}S^-]_3 Fe^{3+}$$

ziram 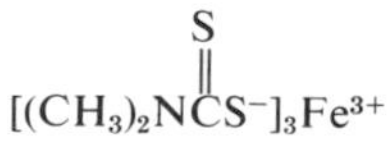 $\qquad\qquad$ ferbam

Similar heavy metal salts of ethylene-bis-dithiocarbamate have been developed as fungicides. These include manganous ethylene-bis-dithio-carbamate or *maneb*, and zinc ethylene-bis-dithiocarbamate or *zineb*

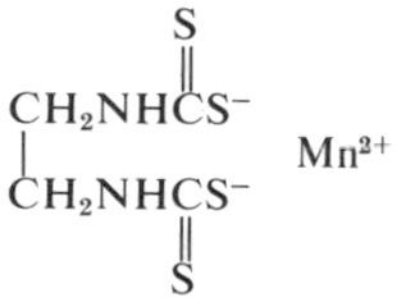

$$\begin{array}{ll} CH_2NH\overset{\overset{\displaystyle S}{\|}}{C}S^- & \\ | & Mn^{2+} \\ CH_2NHCS^- & \\ \overset{\displaystyle\|}{S} & \end{array} \qquad \begin{array}{ll} CH_2NH\overset{\overset{\displaystyle S}{\|}}{C}S^- & \\ | & Zn^{2+} \\ CH_2NHCS^- & \\ \overset{\displaystyle\|}{S} & \end{array}$$

maneb 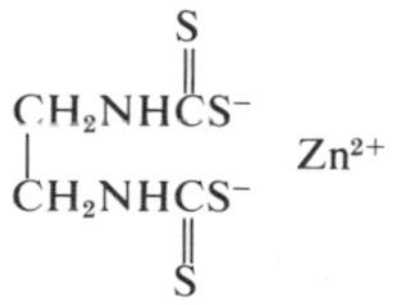 $\qquad\qquad$ zineb

A more recent product is sodium N-methyldithiocarbamate or *metham*.

$$CH_3NH\overset{\overset{\displaystyle S}{\|}}{C}SNa \cdot 2H_2O$$

metham

Thiram can be reduced to form dimethyldithiocarbamic acid which forms salts and complexes with a variety of metals, e.g., ferbam and ziram, and reduced still further to dimethylamine and carbon bisulfide.

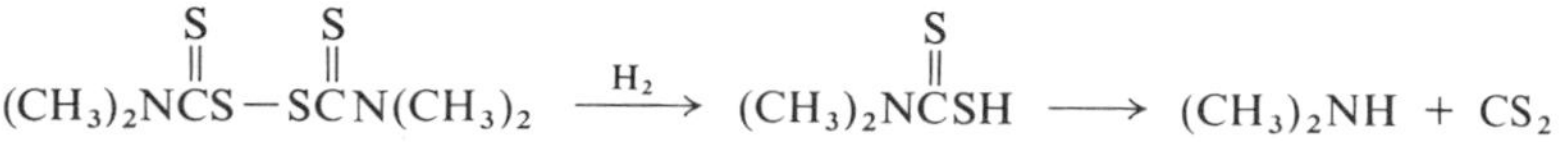

$$(CH_3)_2 N\overset{\overset{\displaystyle S}{\|}}{C}S - S\overset{\overset{\displaystyle S}{\|}}{C}N(CH_3)_2 \xrightarrow{H_2} (CH_3)_2 N\overset{\overset{\displaystyle S}{\|}}{C}SH \longrightarrow (CH_3)_2 NH + CS_2$$

However, nabam and other ethylene-bis-dithiocarbamates form ethy-lenethiuram disulfide and ethylenethiuram monosulfide, which is readily converted to an isothiocyanate.

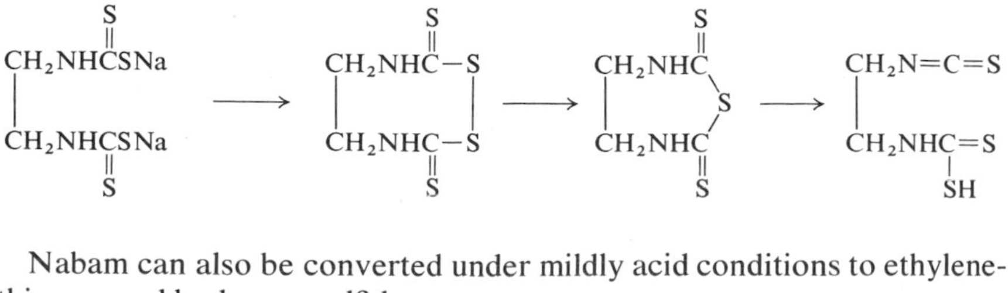

$$\begin{array}{l} CH_2NH\overset{\overset{\displaystyle S}{\|}}{C}SNa \\ | \\ CH_2NHCSNa \\ \overset{\displaystyle\|}{S} \end{array} \longrightarrow \begin{array}{l} CH_2NH\overset{\overset{\displaystyle S}{\|}}{C} - S \\ | \hspace{1.5em} | \\ CH_2NHC - S \\ \hspace{1em}\overset{\displaystyle\|}{S} \end{array} \longrightarrow \begin{array}{l} CH_2NH\overset{\overset{\displaystyle S}{\|}}{C} \\ | \hspace{1.5em}\diagdown \\ \hspace{3em} S \\ CH_2NHC \diagup \\ \hspace{1em}\overset{\displaystyle\|}{S} \end{array} \longrightarrow \begin{array}{l} CH_2N{=}C{=}S \\ | \\ CH_2NHC{=}S \\ \hspace{2em}| \\ \hspace{2em}SH \end{array}$$

Nabam can also be converted under mildly acid conditions to ethylene-thiourea and hydrogen sulfide,

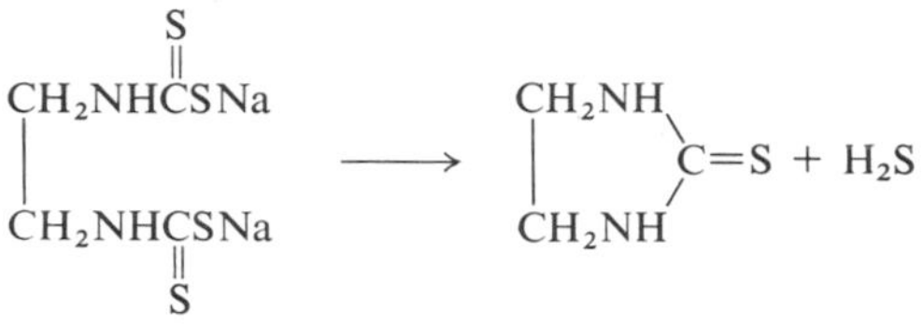

$$
\begin{array}{c}
\overset{\text{S}}{\overset{\|}{\text{CH}_2\text{NHCSNa}}} \\
| \\
\text{CH}_2\text{NHCSNa} \\
\overset{\|}{\underset{\text{S}}{}}
\end{array}
\longrightarrow
\begin{array}{c}
\text{CH}_2\text{NH} \\
| \quad\quad\;\;\text{C}{=}\text{S} + \text{H}_2\text{S} \\
\text{CH}_2\text{NH}
\end{array}
$$

and under more drastic conditions to ethylenediamine and carbon bisulfide (*82*).

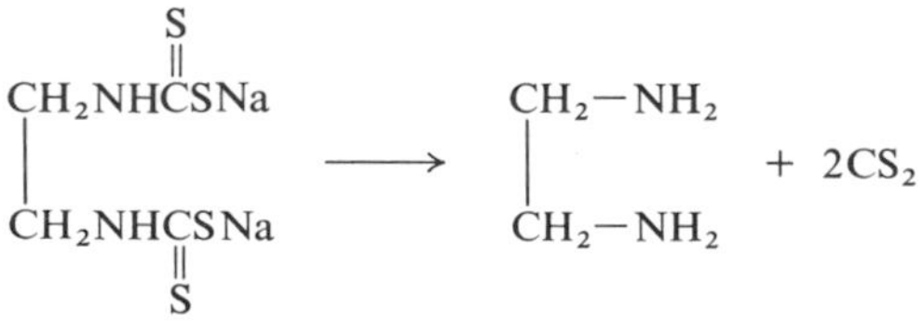

$$
\begin{array}{c}
\overset{\text{S}}{\overset{\|}{\text{CH}_2\text{NHCSNa}}} \\
| \\
\text{CH}_2\text{NHCSNa} \\
\overset{\|}{\underset{\text{S}}{}}
\end{array}
\longrightarrow
\begin{array}{c}
\text{CH}_2{-}\text{NH}_2 \\
| \quad\quad\;\; + \; 2\text{CS}_2 \\
\text{CH}_2{-}\text{NH}_2
\end{array}
$$

Metham also readily forms the highly reactive methylisothiocyanate

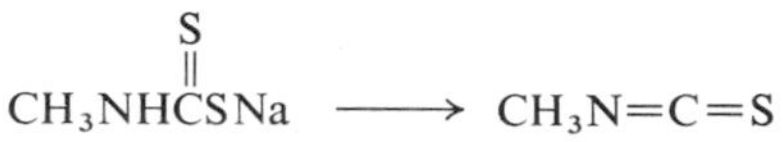

$$
\overset{\text{S}}{\overset{\|}{\text{CH}_3\text{NHCSNa}}} \longrightarrow \text{CH}_3\text{N}{=}\text{C}{=}\text{S}
$$

2. Structure — Activity Relationships. The two types of active dithiocarbamates, i.e., the dimethyldithiocarbamates and the ethylene-bis-dithiocarbamates, have different modes of action, and it is difficult to make generalizations regarding the effects of structural changes. Thus, methyldithiocarbamic acid is a weaker fungicide than dimethyldithiocarbamic acid while in the corresponding ethylene-bis-dithiocarbamates, replacement of the hydrogen atom by a methyl group greatly decreases fungicidal activity (Table 1-2).

Increasing the size of the alkyl groups on the nitrogen atom of the dialkyldithiocarbamates greatly decreases the toxicity. Similarly, increasing the number of methylene groups between the two thiocarbamyl nitrogens of the bis-thiocarbamates also decreases activity. Replacement of C=S by C=O in dimethyldithiocarbamates greatly decreases the activity.

3. Mode of Action. A great deal of study has been given to the mode of action of the dithiocarbamate fungicides. All of the highly active compounds possess the —C(=S)S— grouping. However, there are certain differences in behavior that suggest that the N,N-dimethyldithiocarbamates may have a different mode of action from that of the ethylene-bis-dithiocarbamates. For example, they have different spectra of action on various species of fungi, and histidine has been found to antagonize the

TABLE 1-2
FUNGISTATIC ACTIVITY OF DITHIOCARBAMATES(75)

	Growth inhibiting concentration, ppm			
	Botrytis cinerea	Penicillium italicum	Aspergillus niger	Rhizopus nigricans
$CH_3NHC(S)Na$	10	10	50	200
$(CH_3)_2NC(S)Na$	0.2	0.5	20	2
$(C_2H_5)_2NC(S)Na$	1	2	10	5
$(C_3H_7)_2NC(S)Na$	200	200	200	1000
$(C_4H_9)_2NC(S)Na$	200	200	200	1000
$[(CH_3)_2CH]_2NC(S)Na$	100	50	50	50
$[(CH_3)_2NC(S)S]_3Fe$	0.1	0.2	10	1
$[(CH_3)_2NC(S)S]_2Zn$	0.1	0.1	5	1
$H_2NC(S)$—S—$SC(S)NH_2$	10	10	5000	5000
$CH_3NHC(S)SSC(S)NHCH_3$	5	5	20	100
$(CH_3)_2NC(S)S$—$SC(S)N(CH_3)_2$	0.2	0.2	10	2
$(C_2H_5)_2NC(S)S$—S—$C(S)N(C_2H_5)_2$	0.5	1	5	2
$(CH_3)_2NC(S)SCN(CH_3)_2$	0.5	1	20	10
$(CH_3)_2NC(S)(S)_6C(S)N(CH_3)_2$	0.5	0.5	5	2
$(CH_3)_2NC(S)N(CH_3)_2$	5000	5000	10000	5000
$(CH_3)_2NC(S)SN(CH_3)_2$	0.2	0.5	20	5
$(CH_3)_2NC(S)OC(S)N(CH_3)_2$	0.5	1	10	2
$(CH_3)_2NC(O)SNa$	200	100	1000	500
$[$—$CH_2NHC(S)SNa]_2$	1	0.5	2	10
$[$—$CH_2N(CH_3)C(S)SNa]_2$	50	50	1000	1000
$[$—$CH_2CH_2NHC(S)SNa]_2$	1	1	5	50
$[$—$CH_2CH_2CH_2NHC(S)SNa]_2$	2	2	5	100
$[$—$CH_2CH_2CH_2CH_2NHC(S)SNa]_2$	5	2	10	500
$[$—$CH_2(CH_2)_4NHC(S)SNa]_2$	10	10	100	1000

fungicidal action of thiram and sodium dimethyldithiocarbamate but not of nabam.

The reactive hydrogen of nabam can lead to the formation of an isothiocyanate through the formation of ethylenethiuram disulfide as demonstrated by Ludwig and Thorn(82). Such isothiocyanates readily react with —SH groups such as those of essential enzymes in fungus cells:

$$RN{=}C{=}S + HS{-}\text{enzyme} \rightleftharpoons RNHCS{-}\text{enzyme}$$
$$\underset{S}{\overset{\|}{}}$$

Therefore, it has been suggested that the oxidative pathways of carbohydrate metabolism may be involved since ethylenethiuram monosulfide at 10^{-5} M produced 77% inhibition of glucose-6-phosphate dehydrogenase and 80% inhibition of 6-phosphogluconate dehydrogenase(26).

Such a plausible explanation is lacking for thiram and the other dimethyldithiocarbamates that do not obviously form isothiocyanates. However, thiram at 10^{-5} M inhibited glucose-6-phosphate dehydrogenase to 77% and 6-phosphogluconate dehydrogenase to 72% being approximately as active in this in vitro test as ethylenethiuram monosulfide(26). Therefore, it is difficult to separate the action of the two types of dithiocarbamates(82). Thiram can be reduced to dimethyldithiocarbamic acid, and it is possible that the latter may react with the —SH groups of essential enzymes to form disulfides(82).

On the other hand, Van der Kerk(145) suggests that the action of thiram and other dimethyldithiocarbamates results from the strong capacity of these compounds for chelating metals such as copper. Dimethyldithiocarbamates form both 1:1 $Cu^+SC(S)N(CH_3)_2{}^+$ and 1:2 $(CH_3)_2NC(S)SCuSC(S)N(CH_3)_2$ complexes with copper, and these may be the toxic agents. Van der Kerk(145) speculates that the inhibition of fungus growth is attributable to the chemical combination of an essential enzyme with 1:1 copper complex

$$[DDCCu]^+ + \text{enzyme} \rightleftharpoons DDC—Cu—\text{enzyme}$$

The antagonistic action of histidine can then be explained by its competition with the enzyme for the toxic 1:1 copper complex. Much more research is obviously needed to fully explain the fungitoxicity of the dithiocarbamates.

4. Biological Activity. The dithiocarbamate fungicides are the most widely used and most versatile of the organic fungicides and are used as foliage sprays, seed protectants, turf treatments; for the treatment of skin diseases of humans and animals; and as textile and paint protectants. Thiram is most widely used as a seed protectant at 1 to 3 oz per 100 lb of seed but is also used as a foliar spray for rusts and blights. Ferbam is a protective fungicide with a wide spectrum of activity against fruit diseases such as apple blotch and scab, apple bitter rot, peach leaf curl, tobacco blue mold, rose black spot, and various rusts. Ziram is most useful for control of vegetable diseases such as tomato anthracnose and tomato and potato early blight.

Nabam is water soluble and has had limited commercial use as a foliar fungicide for fruits and vegetables. Its water-soluble analogs, zineb and maneb, have very wide spectra of activity as foliar sprays for fruits, vegetables, and ornamentals especially for the control of blights, leafspots, blotches, and mildews.

Metham is a soil fungicide, it gives effective control of soil-borne fungi, nematodes, and weeds in potting soils and turf and in vegetable, ornamental, and field crop plantings. Treated soils should be aerated for 2 to 3 weeks before seeding or planting.

5. Toxicology. The dithiocarbamate fungicides are generally of low mammalian toxicity, and the following rat oral LD_{50} values have been recorded: thiram 780, zineb > 5200, ferbam 17,000, ziram 1400, nabam 395, maneb 7500, and metham 820 mg/kg. Their dermal toxicity is correspondingly low although heavy use has caused occasional skin irritation, conjunctivitis, bronchitis, etc. They are also of low chronic toxicity and have been fed to rats for two years at dosages of up to 500 ppm without marked effects(55).

E. Captan and Analogs

1. Introduction and Chemistry. *Captan*, or N-trichloromethylthiotetrahydrophthalimide, a white crystalline compound m.p. 172°C, was developed by Kittleson(73, 74) and has become one of the most versatile and general fungicides for foliar treatment of fruits, vegetables, and ornamentals, of soil and seed treatments, and for postharvest applications. Analogs that have subsequently been developed as foliar fungicides include *folpet*, or N-trichloromethylthiophthalimide m.p. 177°C, and *difolatan*, or N-1,1,2,2-tetrachloroethylthiophthalimide m.p. 160°C.

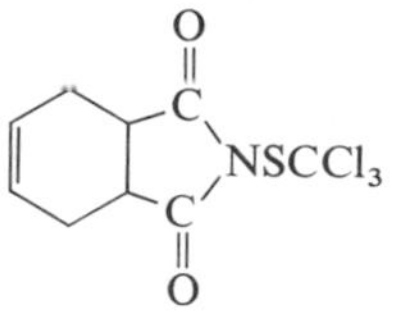

captan 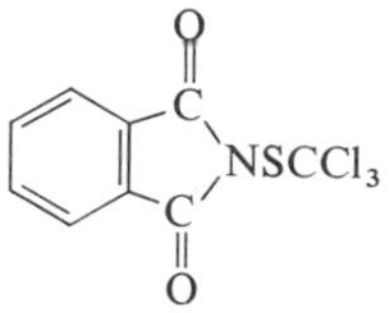   folpet

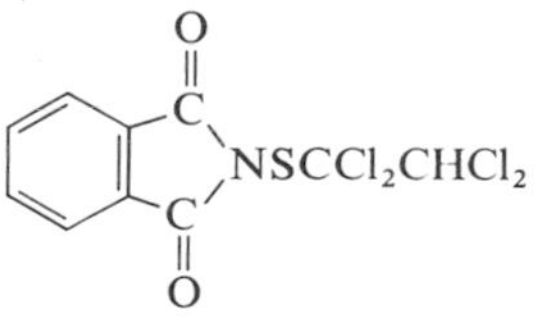

difolatan

2. Structure–Activity Relationships. Kittleson(74) has reported the fungicidal activity of 18 compounds containing the $=NSCCl_3$ group. The most active N-trichloromethylthio derivatives to *Alternaria solani* and *Sclerotinia fructicola* were: tetrahydrophthalimide, phthalimide,

4-nitrophthalimide, and succinimide, 5,5-dimethyl-2,4-oxazolidinedione, 5-methyl, 5-ethyl-2,4-oxazolidinedione, and 5,5-pentamethylene-2,4-oxazolidinedione. These compounds were from 10 to 100 times more effective as fungicides than N-trichloromethylthio-5,5-dimethyl-hydantoin and its 1-nitro- and 1-acetyl derivatives, and as N-trichloromethylthio-N-butylbenzenesulfonamide and N-trichloromethylthio-N-phenyl-benzenesulfonamide. As all of these compounds were fungicidal, Kittleson(74) has attributed their activity to the presence of the =NSCCl₃ group.

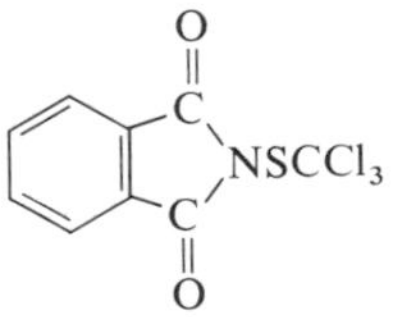

N-trichloromethylthio-phthalimide

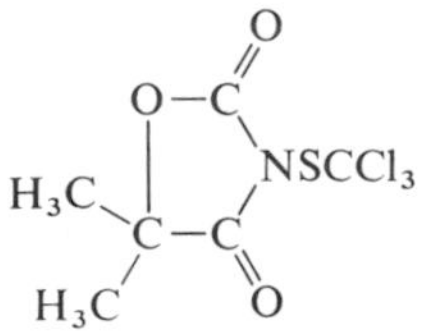

N-trichloromethylthio-5,5-dimethyl-2,4-oxaz-
olidinedione

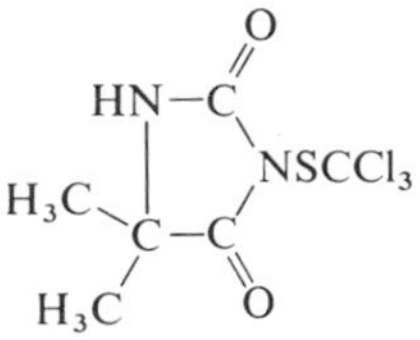

N-trichloromethylthio-5,5-dimethyl
hydantoin

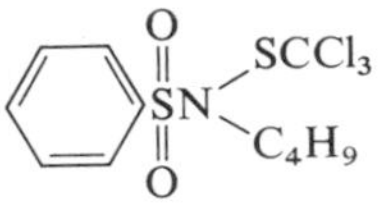

N-trichloromethylthio-N-butyl-benzene
sulfonamide

3. Mode of Action. The action of captan was ascribed by Kittleson(73) to result from the toxophore N—S—CCl₃. Based on the analogy with the fungicide cycloheximide, which has the —CONRCO— grouping of captan but not the —SCCl₃ group, Horsfall and Rich(117) took the opposite view: that the former was the toxophore and the latter the conductophore. It seems probable that both ideas are oversimplifications as Lukens and Sisler(83) have shown that captan interacts with —SH group within fungus cells to produce thiophosgene:

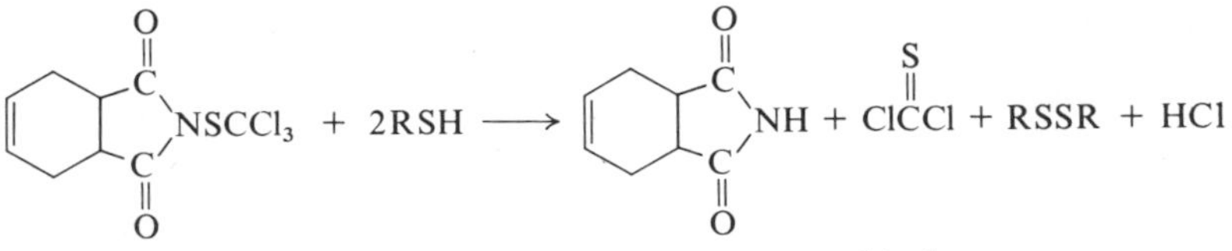

captan   thiophosgene

The highly reactive thiophosgene is suggested to react with free sulfhydryl, amino, and hydroxyl groups of enzymes. Another theory suggests that captan may form the transitory N-chlorotetrahydrophthalimide, which is highly reactive and may combine with reactive groups within the fungal cell(*117*).

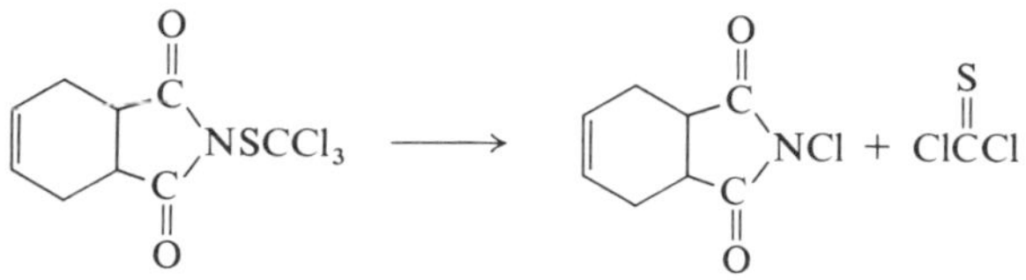

Captan has been found by Hochstein and Cox(*61*) to decrease oxygen uptake of *Fusarium roseum* at concentrations as low as $1 \times 10^{-5} M$. When the fungus was treated with captan, the conidia accumulated large amounts of pyruvate and in in vitro experiments with yeast, captan at concentrations as low as $5 \times 10^{-5} M$ inhibited the decarboxylation of pyruvate by the enzyme pyruvate carboxylase. The inhibition of decarboxylation was reversed by the addition of thiamine, suggesting a competition between captan and cocarboxylase for sites on the apoenzyme.

Additional study of the mode of action of these interesting compounds is urgently needed.

4. Biological Activity. Captan is a very versatile fungicide and has a high degree of safety to plants. It has, therefore, become widely used for the control of plant diseases such as scab, rots, leafspot, blotch, mildew, blights, and damping off of fruits, vegetables, ornamentals, and lawns. Its wide acceptance by commercial fruit growers results largely from its efficiency in plant disease control coupled with lack of adverse effects on fruit appearance and quality. Captan has FDA tolerances of 100 ppm on virtually all of the common deciduous and citrus fruits, avocadoes, berries, grapes, melons, potatoes, tomatoes, and other common vegetables. Captan is also used for postharvest control of fruit rots.

5. Toxicology. Captan is one of the safest of all fungicides with an oral LD_{50} to the rat of $> 15,000$ mg/kg and an intravenous LD_{50} of 50 to 100 mg/kg. It has been fed to rats at 10,000 ppm (1%) in the diet for 2 years without adverse effects. Folpet has an oral LD_{50} to the rat of $> 10,000$ and difolatan of 6000 mg/kg.

F. Chloronitrobenzenes

1. Introduction and Chemistry. The chloronitrobenzenes were introduced as fungicides by I. G. Farbinindustrie in the late 1930's(*139*) when it was found that 1,3,5-trichloro-2,4,6-trinitrobenzene m.p. 195°C, and 1,2,4-trichloro-3,5-dinitrobenzene, m.p. 102.5°C, were highly toxic to specific fungi. The former compound was subsequently abandoned because of its explosive nature. Other closely related compounds that have been found to be effective soil fungicides are 1,2,3-trichloro-4,6-dinitrobenzene, m.p. 93°C, 1,2,3,4,5-pentachloronitrobenzene, or *PCNB*, m.p. 146°C, and 2,3,5,6-tetrachloronitrobenzene, or *TCNB*, m.p. 99°C.

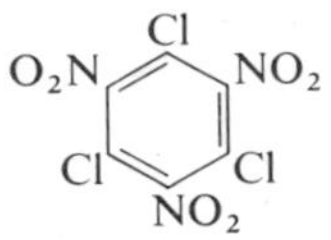

1,3,5-trichloro-2,4,6-
trinitrobenzene

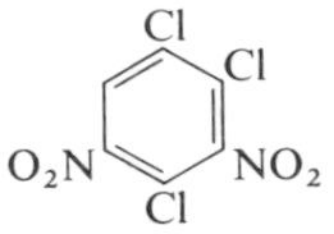

1,2,4-trichloro-3,5-
dinitrobenzene

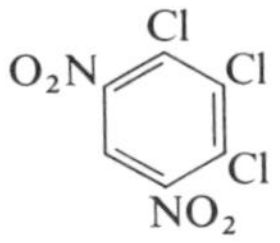

1,2,3-trichloro-
4,6-dinitrobenzene

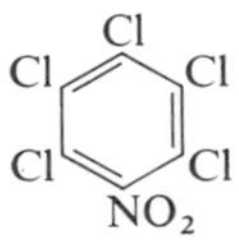

1,2,3,4,5-pentachloro-
nitrobenzene

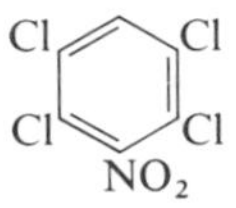

2,3,5,6-tetrachloro-
nitrobenzene

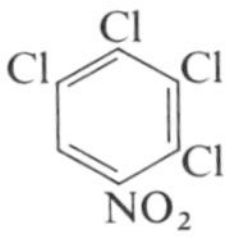

2,3,4,5-tetrachloro-
nitrobenzene

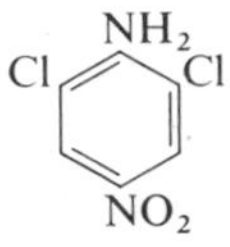

2,6-dichloro-4-
nitroaniline

Pentachloronitrobenzene and 1,2,4,5-tetrachloro-3-nitrobenzene have also been found effective against dry rot of seed potatoes and a newer compound, 2,6-dichloro-4-nitroaniline or BOTRAN, m.p. 194°C, has been developed for the control of postharvest rots of fruits and vegetables. Although it has no nitro-group, *hexachlorobenzene*, m.p. 229°C, should be included in this category. It is effective as a seed treatment for smut control of wheat, sorghum, and onions.

2. Structure–Activity Relationship. Although critical analysis is lacking, the activities of these polychlorobenzene derivatives as soil fungicides seem to be related largely to vapor pressures. Pentachloronitrobenzene with the highest melting point of the nitro derivatives is the most persistent in soil. Brooks(*16*) has reported that 2,3,4,5-tetrachloronitrobenzene is less fungicidal and more phytotoxic than 2,3,5,6-tetrachloronitrobenzene. 2,3,4,6-Tetrachloronitrobenzene was more active than the two other isomers.

3. Mode of Action. Little critical work has been done on the mode of action of the chloronitrobenzenes. Brown(*18*) considers these compounds as vapor fungistats which, however, do not stop hyphal growth. Rich (*117*) states that pentachloronitrobenzene may be a competitive inhibitor of inositol, and essential growth substance for fungi.

4. Biological Activity. The chloronitrobenzenes are fungistats effective principally against soil fungi such as *Botrytis, Rhizoctonia, Mucor, Trichoderma*, and *Fusarium*. They act largely by fumigation and retard germination and colony growth as well as suppressing sporulation. These compounds are also useful in delaying sprouting of stored potatoes(*19*).

5. Toxicology. The oral LD_{50} values to the rat of dinitrotrichlorobenzene 500, and pentachloronitrobenzene of 1650 mg/kg suggest that these compounds are of low to moderate toxicity. Tetrachloronitrobenzene has been fed to rats at 100 mg/kg daily with no apparent ill effects, and rats survived two-year feeding studies with pentachloronitrobenzene at 2500 ppm.

G. Mercury Compounds

1. Introduction and Chemistry. The development of mercurial fungicides was an outgrowth of the usefulness of mercuric chloride as a bactericide. *Mercuric chloride*, $HgCl_2$, was first tested as a seed treatment on cereals by Kellerman and Swingle in 1890(*71*) and by Hiltner in 1915(*59*) who paved the way for the concept of protective seed dressings by observing that mercury treatment would prevent reinfestation by dormant mycelium of *Fusarium* disease of rye. However, the very poisonous nature of mercuric chloride prevented its widespread use until I. G. Farbenindustrie introduced an organic mercurial *"chlorphenol mercury"* $ClC_6H_3OH \cdot Hg \cdot OSO_3Na$ in 1915(*87*), for use as a liquid seed disinfectant. From that time on, a succession of organic mercurials of varying chemical structures have been marketed as shown below. Organic mercury dusts for dry seed treatment were introduced in 1924 with *o-nitrophenol mercury* $HOC_6H_3NO_2 \cdot HgOH$. Dust treatments with organic mercurials are not only unpleasant but also are hazardous, and slurry treatments became popular with the introduction, about 1930, of a new type of organic mercurial, represented by *methoxyethylmercury chloride* $CH_3OCH_2CH_2Hg \cdot Cl$, developed for use as a slurry treatment or its corresponding silicate used as a dry seed treatment.

The organomercury fungicides can be represented by the general formula RHgX where R = aryl-, aryloxy-, alkyl-, or alkoxyethyl-, and X is an anionic group such as chloride, acetate, lactate, urea, carbamate, hydroxy, or related structure as shown in Table 1-3. The nature of the

TABLE 1-3

ORGANOMERCURY FUNGICIDES

Phenyl mercurials

Phenyl mercuric acetate	$C_6H_5HgOCOCH_3$	Seed, turf, foliage, and industrial fungicide
Tolyl mercuric acetate	$CH_3C_6H_4HgOCOCH_3$	Scab treatment on fruits, ornamentals, turf
Phenyl mercuritriethanolammonium lactate	$C_6H_5HgN(C_2H_4OH)_3 \cdot OC(O)CH(OH)CH_3$	Scab treatment on fruits, ornamentals
Phenyl mercurimonoethanolammonium acetate	$C_6H_5HgNH_2C_2H_4OH \cdot OC(O)CH_3$	Seed treatment
Phenyl mercury urea	$C_6H_5HgNHC(O)NH_2$	Slimicide for paper mills and mold retardant for paper
Phenyl mercuric dimethyldithiocarbamate	$C_6H_5HgSC(S)N(CH_3)_{3/2}$	

Phenoxy mercurials

Hydroxy mercurichlorophenol	ClC_6H_4OHgOH	Seed, bulb, turf, foliage treatment
Hydroxy mercurinitrophenol	$O_2NC_6H_4OHgOH$	Disinfectant for potatoes
Methyl mercury 8-hydroxyquinolate	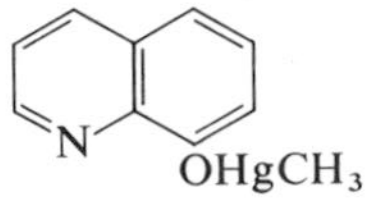	Seed fungicide
Ethyl mercurithiosalicylic acid	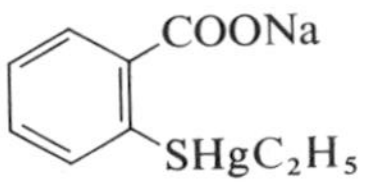	Seed and bulb fungicide

Alkyl mercurials

Methoxyethylmercuric chloride	$CH_3OCH_2CH_2HgCl$	Seed treatment
Ethylmercuric chloride	C_2H_5HgCl	Seed treatment

Ethylmercuric acetate	$C_2H_5HgOC(O)CH_3$	Slimicide for paper mills
1-Acetoxymercuri-2-hydroxyethane	$HOC_2H_4HgOC(O)CH_3$	Seed treatment, turf, soil
Methylmercuric dicyandiamide	$CH_3HgNHC(=NH)NHCN$	Seed treatment
Ethylmercury 2,3-dihydroxypropyl mercaptide	$HOCH_2CH(OH)CH_2SHgC_2H_5$	Seed treatment, slimicide, soil disinfectant, paint preservative
Chloromethoxypropylmercuric acetate	$ClCH_2OC_3H_6HgOC(O)CH_3$	

Miscellaneous mercurials

N-(Ethylmercuri)-p-toluenesulfonamide	$CH_3C_6H_4SO_2N(C_6H_5)HgC_2H_5$	Seed treatment
N-(Ethylmercuri)-1,4,5,6,7,7-hexachloro-bicyclo-[2.2.1]-hept-5-ene-2,3-dicarboximide (also mercurymethyl derivative)	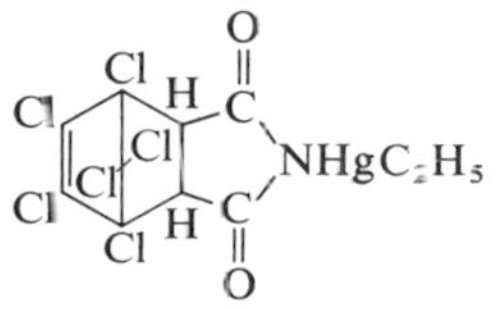	Seed treatment, lawn

organic group R appears to regulate the transport and stability of the compound, and the anionic group determines the solubility. Thus, Booer (*10*) has described ethyl- and methoxyethylmercury radicals as resembling sodium ion in that they are strongly alkaline and form highly ionized salts that are generally water soluble and appreciably volatile. These compounds are quantitatively decomposed by strong acids:

$$CH_3OCH_2CH_2Hg\cdot Cl + HCl \longrightarrow C_2H_4 + CH_3OH + HgCl_2$$

In contrast, the phenyl- and tolyl-mercury radicals resemble silver in giving very stable and generally insoluble salts that can be melted and sublimed or boiled with strong acid without decomposition.

In soil all the organomercury compounds are decomposed to mercury salts or to metallic mercury which are the active fungicides. Booer(*10*) has suggested that this decomposition takes place through base exchange to form organomercury clays which subsequently form mercury salts by further base exchange. These mercuric salts are then reduced to mercurous salts and to mercury. Typical reactions suggested are:

(1) $(C_2H_5Hg)_2$—Clay $\longrightarrow (C_2H_5)_2Hg + Hg\!\!=\!\!Clay$

   $Hg\!\!=\!\!Clay + CaO \longrightarrow Ca\!\!=\!\!Clay + HgO$

(2) $2CH_3OCH_2CH_2HgO\overset{\text{O}}{\overset{\|}{C}}CH_3 + \;\;\underset{H}{\overset{H}{\diagdown \diagup}}Clay \longrightarrow (CH_3OCH_2CH_2Hg)_2\!\!=\!\!Clay +$

2CH₃COOH

   $(CH_3OCH_2CH_2Hg)_2\!\!=\!\!Clay + 2CH_3COOH \longrightarrow Hg\!\!=\!\!Clay + 2CH_3OH +$

$2C_2H_4 + Hg(O\overset{\text{O}}{\overset{\|}{C}}CH_3)_2$

The metallic mercury liberated in the soil is ultimately converted to mercury sulfide by reaction with H_2S liberated by soil microorganisms.

2. Structure–Activity Relationships. As we have seen, the organomercury fungicides can be represented by the general formula RHg·X and the organic radical and anionic group determines the stability, transport, and solubility of the compound. Perhaps the most thoroughly studied series of compounds is the alkoxyalkylmercury salts(*89*). The methoxy- and ethoxyethyl- mercury compounds were of equal effectiveness as fungicides and had the most favorable ratio between "curative dosage for fungi" vs. "tolerated dosage for seed germination." The propoxy-, isopropoxy-, butoxy-, and isobutoxyethyl mercury compounds were also effective fungicides but had less favorable curative vs. tolerated ratios. For the anionic portion of the molecule there was no general difference in effectiveness as dry seed treatments between inorganic salts and salts of carboxylic acids, which were effective at usual dosages in preparations containing 1.5% mercury. When the free valence of mercury is attached to moieties containing N ($CH_3OCH_2CH_2HgN-$) or S ($CH_3OCH_2CH_2HgS-$), activity was decreased so that 2% mercury was required in dry seed treatments. If the free valence of mercury is attached to a second carbon as in $CH_3OCH_2CH_2HgC \equiv CHgCH_2CH_2OCH_3$, activity is still further decreased so that a content of 3% mercury is required for dry seed treatment.

3. Mode of Action. The mechanism of toxicity of mercury is not very specific and the mercurial fungicides all owe their activity to the Hg^{2+} moiety. In the organic mercurials the alkyl or aryl portion serves to conduct the Hg^{2+} to the site of action, by reason of lipoid solubility, and also determines the stability and rate of release of the mercury ion. The biochemical action of the mercury fungicides is related to the affinity of Hg^{2+} for the sulfhydryl groups of essential respiratory enzymes. Thus mercury treatment decreases the oxygen uptake of treated fungus spores, and poisoned spores can be revived by subsequent application of sulfhydryl compounds such as glutathione or cysteine. The specific mercury chelating agent BAL or 2,3-dimercapto-1-propanol is also effective in protecting fungus spores against mercury (91).

4. Biological Activity. The mercury fungicides (see Table 1-3) are generally applied as seed dressings used either as dusts or slurries containing from 1.5 to 3.2% metallic mercury for application to seeds of cotton, rice, wheat, flax, peanuts, safflower, and other crops. They are also used as foliar fungicides for scab of apples, pears, strawberries, and other fruits; for the treatment of gladiolus corms and other bulbs; and as protectants for potato seed pieces; and for control of fungus diseases of turf. Certain of the compounds are very effective slimicides for paper mills, and are mold resistant treatments for paper and paints.

5. Toxicology. The organomercury compounds are of high to moderate acute toxicity to animals with oral rat LD_{50} values of ethylmercuric chloride 30, phenylmercuri triethanolamine lactate 30, methylmercury 8-hydroxyquinolate 72, ethylmercurithiosalicylic acid 75, N-(ethylmercuri)-1,4,5,6,7,7-hexachlorobicyclo-[2.2.1]-hept-5-ene-2,3-dicarboximide 148, methoxyethylmercuric chloride 570, methoxyethylmercuric silicate 1140, and phenyl mercuric acetate 2080 mg/kg. These values compare with the oral LD_{50} of 37 mg/kg for mercuric chloride.

From chronic exposure the mercury compounds are highly toxic and diets of phenyl mercuric acetate as low as 0.5 ppm produced kidney damage in animals. The nature of the organic radical determines to some extent the amount of tissue storage and damage in the liver and kidney: inorganic mercury < aryl mercury < alkyl mercury. The concentration of all three classes increased after daily oral dosage in the following order: brain < blood < liver < kidney. In man, long exposure to organic mercurials causes nervous symptoms: tremors, incoordination, disturbance of hearing, irritability, and severe mental disturbance (55).

Mercury compounds can enter the body through skin absorption, by mouth, and by the respiratory tract. The appreciable volatility of some of the alkyl mercury compounds increases the hazard of inhalation, and the threshold limit for organic compounds in air is 0.01 mg/m^3.

The organic mercury compounds are severe skin irritants and may cause severe dermatitis. Protective clothing should be used when handling these materials and scrupulous personal hygiene is required. Mercury compounds pose a severe environmental hazard through concentration and accumulation in food chains as biologically formed dimethyl mercury.

H. Organo-Tin Compounds

1. Introduction and Chemistry. Investigation of the fungicidal activities of tin compounds was begun in 1950 by Van der Kerk and co-workers (146). It was soon found that although inorganic tin compounds were inactive, certain organo-tin compounds, particularly the trialkyl and triaryl tins, are among the most effective fungicidal agents yet discovered. These organo-tin compounds are very general biocides and are toxic to plants, insects, and various marine organisms. At present *tributyl tin hydroxide* is used as an anti-mildew agent in wood, textiles, and paints; in antifouling paints for marine vessels; for prevention of microbial slime in paper mills; and as a germicide. *Triphenyl tin chloride*, m.p. 105°C, is used in antifouling paints; and *triphenyl tin hydroxide*, m.p. 120°C, and *triphenyl tin acetate*, m.p. 121°C, are used as agricultural fungicides.

$$(C_4H_9)_3SnOH \qquad (C_6H_5)_3SnO\overset{\overset{\textstyle O}{\|}}{C}CH_3$$

tributyl tin hydroxide   triphenyl tin acetate

$$(C_6H_5)_3SnCl \qquad\qquad (C_6H_5)_3SnOH$$

triphenyl tin chloride   triphenyl tin hydroxide

Tin is a group IV element and forms a limited series of organic compounds analogous to those of carbon. In the stable compounds the tin is tetravalent and may be bonded to one, two, three, or four alkyl or aryl groups with the remaining bonds to halogens, hydrogen, or oxygen. The organo-tin compounds are generally prepared by the reaction of tin halides with the Grignard reagent or with organozinc compounds. The organo-tin halides react slowly with water:

$$R_3SnCl + HOH \longrightarrow R_3SnOH + HCl$$

2. Structure–Activity Relationships. Some of the fungicidal properties of simple tin compounds to four species of fungi are shown in Table 1-4. The inorganic tin derivatives are inactive, and highest fungicidal action is found in the triorgano-tin compounds R_3SnX. Although fungitoxicity is little affected by the nature of X, it varies considerably with R. In the aliphatic derivatives, toxicity reaches a maximum with tripropyl and tributyl tins and decreases with longer chains, while triphenyl tins are the most effective aryl derivatives.

3. Mode of Action. The general biocidal action of the tetravalent organo-tin compounds appears to result from their fundamental action as inhibitors of oxidative phosphorylation. Thus these compounds prevent the incorporation of inorganic phosphate into the highly reactive, high-energy ATP or adenosine triphosphate, a reaction which typically takes place in the mitochondria. This inhibitory process, about which little is

TABLE 1-4

FUNGITOXICITY OF SOME ORGANO-TIN COMPOUNDS(146)

	Minimum lethal concentration, ppm			
Compound	Botrytis allii	Penicillium italicum	Aspergillus niger	Rhizopus nigricans
$(C_2H_5)_4Sn$	50	> 1000	100	100
$(C_2H_5)_3SnCl$	0.5	2	5	2
$(C_2H_5)_2SnCl_2$	100	100	500	200
$C_2H_5SnCl_3$	> 1000	> 1000	> 1000	> 1000
$(CH_3)_3SnOCOCH_3$	20	20	200	200
$(C_2H_5)_3SnOCOCH_3$	1	2	5	2
$(C_3H_7)_3SnOCOCH_3$	0.1	0.1	1	1
$(C_4H_9)_3SnOCOCH_3$	0.1	0.1	0.5	0.5
$(C_6H_{13})_3SnOCOCH_3$	1	10	20	100
$(C_8H_{17})_3SnOCOCH_3$	> 100	> 100	> 100	> 100
$(C_6H_5)_3SnOCOCH_3$	5	5	1	10

known, can be depicted as:

$$Pi + ADP \longrightarrow ATP$$
$$\uparrow$$
$$R_3SnX$$

4. Biological Activity. An important problem in the development of organo-tin fungicides has been the high phytotoxicity of many of the compounds. The trialkyl-tin compounds have in general proved too phytotoxic for foliar use and their use is restricted to mildew-proofing, wood preservation, slime control, and in antifouling paints. However the triphenyl-tin compounds are substantially less phytotoxic, and triphenyl tin acetate or BRESTAN has proved effective as a foliar fungicide for the control of potato blight, leaf spot of sugar beet, and leaf spot of celery. It is also used to control fungal diseases of coffee, banana, and sugar cane. The phytotoxic effect in the triphenyl-tins $(C_6H_5)_3SnX$ is related to the nature of X and is highest for Cl and SO_4. Among the safest of all the triphenyl-tins are triphenyl tin sulfide, $(C_6H_5)_3SnSSn(C_6H_5)_3$, and triphenyl-tin disulfide, $(C_6H_5)_3SnS-SSn(C_6H_5)_3$ which, however, are of more limited spectrum of fungicidal activity than the triphenyl-tin acetate(*146*).

5. Toxicology. As mentioned previously the triorgano-tin compounds are general biocides and are inhibitors of oxidative phosphorylation in mammals as well as in plants and fungi. However, triphenyl-tin acetate, rat oral LD_{50} 200♂ and 300♀ mg/kg, and triphenyl-tin hydroxide, rat oral LD_{50} 500 to 600 mg/kg, are not highly toxic. Triphenyl-tin sulfide and disulfide, rat oral $LD_{50} > 1470♀$, are compounds of low acute toxicity.

I. *Miscellaneous Fungicides*

1. 2-Mercaptobenzothiazole. This compound, m.p. 180°C, is insoluble in water but soluble in benzene to 1%, alcohol to 2%, and acetone to 10%. It is a weak acid and forms salts with bases such as monoethanolamine and laurylpyridinium chloride, and with manganese and zinc dimethyldithiocarbamates:

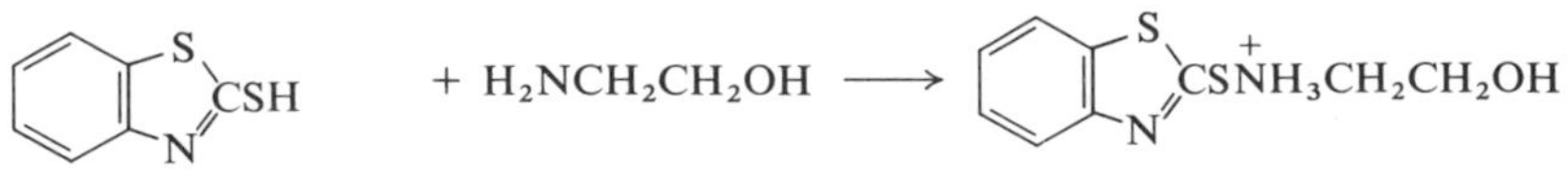

mercaptobenzothiazole   monoethanolamine salt

These salts are effective in mildew-proofing of fabrics, in slime and algae control, and, in the case of the dimethyldithiocarbamate salts, as a seed treatment and foliar fungicide for fruits and vegetables. The mercaptobenzothiazole salts are of low acute toxicity to rats, oral LD_{50} 525 to 3000 mg/kg.

2. 2-Alkylimidazolines. These compounds, also called *glyoxalidines*, are 2-alkyl derivatives of imidazolidine and were first reported as fungicides by Wellman and McCallan(*154*). *Glyodin* or 2-heptadecyl glyoxalidine acetate, m.p. 94°C, is used as a foliar fungicide for fruits and ornamentals particularly against apple scab. In this series of compounds

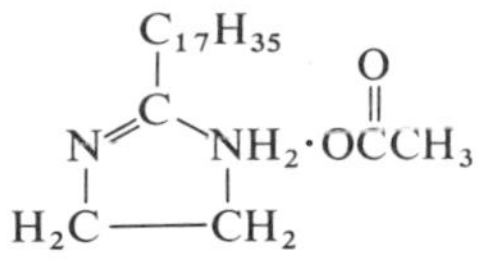

$$\begin{array}{c} C_{17}H_{35} \\ | \\ C \\ N \diagup \; \diagdown NH_2 \cdot OCCH_3 \\ | \qquad\qquad | \\ H_2C \!\!-\!\!-\!\! CH_2 \end{array}$$

glyodin

the unsubstituted imidazoline is almost inactive and fungistatic action increased rapidly with increasing length of the 2-alkyl side chain reaching a maximum at C_{17}. The following LD_{50} values were given against *Sclerotinia fructicola* for 2-substituted 1-hydroxyethylglyoxalidines, C_5H_{11} 880, $C_{11}H_{22}$ 14, $C_{13}H_{27}$ 1.1, $C_{15}H_{31}$ 1.0, $C_{17}H_{35}$ 0.6, $C_{21}H_{43}$ 4.0, and $C_{25}H_{51}$ 116. These results were interpreted by Woodcock(*159*) as indicating the importance of the correct lipophilic–hydrophobic balance (oil–water distribution coefficient) and appear to be an example of the conductophore–toxaphore relationship. Glyodin is slowly hydrolyzed in water, with ring opening, to give N-(2-aminoethyl) stearamide $C_{17}H_{35}$-$CONHCH_2CH_2NH_2$ which is less active. Glyodin has pronounced surface-active properties and is concentrated in fungus spores to as much as 10,000 times the concentration applied(*102*). The fungitoxicity of glyodin is reversed by the purines guanine and xanthine, suggesting that glyodin is a competitive inhibitor of a biosynthetic reaction involving these purines(*155*). Glyodin has a rat oral LD_{50} of 3720 mg/kg.

3. Triazines. Certain substituted triazines closely related to the triazine herbicides (Sec. 1-3, G) have been shown by Schuldt and Wolf(*127*) to be highly fungicidal. The most active compound is *dyrene* or 2,4-dichloro-6-(2-chloroanilino)-1,3,5-triazine, m.p. 159–160°C, which is a foliar fungicide for fruits, vegetables, and turf. The rat oral LD_{50} is 2710 mg/kg.

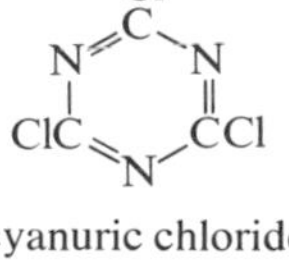

cyanuric chloride 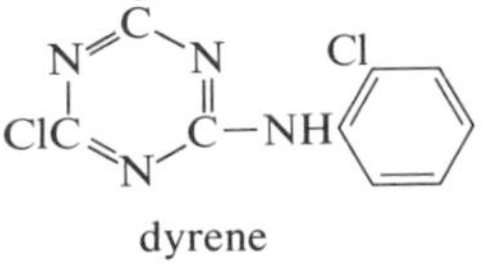   dyrene

In this series of compounds, the parent substance, cyanuric chloride, was nonfungitoxic. However, substitution of a single chlorine by aryl-amino or aryloxy groups produced highly toxic compounds of which dyrene was the most promising. Disubstitution of chlorines reduced activity, and the trisubstituted compounds were nearly inactive. Nuclear substitution of the 6-anilino group had an important effect on fungitoxi-city, as shown by the following ED_{95} values in ppm to *Alternaria solani*: unsubstituted 8000, *o*-CH_3 80, *p*-CH_3 5000, *o*- and *p*-OCH_3 250, *o*-Cl 80, *m*-Cl 750, *p*-Cl 300, 2,4-di-Cl 200, 2,5-di-Cl 2500, *o*-Br 60, *p*-CN > 10,000, *p*-NO_2 > 10,000, *o*-C_6H_5 4000, and *p*-C_6H_5 > 10,000.

4. Quinolinols. 8-Hydroxyquinoline or oxine, m.p. 76°C, has long been used as a bacteriostat and as a chelating agent in analytical chemistry. The compound is almost insoluble in water but is soluble in alcohol, benzene, acetone, and chloroform. Oxine and its copper chelate are used as fungistats in fabric protection, in the chemotherapy of Dutch elm disease, *Ceratostomella ulmi*, and of *Verticillium*, and as wood preserva-tives.

8-hydroxyquinoline   8-hydroxyquinoline copper chelate

Substitution of the oxine nucleus in the 5- or 7- positions with halogen reduces activity and substitution with nitro- groups produced inactive compounds. The methylation of oxine in various ring positions has little effect on fungistatic action. With the introduction of longer alkyl chains in the 5-position, fungistatic action increases, reaching a maximum at *n*-C_5H_{11} to *n*-C_6H_{13} and then decreases, again a reflection of the impor-tance of proper lipophilic–hydrophobic balance(*159*). The fungistatic activity of the 8-hydroxyquinolines is apparently related to their ability to form complexes with essential trace metals such as copper, iron, and zinc within the fungal cell(*163, 164*). Thus, oxine methyl ethers and 5-*n*-amyl-6-hydroxyquinoline, which are incapable of chelation with heavy metals, were of low activity, and the fungistatic action of oxine and its copper chelate to *Aspergillus niger* were found to be reversed by L-histidine or L-cysteine, or by EDTA, all of which compete for copper and liberate the free oxine(*159*). However, there is some evidence that the oxine-copper 1:1 complex is more toxic than oxine alone, and Rich (*117*) has suggested that the copper chelate may be the active moiety that blocks an essential enzyme reaction.

DEXON or sodium *p*-dimethylaminobenzenediazosulfonate, m.p. 200°
d., is a soil fungicide for the control of *Pythium*, *Aphanomyces*, and *Phyto-phthora* attacking field and vegetable crops and ornamentals and turf.
DEXON has a rat oral LD_{50} of 60 mg/kg.

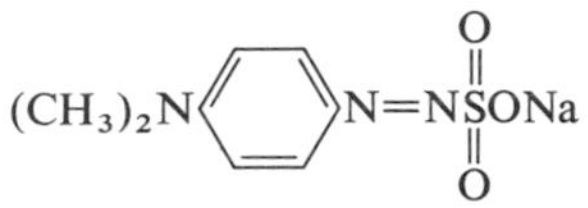

$$(CH_3)_2N\langle\bigcirc\rangle N=N\overset{O}{\underset{O}{\overset{\|}{\underset{\|}{S}}}}ONa$$

DEXON

J. Antibiotics

1. Introduction and Chemistry. Antibiotics are chemicals produced
by living organisms that are selectively toxic to other organisms. In
practice the use of antibiotics has been restricted to chemicals produced
by microorganisms that prevent the growth and development of bacteria
and fungi. More than 340 antibiotics have been identified and more than
100 of these are produced by fungi. It seems clear that such naturally
occurring substances play a role in the biological control of soil pathogens.
However, most of these naturally occurring antibiotics are of complex
structures and are generally highly unstable so that their use as pesticides
has been limited. As most of the antibiotics have been developed for use
in medicine, the exploration of their use as fungicides has been carried
out largely with those substances that were well characterized chemically
and available in substantial quantities. This subject has been thoroughly
reviewed by Goodman(*46, 47*).

Penicillin, the first successful antibiotic, was discovered by Fleming in
1929 and used by Chain and Florey in 1940 for the treatment of human
disease. Its use in plant pathology for controlling *Agrobacterium tume-faciens* in *Bryophyllum* was reported by Brown and Boyle(*17*); however,
penicillin has not been used commercially for plant disease control.

The pronounced fungistatic properties of *gliotoxin* from *Trichoderma
viride* were described by Brian and Hemming(*13*) who found that it in-hibited the growth of *Fusarium* and *Botrytis* spores at 2 to 4 ppm but the
compound (m.p. 221°C, water solubility 0.007% at 30°), is too unstable
to have found application as a soil fungicide. The chemical structure of
gliotoxin has been elucidated by Bell et al. (*9*).

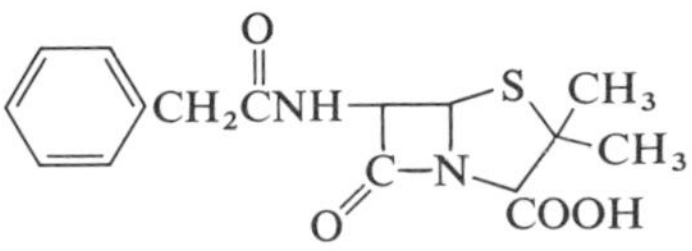

benzyl penicillin   gliotoxin

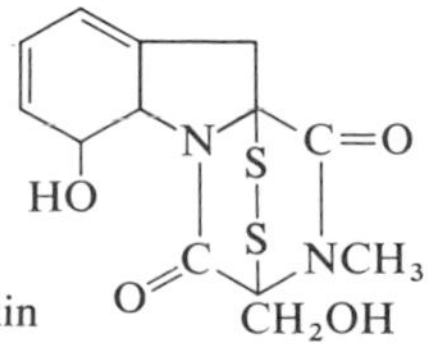

Griseofulvin was first isolated by Oxford et al.(*112*) from *Penicillium griseofulvin* and has been shown to be effective at 0.1 to 10 ppm in producing severe stunting, excessive branching, and abnormal swelling and distortion of the hyphae of fungi that have chitinous cell walls(*12*). It also appears to have systemic fungicidal properties(*30*). The compound m.p. 222°C, is a crystalline solid of very low solubility(*49*).

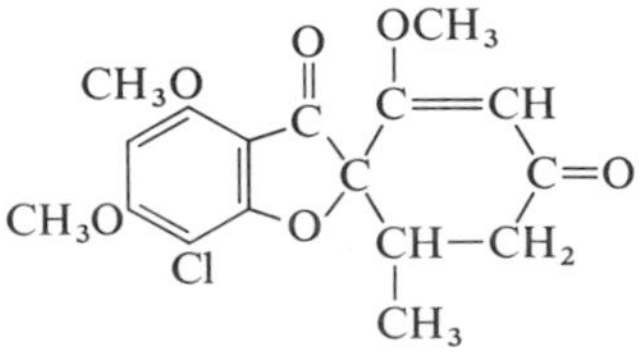

griseofulvin

The antibiotic most widely used as a fungicide is *streptomycin*, which was isolated from *Streptomyces griseus* in 1942 by Wakesman(*151, 152*). Its systemic effectiveness in the control of bacterial pathogens of plants was first reported by Mitchell et al.(*103*). Streptomycin is isolated as the trihydrochloride, m.p. 190–200°C, which is readily soluble in water (> 20%) but is of very low solubility in most organic solvents. Its chemical structure has been established after extensive cooperative research(*151, 152*) as:

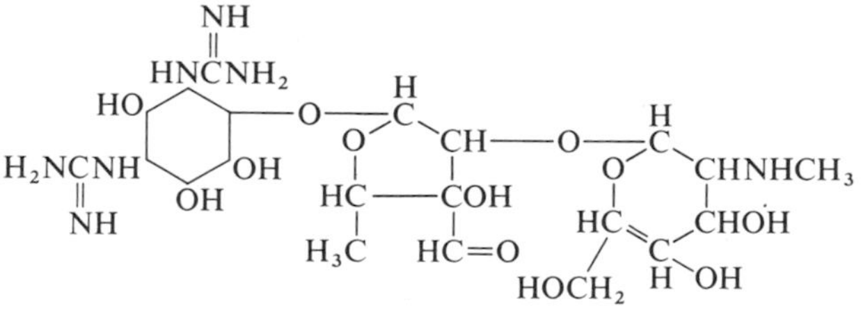

streptomycin

Streptomycin is a basic substance that is irreversibly hydrolyzed under either acidic or alkaline conditions.

Cycloheximide is another antibiotic, in addition to streptomycin, isolated from *Streptomyces griseus* by Whiffen et al.(*156*), which has been shown to have activity against plant pathogenic fungi at concentrations of 1.0 to 5 ppm(*25, 34*). Cycloheximide is a colorless crystalline compound, m.p. 119.5–21°C, that is soluble in water to 2.1% and is soluble in organic solvents (7% in amylacetate). Its structure has been shown by Kornfeld et al. (*77*) to be:

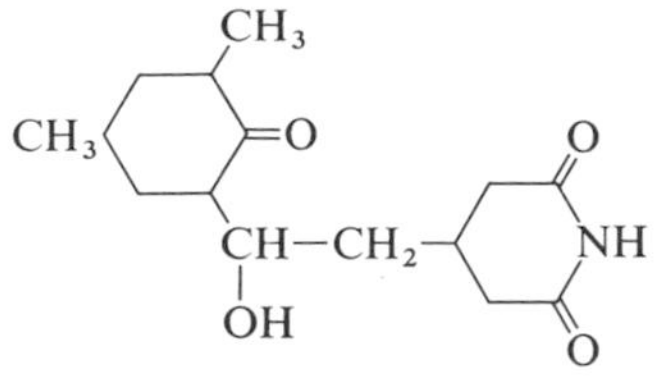

cycloheximide

Cycloheximide is relatively heat stable and is stable under acid conditions but is decomposed by alkali. It has the interesting property of forming stable derivatives such as the acetate, oxime, and semicarbazone that are active as fungicides.

Blastmycin is an antibiotic produced by *Streptomyces blastmycetius* that has particular effectiveness against *Piricularia oryzae* and *P. grisea*, which cause rice stem blast(*153*). Blastmycin, m.p. 168–169°C, is insoluble in water and soluble in acetone, ethyl acetate, benzene, chloroform, methyl isobutyl ketone, and carbon tetrachloride. Its structure has been determined by Yonehara and Takeuchi(*162*):

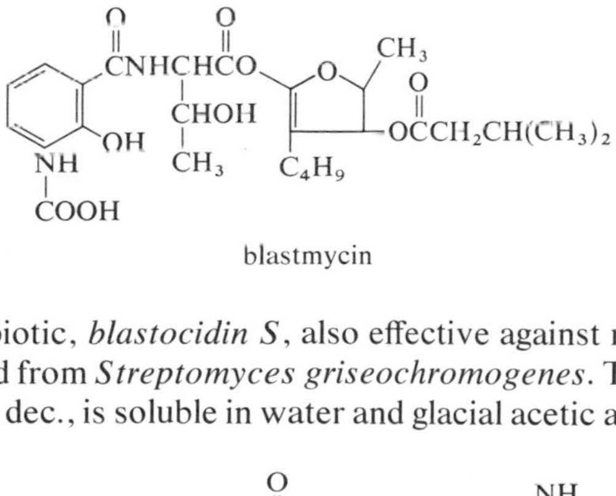

blastmycin

Another antibiotic, *blastocidin S*, also effective against rice stem blast, has been isolated from *Streptomyces griseochromogenes*. The compound, m.p. 252–253°C dec., is soluble in water and glacial acetic acid.

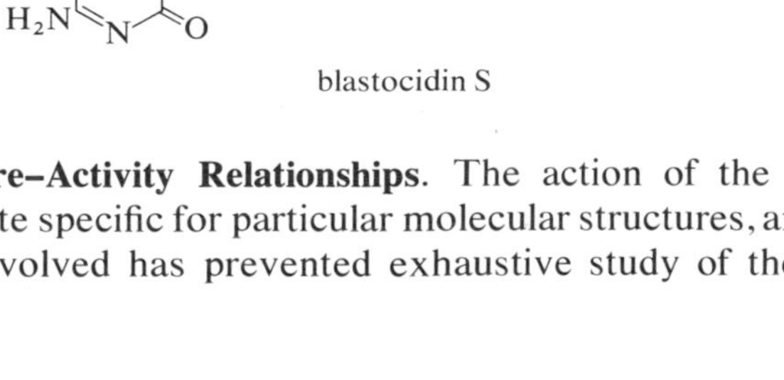

blastocidin S

2. Structure–Activity Relationships. The action of the antibiotics is generally quite specific for particular molecular structures, and the difficult chemistry involved has prevented exhaustive study of the influence of

TABLE 1-5

COMPARATIVE IN VITRO ACTIVITIES OF ANTIBIOTICS TO PLANT PATHOGENIC BACTERIA(*107*)

Bacterium	Sensitivity to, ppm							
	Chlortetra cycline	Neo- mycin	Oxytetra- cycline	Strepto- mycin	Poly- mixin	Strepto- thricin	Vio- mycin	Chloro- mycetin
Agrobacterium tumefasciens	0.025	0.4	0.05	0.4	3.2	3.2	3.2	> 12.5
Erwinia amylovora	0.2	0.25	0.6	0.2	0.13	0.1	1.2	0.9
Pseudomonas pisi	1.6	0.8	6.3	> 12.5	0.4	> 12.5	1.6	> 12.5
Xanthomonas phaseoli	0.1	0.1	0.8	0.1	0.2	0.2	0.8	6.3

structural modifications. The information in Table 1-5, which compares the activities of various antibiotics to representative species of bacterial plant pathogens, gives an indication of the comparative activities of some of the important compounds.

3. Mode of Action. Streptomycin is absorbed both by plant cuticle and by plant roots and is systemic in action in that it is translocated to the aerial portions of the plant following root application. Streptomycin also moves in appreciable quantities from distal to proximal portions of leaves and in the reverse direction, and from leaves to stem and both upward and downward. Translocation in these ways takes place in sufficient amounts to inhibit the growth of plant pathogens (47).

Streptomycin exerts its bacteriostatic effect in plants as the parent compound rather than in modified or metabolized form. Although earlier work suggested that this antibiotic inhibited energy production in the Krebs cycle, the more recent view is that streptomycin inhibits protein synthesis through preventing the proper attachment of messenger RNA to ribosomes.

4. Biological Activity. Although a considerable number of antibiotics have fungicidal properties, only streptomycin and cycloheximide have been used commercially (46, 47). Streptomycin is particularly effective against fireblight of apple and pear caused by the bacterium *Erwinia amylovora* where it is used as a foliar spray at 50 to 100 ppm. Sprays of 200 ppm are effective in controlling bacterial spot of tomato and pepper caused by *Xanthomonas vesicatoria* and tobacco wildfire caused by *Pseudomonas tabaccina*. Dipping potato seed pieces in 200 ppm streptomycin has protected them against seed piece decay (*Erwinia atroseptica* and *Pseudomonas fluorescens*).

Cycloheximide is an antifungal antibiotic that has a wide spectrum of action. It is generally used as an eradicative spray and is most effective against cherry leaf spot, brown rot, and powdery mildew of roses at 1 to 2 ppm. It is also effective against a variety of other powdery mildews, rusts, leafspots, and blights. It is particularly useful in controlling *Helminthosporium* leafspots, and *Puccinia* rusts of grasses along with other turf pests, and for the control of white pine blister rust, *Cronartium ribicola*.

5. Toxicology. Streptomycin has an oral LD_{50} to the rat of 300 mg/kg and cycloheximide of 1750 mg/kg.

K. Postharvest Fungicides

The attacks of fungi and bacteria cause severe damage to fruits and vegetables during shipment and storage after harvest. As examples, potatoes are attacked by late-blight rot *Phytophthora infestans*; onions by onion neck rot *Botrytis allii*; apple and citrus fruits by blue green molds *Penicillium* spp; watermelon by stem-end rot *Diplodia* spp; pineapple by black rot *Thieloviopsis paradoxa*; and pears by *Botrytis* spp. A variety of special fungicidal or fungistatic compounds have been developed to control these diseases and to aid in the marketing of high-quality produce.

Citrus fruits are almost universally protected against blue green mold by the use of wrappers impregnated with *biphenyl* (diphenyl), m.p. 70.5°C, insoluble in water and soluble in organic solvents. The rat oral LD_{50} is 3280 mg/kg. *Sodium o-phenylphenate*, m.p. 57°C (Sec. 1-2, B), is used as a postharvest dip for citrus and deciduous fruits to control blue green molds. It is soluble in water and has a rat oral LD_{50} of 2480 mg/kg.

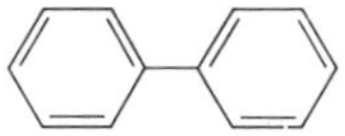

biphenyl 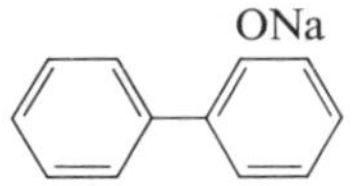   sodium *o*-phenylphenate

A newer compound, 2,6-dichloro-4-nitroaniline, BOTRAN m.p. 194°C (Sec. 1-2, F), has given exceptionally promising results in the control of postharvest rots of fruits and vegetables.

Dehydroacetic acid, m.p. 109°C, is used as a mold preservative on processed foods, bananas, strawberries, and squash. It is soluble in water to 0.1% at 25°C and in acetone to 22%, benzene to 18%, and ethanol to 3% (Merck Index, Ref. *93*), and has a rat oral LD_{50} of 1000 mg/kg.

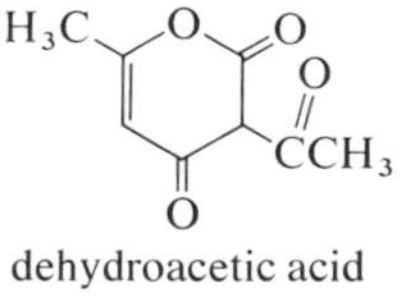

dehydroacetic acid   calcium propionate

Benzoic acid has been used to control black rot of pineapple and *copper sulfate* to control stem end rot of watermelon. *Calcium propionate* is widely used as a fungistat in bread and other foods and on tobacco.

1-3 HERBICIDES

A. Inorganic Herbicides

Inorganic arsenicals (see Sec. 1-4, A) such as *arsenic trioxide* As_2O_3, *sodium arsenite* $NaAsO_2$ and Na_2HAsO_3, and *calcium arsenate* Ca_3-

$(AsO_4)_2$ have been used for many years as soil sterilants and nonselective herbicides. These compounds are the most persistent soil sterilants and may remain effective for periods as long as five to eight years especially in areas of low rainfall. Sodium arsenite is also widely used to kill potato vines, to defoliate cotton, and for aquatic weed control. These inorganic arsenicals are highly toxic to man and animals, and sodium arsenite in particular is attractive because of its salty taste and is hazardous to livestock. Therefore, several newer organic arsenicals have been developed that are much more selective and less poisonous. These include *dimethylarsinic acid* $(CH_3)_2As(O)OH$ and *methylarsonic acid*, CH_3-$As(O)(OH)_2$, used as the disodium or diamine salts. These compounds are used for the control of crabgrass in lawns and are effective in pasture renovation and sod elimination. Dimethylarsinic acid is also effective as a preharvest dessicant. *Arsenic acid*, $As(=O)\cdot(OH)_3$, is also used as a cotton dessicant and defoliant.

The inorganic arsenicals are highly irritant to skin and may cause severe burning. The oral LD_{50} values to the rat are: sodium arsenite ca. 30 mg/kg, arsenic trioxide 138 mg/kg, disodium methylarsonate 750 mg/kg, and dimethylarsinic acid 1350 mg/kg (see Sec. 1-9, A).

Sodium cyanate, $NaOC\equiv N$, and *potassium cyanate*, $KOC\equiv N$, are water-soluble salts used as postemergent contact herbicides to control crabgrass in turf.

Ammonium sulfamate, $H_2NS(O)_2ONH_4$, m.p. 128°C, is a water-soluble compound used to kill brush, poison ivy, and other woody plants especially along ditches and right-of-ways. It acts as both a contact and translocated herbicide and has an oral LD_{50} to the rat of 3900 mg/kg.

Sodium tetraborate or borax, $Na_2B_4O_7$, water solubility 2.7% at 20°C, is used as a nonselective herbicide and soil sterilant. It may remain effective in the soil for a year or more and has a rat oral LD_{50} of 5330 mg/kg.

Sodium chlorate, $NaClO_3$, has been widely used as a soil sterilant. It is soluble in water to 75% and is most persistent in areas of low rainfall where it may remain toxic as long as 5 years. It is readily absorbed through both roots and leaves and may be translocated throughout the plant. The mode of action is not well understood and may be related to the high oxidizing capacity of the chlorate ion.

Magnesium chlorate, $Mg(ClO_3)_2\cdot6H_2O$ is used as a defoliant and dessicant.

The chlorates are unstable molecules and are strong oxidizing agents. They are highly flammable when in contact with organic materials and sometimes ignite spontaneously. Great care must, therefore, be used in their storage and application. Although the oral LD_{50} of sodium chlorate is 5000 mg/kg, the compound has a salty taste and cattle may consume enough to be poisoned.

Calcium cyanamide, $Ca=N-C\equiv N$, has been used as a fertilizer, defoliant, and herbicide where it is used in plant beds. The calcium cyanamide reacts in soil to successively form urea, ammonia, nitrite, and nitrate. Soil moisture and acidity affect the rate of decomposition that takes place in a few days under usual conditions. The rat oral LD_{50} is 1000 mg/kg.

B. *Chlorophenoxy and Chlorobenzoic acids*

1. Introduction and Chemistry. The discovery of the auxin-like or growth regulating properties of the chlorinated phenoxyacetic acids by Zimmerman and Hitchcock[165] and their subsequent employment as herbicides by Hamner and Tukey[54] and Marth and Mitchell[86] began the modern era of selective chemical weed control. These compounds were cheap and possessed unique properties of plant translocation. They were selective to broad-leaved weeds in cereals and could be absorbed from soil as pre-emergent herbicides. Moreover, the growth-regulating action was shared by hundreds of related molecules so that to a large extent, herbicides could be tailormade for employment in a wide variety of herbicidal operations.

The parent substance in this group of compounds is 2,4-dichloro-phenoxyacetic acid or *2,4-D*, a white crystalline substance (m.p. 138–140°C) soluble in water to 0.06% at 22°C, and only slightly soluble in petroleum oils. The compound is a typical organic acid, and readily forms sodium, potassium, and ammonium salts that are water soluble (sodium salt to 3.5%).

An isostere of 2,4-D, 2-methyl-4-chlorophenoxyacetic acid, *MCPA*, m.p. 99°C, soluble in water to 0.16% at 25°C, has chemical and herbicidal properties very similar to those of *2,4-D*. It is used extensively in Europe for weed control in small grains[132].

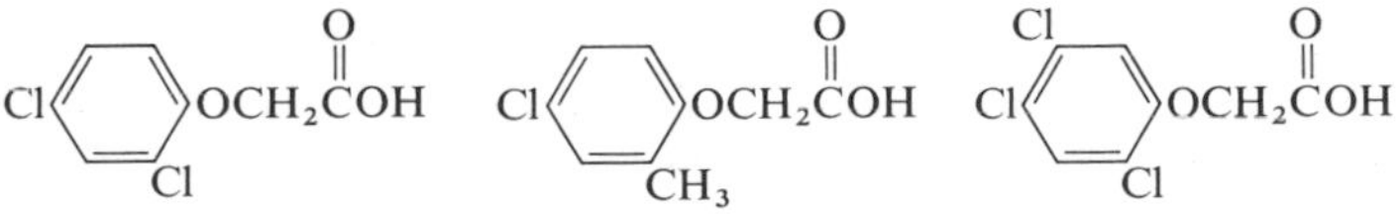

2,4-D   MCPA   2,4,5-T

The closely related 2,4,5-trichlorophenoxyacetic acid or *2,4,5-T* m.p. 153°C, is also of considerable importance as an herbicide for the control of woody species resistant to 2,4-D. 2,4,5-T is soluble in water to 0.028% at 25°C.

a. OTHER ACIDS. In addition to the chlorophenoxyacetic acids, other organic acids and related groups have valuable herbicidal properties.

2-(2,4-Dichlorophenoxy)-propionic acid or *2,4-DP* (m.p. 117°C water solubility 0.018%) is useful for brush control on rangelands and right-of-ways and for aquatic weeds.

2-(2,4-5-Trichlorophenoxy)-propionic acid or *silvex* (m.p. 181°C, water solubility 0.014% at 25°C) is suitable for the control of plant species such as oaks, maples, aquatic weeds, and others resistant to 2,4-D and 2,4,5-T. It is formulated into the same salts and esters. 2-(2-Methyl-4-chlorophenoxy)-propionic acid or *MCPP* (m.p. 94–95°C water solubility 0.0895% at 25°C) is also of importance.

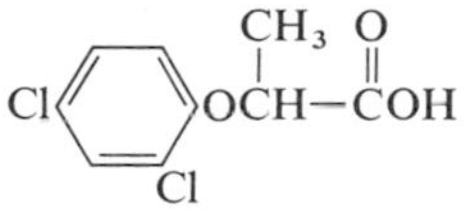

2,4-DP 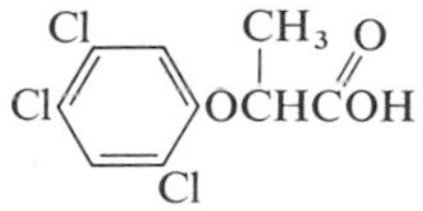   silvex 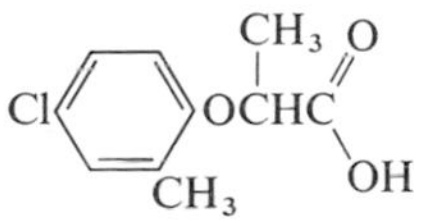   MCPP

4-(2,4-Dichlorophenoxy)-butyric acid, *2,4-DB* (m.p. 121°C, water solubility 0.0053%), is an herbicide cleverly designed to take advantage of lethal synthesis within the plant by means of β-oxidation. By this reaction, a specific β-oxidase enzyme in plant tissues attacks the beta-carbon of the organic acid and oxidizes it, removing two carbon atoms from the chain. Thus, as shown below, the inactive 2,4-dichlorophenoxy-butyric acid is oxidized to form the active 2,4-dichlorophenoxyacetic acid.

2,4-DB

2,4-D

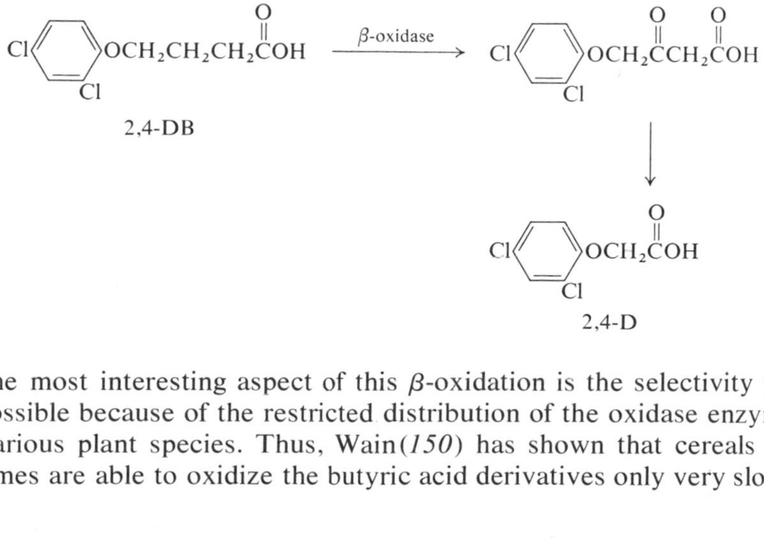

The most interesting aspect of this β-oxidation is the selectivity that is possible because of the restricted distribution of the oxidase enzymes in various plant species. Thus, Wain(*150*) has shown that cereals and legumes are able to oxidize the butyric acid derivatives only very slowly

and thus escape the herbicidal effects while most weeds present in these crops carry out the β-oxidation very readily and are killed by the formation of the 2,4-dichlorophenoxyacetic acid. Selectivity of this type is also practicable with the 4-(2-methyl-4-chlorophenoxy)-butyric acid, m.p. 102–103°C, water solubility 0.0048%, and with 4-(2,4,5-trichlorophenoxy)-butyric acid, m.p. 114–115°C, water solubility 0.0042%.

Sodium 2,4-dichlorophenoxyethyl sulfate, *sesone* (m.p. 170°C, water solubility 20% at 20°C), has a unique selectivity of action based upon its ready hydrolysis at pH values below 5.5 to form 2,4-dichlorophenoxy-

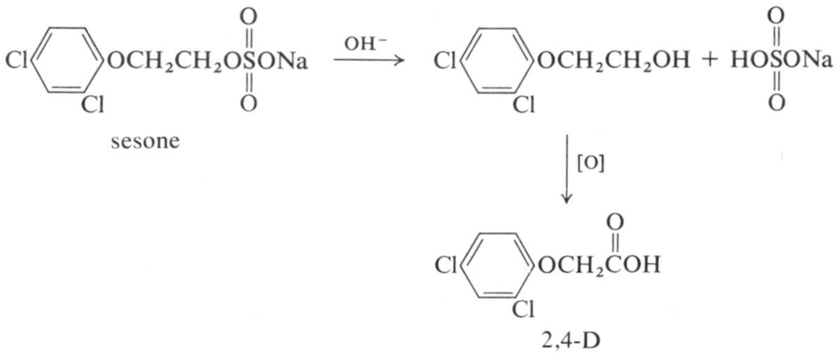

$$Cl\text{—}\langle\bigcirc\rangle\text{—}OCH_2CH_2O\overset{O}{\underset{O}{\overset{\|}{\underset{\|}{S}}}}ONa \xrightarrow{OH^-} Cl\text{—}\langle\bigcirc\rangle\text{—}OCH_2CH_2OH + HO\overset{O}{\underset{O}{\overset{\|}{\underset{\|}{S}}}}ONa$$

sesone

$$\Big\downarrow [O]$$

$$Cl\text{—}\langle\bigcirc\rangle\text{—}OCH_2\overset{O}{\overset{\|}{C}}OH$$

2,4-D

ethanol, which is then further oxidized by soil microorganisms to form 2,4-D. The parent compound, sesone, is herbicidally inactive when applied directly to the plant but when leached into the soil it is converted to 2,4-D and enters the roots as an active weed killer. The decomposition rate in the soil is aided by microorganisms such as *Bacillus cereus* var. *mycoides* and takes place within a few years. Sesone has been used for weed control in peanuts, strawberries, potatoes, evergreens, shrubs, and lawns. The related 2,4,5-trichlorophenoxyethyl sulfate, *natrin*, and 2-methyl-4-chlorophenoxyethyl sulfate (*methin*) behave in exactly analogous fashion. A slightly different type of action is obtained from 2,4-dichlorophenoxyethyl benzoate, *sesin*, which is also readily hydrolyzed to 2,4-dichlorophenoxyethanol. These three compounds have somewhat different patterns of selectivity from *sesone*.

Tris-(2,4-dichlorophenoxyethyl) phosphite, *2,4-DEP*, is a viscous oil, d. 0.96, nearly insoluble in water but very soluble in most organic solvents and is a pre-emergent herbicide also incorporating the principle of hydrolysis to 2,4-dichlorophenoxyethanol and phosphorous acid, with subsequent oxidation of the former to 2,4-dichlorophenoxyacetic acid.

$$(Cl\text{—}\langle\bigcirc\rangle\text{—}OCH_2CH_2O)_3P$$

2,4-DEP

2-(2,4,5-Trichlorophenoxy)-ethyl-2,2-dichloropropionate, *erbon*, combines in one molecule the chlorophenoxyacetic acid type of herbicidal action with the dichloropropionic acid moiety of dalapon (Sec. 1-3, C). It may be expected to have a dual action as a residual nonselective systemic herbicide after hydrolysis in the soil and oxidation of the 2,4,5-trichlorophenoxyethanol produced. Erbon has a broad spectrum of activity because the 2,4,5-trichlorophenoxyethanol fragment is toxic to broad-leaved plants and the 2,2-dichloropropionic acid fragment is specifically toxic to grasses.

2,3,6-Trichlorophenylacetic acid, or *fenac*, m.p. 161°C, water solubility 0.02%, is an effective pre-emergent herbicide for perennial weeds. It can be formulated in any of the ways described for the chlorophenoxy acetic acids and resists decomposition in the soil much longer than the latter.

A related compound is diphenylacetonitrile or *diphenatrile*, m.p. 74°C, water solubility 0.022%, which is readily oxidized to diphenylacetic acid. It is effective against crabgrass in turf.

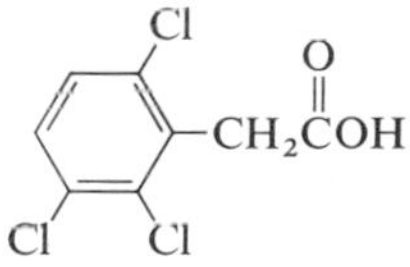

fenac 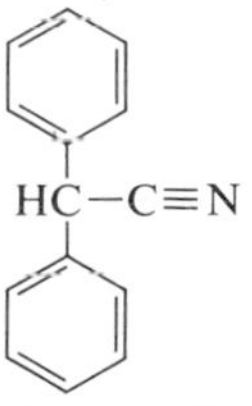   diphenatrile

Two other growth regulating agents belonging to this general group are *1-naphthaleneacetic acid*, m.p. 130°C, and *1-naphthaleneacetamide*, m.p. 183°C, both of which are used for thinning apples.

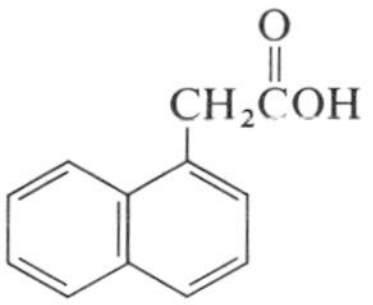

naphthaleneacetic acid 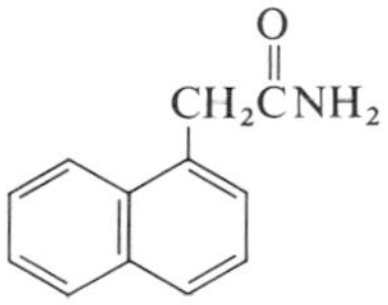   naphthaleneacetamide

b. CHLOROBENZOIC ACIDS. The growth regulating properties of the chlorinated benzoic acids were first recognized by Zimmerman and Hitchcock in 1942(*165*). The activity of these compounds as auxins resembles that of the phenoxyacetic acids but the benzoic acids are much more persistent in the soil and are especially useful for the control of deep-rooted perennial weeds. They may be formulated as salts or esters similarly to 2,4-D.

2,3,6-trichlorobenzoic acid or *2,3,6-TBA*, m.p. 125°C, is water soluble to 0.77% at 22°C.

2-methoxy-3,6-dichlorobenzoic acid, or *dicamba*, m.p. 114–116°C, is useful for weeds in small grain and in turf. The 2-methoxy-3,5,6-trichlorobenzoic acid, or *tricamba*, m.p. 137–139°C, has similar activity.

2,5-dichloro-3-nitrobenzoic acid or *dinoben*, m.p. 220°C, is a nonselective herbicide, and 2,5-dichloro-3-aminobenzoic acid or *amiben*, m.p. 202°C, water solubility 0.06% at 24°C, is a pre-emergent herbicide. An interesting variation of this type of compound is 4-amino-3,5,6-trichloropicolinic acid or *picloram*, m.p. 209.5°C, water solubility 0.0403%. This is the most persistent herbicide of the group.

2,6-dichlorobenzonitrile or *dichlobenil*, m.p. 144°C, water solubility 0.002% at 25°C, is a selective pre-emergent herbicide for orchards, ornamentals, transplanted crops, and aquatic weeds. It is oxidized to the corresponding acid in vivo.

A related herbicide is dimethyl-2,3,5-6-tetrachlorophthalic acid, or DACTHAL, m.p. 156°C, a pre-emergence herbicide for vegetables and for crabgrass.

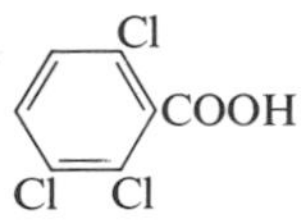

2,3,6-TBA 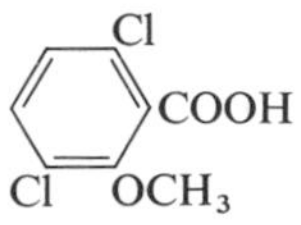   dicamba 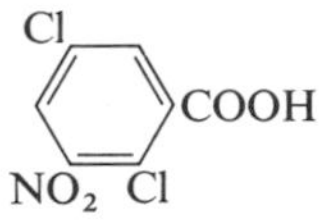   dinoben 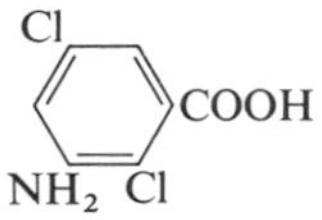   amiben

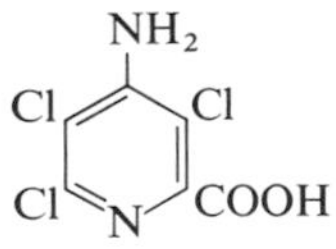

picloram 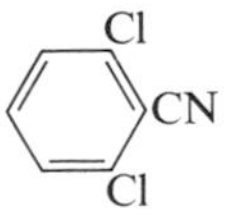   dichlobenil 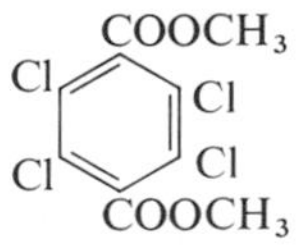   DACTHAL 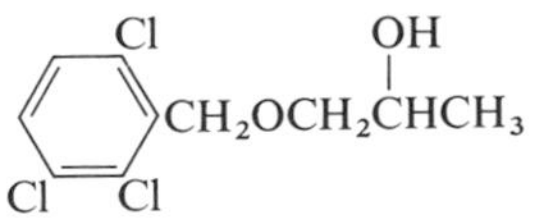   TRITAC

2,3,6-Trichlorobenzyloxypropanol or TRITAC, b.p. 121–124°C/0.1, water solubility 0.0073%, is apparently slowly oxidized to the corresponding 2,3,6-trichlorobenzoic acid by soil microorganisms.

c. SALTS. In order to permit the proper application and formulation of the chlorophenoxyacetic acids, a number of salts and esters have been used commercially. The sodium potassium and ammonium salts have already been mentioned. They are of limited water solubility and are nonvolatile. The amine salts are liquids soluble in water and insoluble in petroleum oils and have generally replaced the sodium and ammonium salts because of better water solubility. The amines commonly employed are di- and tri- ethanolamine, and dimethyl, trimethyl, or triethylamine. All of these salts ionize in water to produce the chlorophenoxyacetic anion and the appropriate cation.

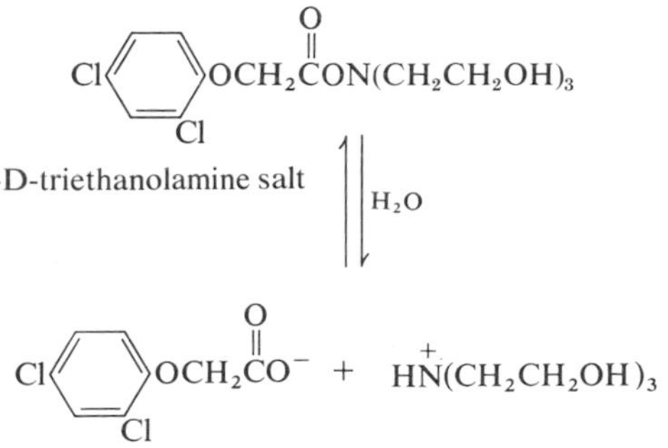

$$Cl\text{-}C_6H_3(Cl)\text{-}OCH_2\overset{\displaystyle O}{\overset{\|}{C}}ON(CH_2CH_2OH)_3$$

2,4-D-triethanolamine salt $\quad\Big\updownarrow\; H_2O$

$$Cl\text{-}C_6H_3(Cl)\text{-}OCH_2\overset{\displaystyle O}{\overset{\|}{C}}O^- \;+\; \overset{+}{H}N(CH_2CH_2OH)_3$$

anion   cation

d. ESTERS. The esters of the chlorophenoxyacetic acids are liquids that are water insoluble but are soluble in organic solvents and readily emulsifiable in water. The esters commonly employed as herbicides are:

methyl ester	$RCH_2OCOOCH_3$
isopropyl ester	$RCH_2OCOOCH(CH_3)_2$
butyl ester	$RCH_2OCOOC_4H_9$
octyl ester	$RCH_2OCOOC_8H_{17}$

These range from very to moderately volatile and under hot, humid conditions the chemical will vaporize and often injure or kill susceptible plants located adjacent to treated areas. In order to decrease this volatility longer chain esters, especially those containing ether linkages, have been developed. These are:

butoxyethyl ester $\qquad\qquad\qquad\qquad RCH_2COOC_2H_4OC_4H_9$

propyleneglycolbutylether ester $\qquad\qquad RCH_2COOCH_2CH(OH)CH_2OC_4H_9$

tetrahydrofurfuryl ester $\qquad\qquad RCH_2COOHC\overset{\displaystyle O}{\underset{\displaystyle H_2C\text{---}CH_2}{\diagdown\, CH_2}}$

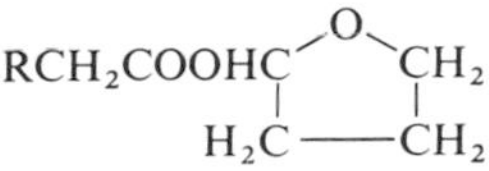

Because of the surface active properties of the ester, the esters of the chlorophenoxyacetic acids are, in general, the most active herbicides. They wet the plant cuticle and aid penetration of the stomata, and the lipoid solubility of the apolar molecules permits more ready penetration of the cuticle waxes.

2. Structure–Activity Relationships. Thousands of aromatic substituted organic acids are known to have auxin-activity, and detailed discussion of their structure–activity relationships have been developed by Veldstra(*149*) and Wain(*150*). With regard to 2,4-D analogs, the number and position of the Cl atoms is of critical importance as illustrated in Table 1-6. Activity is maximal with the 2,4- and 2,5-dichlorophenoxy acetic acids. The high activity of MCPA indicates that the 2-Cl atom can be replaced by CH_3 without much change in activity. However, the 2,4-dimethylphenoxyacetic acid has substantially lower activity. Replacement of chlorine with the other halogens decreases activity in the order Cl > Br > I > F(*142*).

In addition to the phenoxyacetic acids, phenoxypropionic acids are also effective. These exist in optically active (+) and (−) isomers of which the former are active and the latter inactive. Wain(*150*) has concluded that all active compounds of the phenoxy acids must have a free α-hydrogen atom and suggests that activity depends upon the presence in the molecule of three essential structures (a) an unsaturated ring system, (b) an α-hydrogen atom, and (c) a —COOH group or a group that can be biochemically converted to it. These three centers must make contact with three specifically placed receptor cells at a critical boundary surface within the plant.

TABLE 1-6

HERBICIDAL ACTIVITIES OF CHLORINATED PHENOXYACETIC ACIDS(*79*)

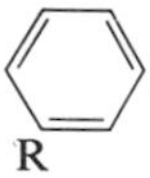

R (OCH₂COOH)	Tomato test, minimum threshold concentrations, mg per gram of lanolin		
R =	Cell elongation	Formative effects	Rooting on stems
2-Cl	0.5	0.25	2.5
3-Cl	0.1	inactive	2.0
4-Cl	0.1	0.06	1.0
2,3-diCl	1.0	inactive	5.0
2,4-diCl	0.015	0.003	0.32
2,5-diCl	0.01	inactive	0.25
2,6-diCl	inactive	20.0	inactive
3,4-diCl	0.05	0.50	2.50
3,5-diCl	inactive	5.0	inactive
2,3,4-triCl	1.0	1.0	inactive
2,3,5-triCl	inactive	inactive	inactive
2,3,6-triCl	inactive	inactive	inactive
2,4,5-triCl	0.025	inactive	0.1
2,4,5-triCl	inactive	active	inactive
3,4,5-triCl	inactive	5.0	inactive

More recent study(*80*) has suggested that the active auxin-like molecule must possess a critical receptor distance of about 5.5 Å between the OH group of the carboxylic acid moiety and an atom on the aryl ring (or a corresponding radical) able to accept a positive charge:

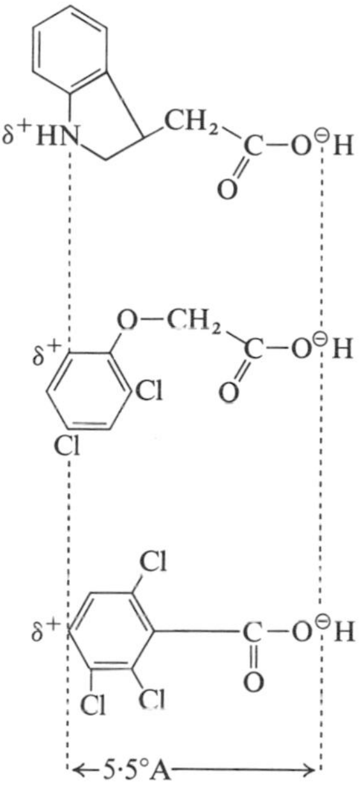

3. Mode of Action. 2,4-D and related phenoxy acids are auxins and cause growth reactions similar to those produced in plants by the naturally occurring indole auxins. The introduction of saturation levels of such artificial auxins into the delicate hormonal balance of the growing plant causes profound changes in the normal growth pattern. As a result, a plant is produced having little or no root elongation, with abnormal growth in the basal stem, and with leaves that do not expand properly and are deficient in chlorophyll. The physiological functions of salt and water uptake by the roots, phloem transport, and photosynthesis are all reduced, contributing to the death of the plant(*148*).

At the cellular level it has been shown that 2,4-D herbicides cause an increase in the rate of RNA production which may result in an auxin–kinin imbalance, producing induced growth with abnormal cellular divisions. Despite an immense amount of study, the exact site of interaction between the auxin molecule and plant protoplasm remains obscure. Sites that have been suggested include interactions at the colloidal protoplasmic boundary of the cell whereby permeability and the flow of essential metabolites are increased, and in the coenzymic activity in nucleic acid metabolism(*150*). In any event, the relative activities of the various phenoxy and benzoic acids are related to their absorption, stability, and rate of transport in the plant tissues.

4. Biological Activity. It has already been pointed out that the chlorophenoxy acids have the widest types of herbicidal action. They are active by contact and by translocation from leaves to roots of perennial weeds and as pre-emergent applications to the soil for control of young seedlings. They are effective for aquatic weed control and for the elimination of unwanted vegetation. The chlorophenoxy compounds are selective against many broad-leaved annual weeds in cereal and grass crops.Through the utilization of beta-oxidation with 4-(2,4-DB), selectivity can be further refined for postemergent control of weeds in legumes. The employment of amide and nitrile derivatives that are slowly metabolized into the active carboxylic acids provides yet other avenues of diversified use. The 2,4,5-trichlorophenoxy acids are exceptionally effective for control of brush on rangeland, pine tree stands, and right of ways. The newer chlorobenzoic acids offer additional opportunities for selective weed control in field, orchard, and turf.

In addition to these avenues of herbicidal activity, the chlorophenoxy acids and their relatives have found important uses in related fields such as thinning of fruit, prevention of preharvest drop, fruit setting, promotion of rooting, and postharvest decay prevention.

The ease with which these organic acids form water-soluble salts and a wide variety of esters with varying degrees of volatility has permitted their use in a great variety of ways. 2,4-D itself is so active an auxin that drift of dusts or sprays has seriously damaged susceptible plants such as cotton miles away from the intended treatment area. Therefore, great care must be exercised in the use of these potent compounds to select suitable methods of application, and appropriately nonvolatile formulations to prevent undesirable plant damage.

2,4-D is readily detoxified by soil microorganisms and at low dosages is normally decomposed in 1 to 4 weeks. However, the rates of disappearance depend upon soil type, moisture, temperature, and aeration; the various analogs are decomposed at widely differing rates.

5. Toxicology. The chlorophenoxy and chlorobenzoic acids have moderate to low acute toxicity to mammals. The acute oral LD_{50} values to the rat for some of the compounds are: 2,4-D acid 375, sodium salt 666–805, and isopropyl ester 700 mg/kg. Additional LD_{50} values are: 2,4,5-T 500, MCPA 700, and sesone 1400. For the chlorobenzoic acids the oral LD_{50} values to the rat are: 2,3,6-TBA 750, amiben 3500, BANOL D 1040, DACTHAL > 3000, dichlorobenil 2710, and dinoben 3500 mg/kg.

Chronically these compounds are of low toxicity, and 2,4-D can be ingested by animals and man in daily dosages approaching those which produce acute toxic effects when given only once. Thus, cumulative effects are minimal(55).

C. Chloroaliphatic Acids

1. Introduction and Chemistry. *Trichloroacetic acid* or *TCA* has been used for many years as a laboratory reagent for denaturing proteins, and its sodium salt has been used for several decades as a soil treatment for the control of grasses. TCA, m.p. 57°C, b.p. 196°C, $d._4{}^{61}$ 1.629, is one of the strongest organic acids, K_a 2×10^{-1} at 18°C. It is soluble to about 10% in water at 25°C and is very soluble in alcohol and ether. TCA readily forms salts, and the sodium salt is soluble in water to 120 g/100 ml at 25°C.

$$
\begin{array}{cc}
\overset{\displaystyle Cl}{\underset{\displaystyle Cl}{Cl-\overset{|}{\underset{|}{C}}-\overset{O}{\overset{\|}{C}}OH}} & \overset{\displaystyle Cl}{\underset{\displaystyle Cl}{CH_3-\overset{|}{\underset{|}{C}}-\overset{O}{\overset{\|}{C}}OH}} \\
TCA & dalapon
\end{array}
$$

2,2-Dichloropropionic acid is an isostere of TCA in which one Cl has been replaced by a CH_3 group. It is also a strong organic acid, $K_a 2 \times 10^{-2}$. The sodium salt, *dalapon*, m.p. 174–176°C, water solubility 90 g to 100 ml at 25°C, is used as an herbicide for the control of perennial grasses where it is more effective than TCA especially following foliage application(69).

Dalapon slowly hydrolyzes in water at 25°C to form pyruvic acid, and should be used within 24 hr after mixing with water:

$$CH_3CCl_2COONa + H_2O \longrightarrow CH_3C(O)COOH + HCl + NaCl$$

2. Mode of Action. The chlorinated aliphatic acids are strong acids that persist in plants for many weeks without change. They inhibit the growing points of the shoot far more than that of the roots. It has been calculated that a dalapon concentration of about $10^{-4} M$ is required for 50% inhibition of the growth of the most susceptible plant species which is 100-fold that required for such herbicides as diuron and simazine. This suggests that dalapon has no acute toxic action but rather produces protein or enzyme denaturation and consequent disturbances of metabolism accompanied by the unusual accumulation of metabolite(148).

Hilton et al. (*60*) suggested that dalapon and other chlorinated aliphatic acids were competitive inhibitors of the biosynthesis of pantothenic acid, but Anderson et al. (*1, 2*) could find no evidence to support this as a mode of herbicidal action.

3. Biological Activity. Salts of TCA are temporary soil sterilants when used at 80 to 100 lb/acre and are useful on perennial grass weeds such as Johnsongrass, Bermudagrass, and quackgrass. Dosages of 5 to 10 lb/acre are effective in selective grass control in sugar beets, cane, and cabbage. TCA generally disappears from soil in 30 to 90 days (*69*).

Dalapon is more effective as a grass killer than TCA and is used to control perennial grasses at 5 to 10 lb/acre for several applications. Lower dosages are used to selectively control annual grasses in some field and vegetable crops and in deciduous fruit orchards. Dalapon disappears from soil more rapidly than TCA and moderate dosages decompose within 2 to 4 weeks while heavier dosages may persist for several months.

4. Toxicology. The oral LD_{50} for TCA to rats is 3300 mg/kg and that of dalapon is 3860 mg/kg. Sodium TCA has been fed to cattle at 1000 mg/kg for three weeks without serious adverse effects. Thus, these compounds have low chronic and acute toxicity. However, both TCA and dalapon are strong acids and concentrated solutions may severely irritate skin and eyes.

D. Amide Herbicides

1. Introduction and Chemistry. The herbicidal activity of a variety of simple amides appears to be closely related to that of the ureas. Hamm and Speziale (*53*) demonstrated that a number of aliphatic amides were effective for pre-emergence weed control of grasses and small broad leaved weeds. The α-chloroacetamides seem to be the most effective and have been developed as N,N-diallyl α-chloroacetamide, *CDAA* or RANDOX. More recent developments have centred around the anilides of which N-(3,4-dichlorophenyl)-propanilide or *propanil* is a good example.

$$\underset{\text{CDAA}}{ClCH_2\overset{\overset{\textstyle O}{\|}}{C}N(CH_2CH{=}CH_2)_2}$$

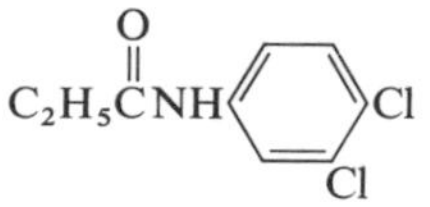

propanil

These compounds are derivatives of the simple chemical acetamide $CH_3C({=}O)NH_2$. The structures and properties of the commercially important compounds are given in Table 1-7. The amide herbicides,

TABLE 1-7
PROPERTIES OF AMIDE HERBICIDES

$$\begin{array}{c} O \\ \| \\ R_1CN \diagdown \begin{array}{l} R_2 \\ R_3 \end{array} \end{array}$$

	R_1	R_2	R_3	m.p., °C	H_2O solubility, %25°C	rat oral LD_{50}, mg/kg
Diphenamide	$CH(C_6H_5)_2$	CH_3	CH_3	135°	0.026	700
Propanil	C_2H_5	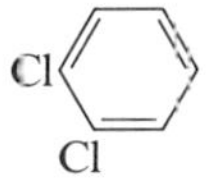 (3,4-dichlorophenyl)	H	85–90°	0.050	1,384
Solan	$C_3H_7CH(CH_3)$	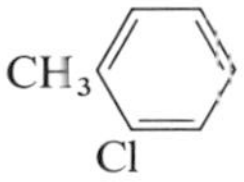 (2-methyl-... Cl phenyl)	H	85–86°	0.0009	> 10,000
KARSIL	$C_3H_7CH(CH_3)$	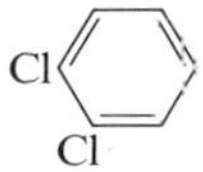 (dichlorophenyl)	H	108–109°		
Dicryl	$CH_3-C{=}CH_2$	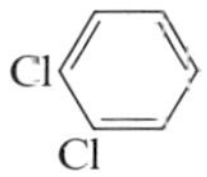 (dichlorophenyl)	H	127–128°	0.0009	3,160
NPA (naptalam)	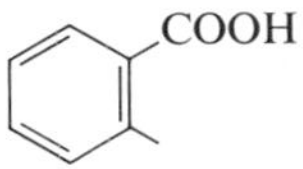 (o-carboxyphenyl, COOH)	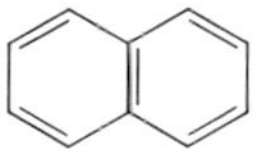 (naphthalene)	H	185°	0.020	> 8,200
CDAA, RANDOX	$ClCH_2-$	$CH_2{=}CHCH_2-$	$CH_2{=}CHCH_2-$		1.97	700
CDEA	$ClCH_2-$	C_2H_5-	C_2H_5-		8.55	500

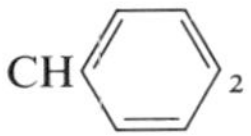

particularly those containing the 2-chloroacetamide structure, are relatively unstable and are hydrolyzed to form amines and glycollic acid.

$$R_2NCCH_2Cl \longrightarrow R_2NH + HOCCH_2OH + Cl^-$$

2. Structure–Activity Relationships. The effects of systematic variations in the structure of α-chloroacetamide on herbicidal activity have been studied extensively by Hamm and Speziale(53). The activity of α-chloroacetamide was negligible and a very large increase was brought about by N-substitution with methyl. At a dosage of 5 lb/acre, the following relative activities were recorded for N-alkyl derivatives – ethyl 1, isopropyl 0, tert-butyl 0, propyl 3, allyl 3, isobutyl 4, sec-butyl 3, isoamyl 3, butyl 1, and amyl 1; and N,N-dialkyl derivatives – dimethyl 4, methyl alkyl 4, diisopropyl 0, dipropyl 4, diallyl 4, and dibutyl 2.

3. Mode of Action. Like the ureas and carbamates, the amide herbicides inhibit the photolysis of water (Hill reaction). The I_{50} for N-phenyl-2-methylpentanamide is 4.5×10^{-5} M.

It has been suggested by Van Overbeek(148) that compounds such as KARSIL owe their biological activity to their ability to form hydrogen bonds through the —NH and —C=O groups with the protein of an enzyme involved in the oxidation of water. There is a remarkable structural resemblance between the highly active amide, urea, and carbamate herbicides as shown below:

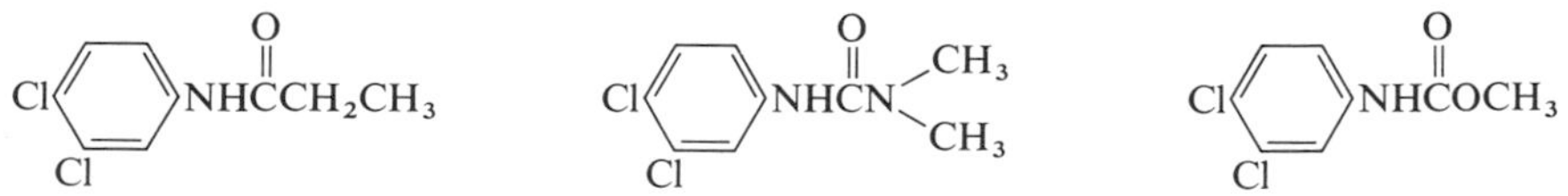

propanil   diuron   swep

This strongly suggests a common mode of action and important structural groups for enzyme interaction.

4. Biological Activity. The amide herbicides are used in a variety of herbicidal applications. Naphthalene acetamide is effective for apple thinning.

The chloroacetamides CDAA and CDEA are effective pre-emergence herbicides for annual grasses and annual broad-leaved weeds but also have foliar contact activity. They are relatively short-lived in soil, decomposing within 3 to 5 weeks under average conditions.

Diphenamide is effective for the pre-emergence control of seedling grasses and is active only upon absorption through the roots.

The substituted N-phenyl amides such as propanil and solan are postemergence herbicides. Propanil has proved especially effective for weed control in rice fields. Naptalam is a selective pre-emergence herbicide for a variety of crops.

5. Toxicology. As shown in Table 1-7, the amide herbicides are generally of low acute toxicity to mammals.

E. Urea Herbicides

1. Introduction and Chemistry. The first commercial urea herbicide was dichloral urea, or DCU, N,N'-di-(2,2,2-trichloro-1-hydroxyethyl) urea introduced in 1950 for pre-emergent control of grasses. The phenyl ureas are the most important urea herbicides and *monuron* or 3-(*p*-chlorophenyl)-N,N-dimethyl urea was first described by Bucha and Todd in 1951(*21*) as effective for the control of annual and perennial grasses. Subsequently, it has proved a very effective soil sterilant, and a number of derivatives have been developed as commercial herbicides.

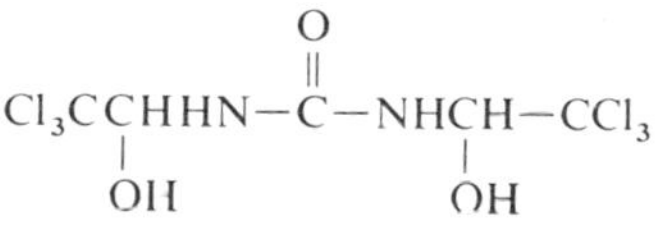

$$Cl_3CCHHN-\overset{\overset{\displaystyle O}{\|}}{C}-NHCH-CCl_3$$
$$\underset{OH}{|} \qquad \underset{OH}{|}$$

DCU

$$Cl\langle \rangle NH\overset{\overset{\displaystyle O}{\|}}{C}N(CH_3)_2$$

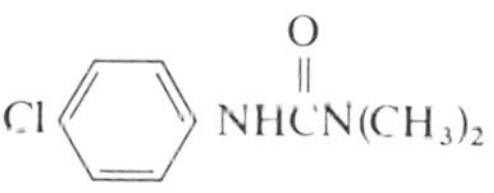

monuron

The substituted urea herbicides may be considered as derivatives of urea, $H_2NC(=O)NH_2$, the first synthetic organic chemical, which is soluble in water to 100%. The water and lipoid solubilities of the ureas are greatly changed by the successive substitution of the hydrogens on the nitrogen groups by a variety of substituents as shown in Table 1-8. Variations in these properties, along with electronic factors affecting adsorption on soil colloids, are important factors in determining persistence of the urea herbicides in pre-emergent applications. The urea herbicides are all crystalline compounds with low water solubility and vapor pressure (Table 1-8).

2. Structure–Activity Relationships. Herbicidal activity seems to be a general property of the phenyl N,N-dimethyl urea structure, and contact toxicity is decreased by increasing the size of the N-alkyl substituent from methyl to ethyl or by removal of the alkyl groups. The phenyl nucleus can be substituted in a variety of ways with chloro or nitro groups

TABLE 1-8
PROPERTIES OF UREA HERBICIDES

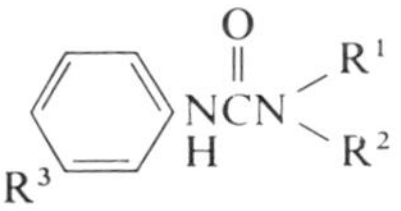 R^1	R^1	R^2	R^3	m.p.,°C	Water sol., ppm 25°C	Oral LD_{50} rat, mg/kg
Fenuron	CH_3	CH_3	H	133–134°	0.3850	7,500
Monuron	CH_3	CH_3	4-Cl	170.5°	0.0230	3,600
Diuron	CH_3	CH_3	3,4-diCl	158–159°	0.0042	3,400
Neburon	CH_3	C_4H_9	3,4-diCl	102–103°	0.00048	> 11,000
Linuron	CH_3	CH_3O	3,4-diCl	93–94°	0.0075	1,500
DCU	N,N-Di-(2,2,2-trichloro-1-hydroxyethyl) urea			191°	0.0074	6,680
Cyclouron	3-Cyclooctyl-1,1-dimethyl urea			138°	0.015	1,500

and activity, which is highest with ortho or para substitution, retained (4).

In a comparison of the growth inhibition of oats in water culture and in clay loam, Sheets and Crafts(136) found the data shown in Table 1-9. The inherent toxicity as expressed by the water culture results of the un-substituted compound (fenuron) is increased about 4 times by p-Cl substitution (monuron) and appreciably further by the second Cl (diuron). The monomethyl urea analog was as toxic as diuron. However, in soil the order of toxicity was revised with fenuron about as active as monuron. Crafts(28) considers that the N-methyl groups affect soil adsorption while the phenyl and chlorophenyl substituents are involved in reactivity with cellular components.

TABLE 1-9
COMPARATIVE TOXICITIES OF UREA HERBICIDES TO OATS IN WATER CULTURE AND SOIL(130)

	I_{50} water	I_{50} soil	ratio S/W
3-(3,4-Dichlorophenyl)-1-methyl urea	1.8×10^{-5} g	130	72
3-(3,4-Dichlorophenyl)-1,1-dimethyl urea (diuron)	1.9	53	28
3-(4-Chlorophenyl)-1,1-dimethyl urea (monuron)	3.0	20	7
3-Phenyl-1,1-dimethyl urea (fenuron)	12.1	23	2

3. Mode of Action. The substituted ureas are readily absorbed by plant roots and translocated to accumulate in leaves where they cause a collapse of the parenchyma vessels. However, they do not penetrate readily through leaves. In the leaves they inhibit photosynthesis (monuron had an I_{50} in *Scenedesmus* of $10^{-7} M$), and are powerful inhibitors of the oxidation of water to oxygen (Hill reaction), monuron at $5 \times 10^{-7} M$ and diuron at $1.2 \times 10^{-7} M$ (52). It has been suggested that monuron blocks photosynthesis at the site of electron transfer by flavin mononucleotide, which has a protective action on monuron-treated plants. The fundamental lesion may be caused by formation of hydrogen bonds between the —NH— and C=O groups and the protein of the enzyme involved (148).

4. Biological Activity. As has been discussed, the herbicidal activities of the phenyl ureas are profoundly affected by their water solubilities. Fenuron, the most water soluble, is used as a soil treatment to kill brush and other woody plants. Neburon, the least soluble, is used for selective weed control in nurseries for turf and ornamentals. Monuron and diuron are used at high dosages as soil sterilants and at low dosages as a selective pre-emergent herbicide for a variety of tolerant crops. Linuron is much more lipoid soluble and has both pre- and post-emergent herbicidal action against annual weeds in tolerant crops.

Monuron and diuron are the most persistent of all herbicides in soil and may remain effective for several years as they are extremely resistant to attack by microorganisms. They must be used with great care to prevent unwanted sterilization of soils devoted to plant growth.

Urox, the trichloroacetate salt of monuron, m.p. 78–81°C, rat oral LD_{50} 2300 mg/kg, was introduced as a soil sterilant; and urab, the trichloroacetate of fenuron (m.p. 65–68°C, rat oral LD_{50} 4000 mg/kg), as a nonselective herbicide for perennial weeds and brush. The oil soluble compound 1-chloro-N-3,4-dichlorophenyl N,N-dimethyl formamidine can be applied in oil spray and in the presence of moisture is rapidly hydrolyzed to diuron:

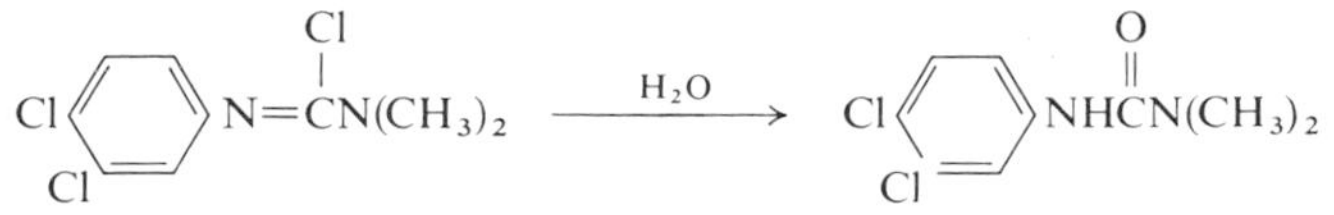

5. Toxicity. As shown in Table 1-8, the urea herbicides are of very low toxicity to warm-blooded animals and do not appear to present any appreciable toxicological hazard.

F. Carbamate Herbicides

1. Introduction and Chemistry. Carbamates are esters of carbamic acid, NH_2COOH, and compounds of this type are also used as fungicides and insecticides (Secs. 1-2, D; 1-4, N). The first observation of the effects of these compounds on plants was made by Friessen(*36*), and Templeman and Sexton(*140*) found that isopropyl N-phenylcarbamate or *IPC*, m.p. 86.5–87.5°C, was an effective pre-emergence herbicide effective against many monocotyledonous plants and relatively inactive to dicotyledonous plants. Subsequently isopropyl-N-(3-chlorophenyl)-carbamate or *CIPC*, m.p. 38–40°C, was found to be more effective.

The carbamate herbicides include a rather miscellaneous group of esters of carbamic and thiocarbamic acids with a variety of N-substituents. They are readily hydrolyzed in soil or in aqueous media:

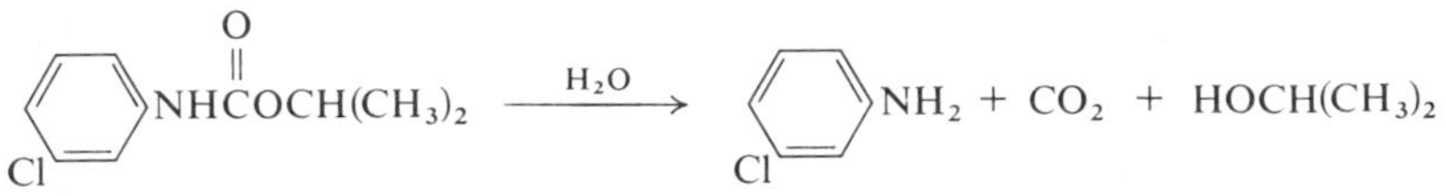

$$\underset{\text{CIPC}}{\underset{\text{Cl}}{\bigcirc}NH\overset{\overset{\displaystyle O}{\|}}{C}OCH(CH_3)_2} \quad\xrightarrow{H_2O}\quad \underset{\text{Cl}}{\bigcirc}NH_2 + CO_2 + HOCH(CH_3)_2$$

Variations in the alcoholic portion of the ester have produced a variety of active compounds as shown in Table 1-10.

The thiocarbamates and dithiocarbamates represent a somewhat different group of compounds that are related to the dithiocarbamate fungicides (whose properties are described in Sec. 1-2, D). Sodium N-methyldithiocarbamate, *metham* or VAPAM, is a soil fumigant for fungi, bacteria, nematodes, insects, weed seeds, and weeds and is active because it releases methylisothiocyanate. A derivative ethyl-N,N-di-*n*-propylthiolcarbamate *EPTC* or EPTAM, b.p. 232°C, is also a soil fumigant.

$$\underset{\text{metham}}{CH_3NH\overset{\overset{\displaystyle S}{\|}}{C}SNa} \qquad\qquad \underset{\text{EPTC}}{(C_3H_7)_2N\overset{\overset{\displaystyle O}{\|}}{C}SC_2H_5}$$

2-Chloroallyl diethyldithiocarbamate, *CDEC* or VEGEDEX, b.p. 128–130°C, is a more stable pre-emergent herbicide introduced in 1954. Several related thiocarbamate herbicides have also been developed for commercial use as shown in Table 1-10 (see page 54).

$$\underset{\text{CDEC}}{(C_2H_5)_2N\overset{\overset{\displaystyle S}{\|}}{C}SCH_2CCl{=\!=}CH_2}$$

2. Structure–Activity Relationships. The herbicidal activity of an extensive series of a number of N-phenylcarbamates and related materials has been evaluated by Shaw and Swanson(*129*) who have given each chemical a phytotoxicity rating based on its average toxicity to eight species of plants as summarized in Table 1-11. Greatest activity was found with the isopropyl and *sec*-butyl carbamates regardless of the aryl substitution. Substitution in the aryl ring produced activity in the order $m > o > p$, with the effectiveness of the substituents decreasing in the order Cl $>$ CH$_3$ $>$ CH$_3$O $>$ CF$_3$ $>$ NO$_2$ $>$ C($=$O)CH$_3$ $>$ OH. In dichlorosubstituted isopropyl N-phenylcarbamates the order of activity was 3,6 $>$ 3,4 $>$ 2,4 $>$ 3,5.

3. Mode of Action. The carbamates are well-known inhibitors of cell growth and Templeman and Sexton(*140*) showed that IPC inhibits the growth of both mono- and dicotyledonous plants by producing a cessation of cell division in the meristems of shoots and roots. The behavior of these compounds resembles colchicine and concentrations as low as 1 ppm inhibit the cell division of barley and higher concentrations produce an interrupted mitotic cycle, blocked metaphases, multinucleate cells, giant nuclei, and an increasing number of partially contracted chromosomes(*33*). The carbamates are also inhibitors of the photolysis of water, the Hill reaction, and it appears that they have a complex mode of action, interfering with both growth and photosynthesis.

The biochemical reactions and possible modes of action of the thiocarbamates are described under Fungicides (Sec. 1-2, D).

4. Biological Activity. The carbamate herbicides such as IPC and Chloro-IPC are especially effective as pre-emergence treatments for grass seedlings and are also used for weed control in cotton, cole crops, and other vegetables. They also prevent sprouting of potatoes in storage when applied as 0.25–0.5% solution by dipping or spraying. These compounds are normally decomposed by soil microorganisms in 3 to 5 weeks.

The most selective of the carbamates is barban, which is widely used as a postemergent herbicide to control wild oats in small grain, etc.

Of the thiocarbamate herbicides, metham is a soil fumigant used in nursery plant beds, potting soil, and turf. EPTC is a pre-emergence herbicide effective against many germinating seeds. CDEC is a pre-emergence herbicide for vegetable crops and nursery stock where it controls many annual grasses and other common weeds. Diallate is used for the pre-emergence control of wild oats in small grains and other crops.

5. Toxicology. The oral LD$_{50}$ values to the rat for the carbamate herbicies are listed in Table 1-10. These compounds are of moderate to low

TABLE 1-10
CARBAMATE HERBICIDES

$$R_1{\diagdown}N{-}CO(S)R_3 \quad \overset{O(S)}{\underset{R_2}{\|}}$$

	R_1	R_2		R_3	m.p., °C	H_2O solubility %, 20°C	Rat oral LD_{50}
IPC	phenyl	H	$-N\overset{O}{\underset{\|}{C}}O-$	$CH(CH_3)_2$	86.5	0.025	> 4420
CIPC	Cl-phenyl	H	$-N\overset{O}{\underset{\|}{C}}-O-$	$CH(CH_3)_2$	38	0.0108	6000
CEPC	Cl-phenyl	H	$-N\overset{O}{\underset{\|}{C}}-O-$	CH_2CH_2Cl			
BIPC	Cl-phenyl	H	$-N\overset{O}{\underset{\|}{C}}-O-$	$CH(CH_3)C\equiv CH$	45	0.054	2500
barban	Cl-phenyl	H	$-N\overset{O}{\underset{\|}{C}}-O-$	$CH_2C\equiv CCH_2Cl$	75	0.0015	1350
swep	Cl,Cl-phenyl	H	$-N\overset{O}{\underset{\|}{C}}O-$	CH_3	113		550

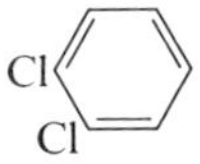

EPTC	C_3H_7	C_3H_7	$\overset{O}{\underset{\|\|}{-NCS-}}$	C_2H_5	b.p. 232	0.0375	1630
Diallate	$(CH_3)_2CH$	$(CH_3)_2CH$	$\overset{O}{\underset{\|\|}{-NCS-}}$	$CH_2CCl{=}CHCl$	b.p. 150/9 mm	0.004 (25°)	395
PEBC	C_4H_9	C_2H_5	$\overset{O}{\underset{\|\|}{-NCS-}}$	C_3H_7	b.p. 142/20 mm	0.0092	1120
CDEC	C_2H_5	C_2H_5	$\overset{S}{\underset{\|\|}{-NCS-}}$	$CH_2CCl{=}CH_2$	b.p. 128	0.01	850
Metham (SMDC)	CH_3	H	$\overset{S}{\underset{\|\|}{-NCS-}}$	Na		0.0072	820
VERNAM	C_3H_7	C_3H_7	$\overset{O}{\underset{\|\|}{-NCS-}}$	C_3H_7	b.p. 150/30 mm	<0.01	1780

TABLE 1-11
RELATION OF STRUCTURE OF N-PHENYLCARBAMATES TO
PHYTOTOXICITY(*129*)

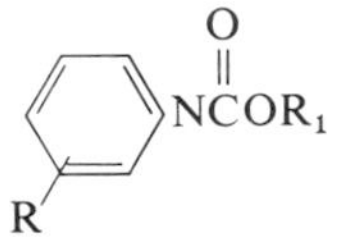

R	R_1	Phytotoxicity[a]
H	C_2H_5	520
H	C_3H_7	690
H	$CH(CH_3)_2$	660
H	C_4H_9	1000
H	$CH(CH_3)C_2H_5$	1330
H	$CH_2CH_2CH_2Cl$	370
H	$CH_2CH(C_2H_5)C_4H_9$	690
H	$C_{12}H_{25}$	680
2-Cl	$CH(CH_3)_2$	990
3-Cl	$CH(CH_3)_2$	1520
4-Cl	$CH(CH_3)_2$	600
2,4-diCl	$CH(CH_3)_2$	730
3,4-diCl	$CH(CH_3)_2$	770
3,5-diCl	$CH(CH_3)_2$	570
3,6-diCl	$CH(CH_3)_2$	1180
2-CH_3	$CH(CH_3)_2$	230
3-CH_3	$CH(CH_3)_2$	1180
4-CH_3	$CH(CH_3)_2$	190
2-CH_3O	$CH(CH_3)_2$	810
3-OH	$CH(CH_3)_2$	80
3-NO_2	$CH(CH_3)_2$	—
4-NO_2	$CH(CH_3)_2$	—
4-$(CH_3)_2N$	$CH(CH_3)_2$	—
4-$(CH_3)_3N^+I$	$CH(CH_3)_2$	180
4-$\overset{\displaystyle O}{\overset{\|}{C}}CH_3$	$CH(CH_3)_2$	176
3-CF_3	$CH(CH_3)_2$	350
3-$\overset{\displaystyle O}{\overset{\|}{C}}CH_3$	$CH(CH_3)_2$	390
2-Cl, 6-CH_3	$CH(CH_3)_2$	1100
3-Cl, 6-CH_3O	$CH(CH_3)_2$	1280

[a]Average activity to cotton, soybeans, corn, wheat, mustard, pigweed, ryegrass, and crabgrass.

acute toxicity. The carbamates appear to present no particular hazard in normal use. However, the thiocarbamate insecticides such as CDEC may irritate skin and eyes and should be immediately washed off with soap and water.

G. Triazine Herbicides

1. Introduction and Chemistry. Although the compound now known as simazine was first synthesized by Hoffman in 1885, the triazine herbicides were introduced by Gast, Knüsli, and Gysin in 1955(41). They have proved to have a wide variety of herbicidal actions and to be valuable as post- and pre-emergent compounds. They are effective, at low dosages, in killing broadleaved weeds in corn and other crops and, in high dosages, as soil sterilants.

The s-triazines are produced from the trimerization of cyanogen chloride CNCl to form the cyclic cyanuric chloride. This is triple imide chloride in which the chlorine atoms are highly reactive and easily substituted with alkyl, alkoxy, alkylthio, and hydroxy groups.

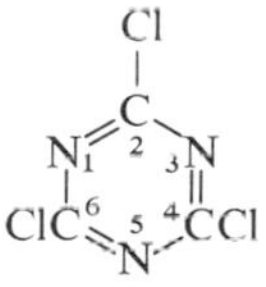

cyanuric chloride 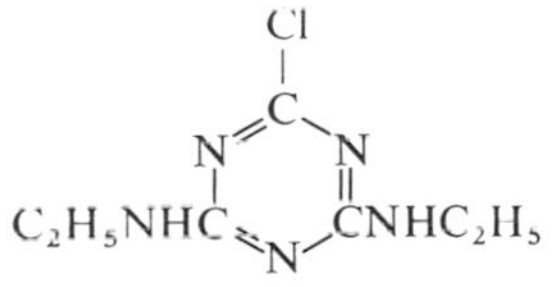   simazine

The resulting compounds have a wide range of water and lipoid solubilities and are of widely varying stabilities and biodegradability. The important compounds are described in Table 1-12.

2. Structure–Activity Relationships. The 2-chloro-4,6-bis-(alkylamino)-S-triazines are the most widely used because of their pre- and post-emergent properties. *Simazine* the 4,6-bis-(ethylamino)-derivative, m.p. 226°C, has a very low water solubility of 0.0005% at 22°C and is very persistent in soils. It is primarily absorbed by roots and is effective as a pre-emergent herbicide. *Atrazine*, the N-ethyl, N'-isopropyl analog, m.p. 173–175°C, is more water soluble and is absorbed through both roots and foliage having both pre- and post-emergent action. It is more effective than simazine where soil moisture is limited.

Replacement of the 2-chloro-group by an alkoxy-group produces compounds of higher water solubility and *prometone*, m.p. 212–214°C, is effective both through leaves and roots as a nonselective herbicide. Substitution of the 2-chloro-group by methylthio- gives compounds

TABLE 1-12
PROPERTIES OF TRIAZINE HERBICIDES

$$N_1 \overset{C}{\underset{C_6}{\overset{2}{\underset{5}{\parallel}}} \overset{N}{\underset{N}{\overset{3}{\underset{4}{}}} C}$$

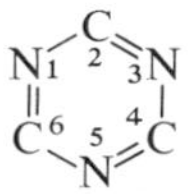	2	4	6	m.p., °C	H_2O solubility %, 22°C[a]	Relative I_{50} for Hill reaction[a]	ED_{50} weeds, ppm	Selectivity index[b]	rat oral LD_{50}, mg/kg
Simazine	Cl	NHEt[c]	NHEt	225–227	0.0005	$1.0(7 \times 10^{-7} M)$	0.18	4.00	> 5000
Propazine	Cl	NHiPr[d]	NHiPr	212–214	0.00086	1.03–3.0	0.92	3.51	> 5000
Atrazine	Cl	NHEt	NHiPr	173–175	0.0070	1.8–2.1	0.31	4.52	3080
Trietazine	Cl	NHEt	N(Et)$_2$	102–103	0.002	0.0025–0.008	0.81	6.85	3750
Chlorazine	Cl	N(Et)$_2$	N(Et)$_2$		0.001	0.021–0.04	0.31	0.38	1750
Ipazine	Cl	NHiPr	N(Et)$_2$		0.004	0.013–0.048	3.49	0.93	2980
Simetone	CH$_3$O	NHEt	NHEt		0.320	1.3–1.5	0.23	0.35	535
Prometone	CH$_3$O	NHiPr	NHiPr		0.075	0.44–0.62	0.17	0.40	2980
Atratone	CH$_3$O	NHEt	NHiPr		0.180	1.3–1.5	0.15	0.77	2400
Simetryne	CH$_3$S	NHEt	NHEt		0.045	5.4–5	1.24	0.79	750
Prometryne	CH$_3$S	NHiPr	NHiPr	118–120	0.0048	6.1–9.1	0.05	0.73	3750

[a] Gysin and Knüsli(52).

[b] ED_{20} for corn/ED_{80} for weeds(92).

[c] Et = ethyl CH$_3$·CH$_2$—.

[d] iPr = isopropyl $\begin{smallmatrix} CH_3 \\ CH_3 \end{smallmatrix} \!\!> \! CH—$.

readily oxidizable to sulfoxide and sulfone and hence more biodegradable and of shorter persistence. *Prometryne* is widely used as a pre-emergent herbicide in potatoes. The structures, water solubilities, and indications of the selectivity of these and some of the many other potentially useful compounds are shown in Table 1-12. The controlling properties are water and lipoid solubility, and biodegradability by soil microorganisms. The activities of a large number of triazines are discussed by Gysin and Knüsli(*52*).

3. Mode of Action. The biological action of the triazine herbicides is not phytohormonal but rather the inhibition of photosynthesis, and in the bean plant this is blocked by exposure of the roots to concentrations of 0.25 to 1.0 ppm.

As a result CO_2 fixation is blocked, sugars are not formed, starch does not accumulate, and the plants become chlorotic. At the molecular level it has been shown that simazine at $7 \times 10^{-7} M$ (I_{50}) inhibits the formation of oxygen from water by isolated chloroplasts in the presence of Fe^{3+} (the Hill reaction), and this is apparently the primary site of action(*52*).

The selectivity of simazine and atrazine for broad-leaved weeds in corn and sugar cane is one of the most interesting features of these compounds and accounts for a very large scale usage. It has been demonstrated that corn contains a "sweet substance" 2,4-hydroxy-7-methoxy-1,4-benzoxazine-3-one that makes a nucleophilic attack upon the chlorine atom at C-2 of simazine replacing it with an OH group(*123*):

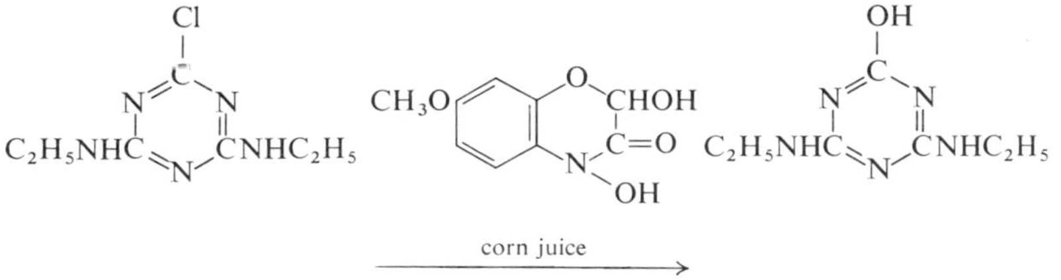

corn juice

The resulting hydroxy simazine is both inactive and unstable and further decomposes in the plant to amines and CO_2.

4. Biological Activity. Simazine is the most versatile of the triazine herbicides because of its very low water solubility and high binding energy to colloidal soil particles. It is especially useful as a pre-emergent herbicide on corn where it provides season-long weed control without crop damage. It can also be applied safely to sugar cane, pineapple, apple, pear, grape, asparagus, ornamentals, shrubs, and forest trees. At high dosages it is an effective soil sterilant. Atrazine is used for pre- and post-emergence applications to corn and sugar cane, and is best suited for very dry conditions. Prometone has been used for the control of deep-rooted perennial grasses such as Bermuda. Prometryne is an effective herbicide for use with potatoes but has caused severe injury under some conditions. Ipazine is a postemergence herbicide for cotton. Trietazine is a pre-emergence herbicide for peas, potatoes, tobacco; and propazine is a selective postemergence herbicide for carrots, celery, and fennel (52).

5. Toxicology. Apart from their herbicidal activity the triazine herbicides are biologically very inactive. The acute oral LD_{50} values to rats are: simazine and propazine > 5000, atrazine 3080, chlorazine 1750, prometone 2980, and simetryne 750 mg/kg. Two years of chronic feeding to rats at 100 ppm in the diet produced no evidence of toxicity(52).

H. Bipyridylium Herbicides

1. Introduction and Chemistry. These compounds were developed as the result of observations that quaternary ammonium germicides such as cetyl trimethylammonium bromide would dessicate young plants. Subsequent researches showed that *diquat*, 2,2'-bipyridylium-1,1'-ethylene dibromide, or 9,10-dihydro-8(a), 10(a)-diazoniaphenanthrene dibromide, m.p. 335–340°C, water solubility 0.0067% at 20°C, was about 120 times more active showing toxicity in greenhouse tests at 125 g/hectare or 2 oz/acre(15).

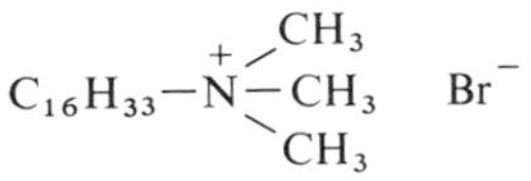

cetyl trimethylammonium
bromide

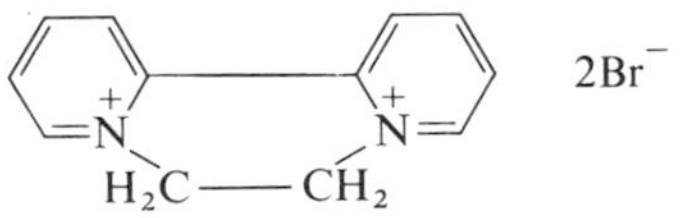

diquat

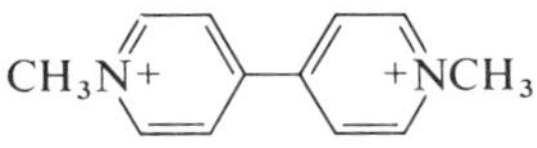

paraquat

The related herbicide *paraquat* or 4,4'-bipyridylium 1,1-dimethyl dichloride was also found to be highly active with a somewhat different spectrum of activity.

The bipyridylium herbicides are formed through alkylation of the bipyridyls with methyl bromide, dimethyl sulfate, or ethylene dibromide, converting them to the respective quaternary ammonium salts with formal positive charges. These compounds are thus the salts of weak bases, and are soluble in water (diquat to 70 g per 100 ml at 20°C). They are stable in acid and neutral solutions.

These compounds readily accept single electrons and are reduced to form free radicals, as is shown by the formation of colored complexes in alkaline solution:

$$CH_3\overset{+}{N} \hspace{-2pt} \text{—} \hspace{-2pt} \overset{+}{N}CH_3 \longrightarrow CH_3N \hspace{-2pt} \text{—} \hspace{-2pt} \overset{+}{N}CH_3$$

2. Structure–Activity Relationships. Herbicidal activity in this group of compounds is restricted to quaternary salts of 2,2'-, 2,4'-, and 4,4'-dipyridyls. All of the quaternary salts of 2,3'- and 3,3'-dipyridyls are inactive. The nature of the quaternary groups greatly affects activity, possibly because of the effect on absorption by the plant cuticle. Thus in the 4,4'-dipyridyl series, the di-*n*-aryl quaternary salt is about 0.1 as active as the di-methyl quaternary (paraquat) and the 1,1'-dibenzyl quaternary is inactive. Some comparisons of the effective threshold concentrations for active compounds are shown in Table 1-13. In the 2,2'-bipyridyls, the activity was maximal for the two carbon bridge, while compounds with larger carbon bridges or the unbridged N,N'-dimethyl-2,2'-dipyridyl dibromide were inactive(*11*).

These structural peculiarities suggest the necessity for the bipyridyl molecule to have coplanar rings in order to be active as a herbicide. Thus, diquat is shown by molecular models to be planar but its inactive N,N-dimethyl analog is not. Many of the quaternary 4,4'-dipyridyls are active and have coplanar rings, but methyl substitution of any of the four ortho-positions adjacent to the inter-ring bond results in inactive compounds that are unable to assume the coplanar ring structure because of steric hinderance(*11*).

3. Mode of Action. The structural peculiarities necessary for herbicidal activity, as discussed in Sec. 1-3, H, 2, suggest that a requisite is the necessity for coplanarity of the two rings. Another important requisite for activity is the capability of forming a resonance stabilized free radical, which can only occur in the 2,2'-dipyridyl and 4,4'-dipyridyl quaternary salts and does not occur with the 2,3'- and 3,3'-dipyridyls or with *ortho* substituted 4,4'-dipyridyls or in 2,2'-dipyridyls without the two-carbon bridge. Thus herbicidal activity is positively correlated with ease of reduction to a stable, water soluble, free radical. This is illustrated in Table 1-13.

Light is essential for the rapid kill of plants with diquat or paraquat, and the presence of chlorophyll and of oxygen is also necessary. Thus it

TABLE 1-13
HERBICIDAL EFFECTIVENESS OF BIPYRIDYLIUM COMPOUNDS[a]

Compound	Eo mV	Mustard		Tomato	
		T	FR	T	FR
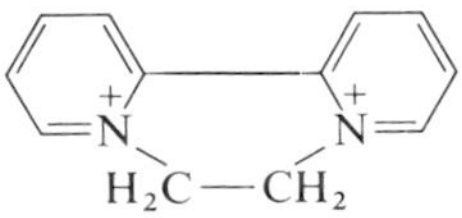 2Br	−348	1.5	1.16	1.1	0.85
 HOCH₂CH₂N⁺ — ⁺NCH₂CH₂OH 2Br	−408	10	2.54	3.5	0.89
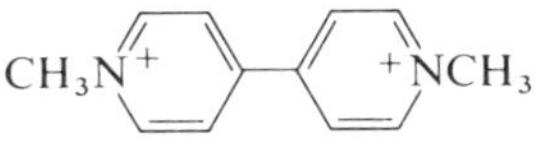 CH₃N⁺ — ⁺NCH₃ 2I	−446	30	2.21	7.5	0.55
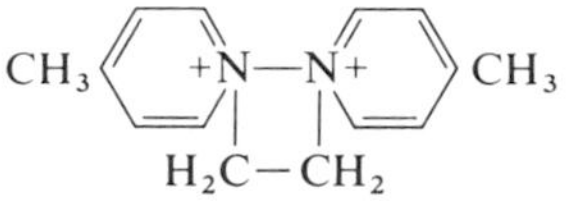 CH₃ +N—N+ CH₃ 2Br	−479	70	1.53	30	0.66
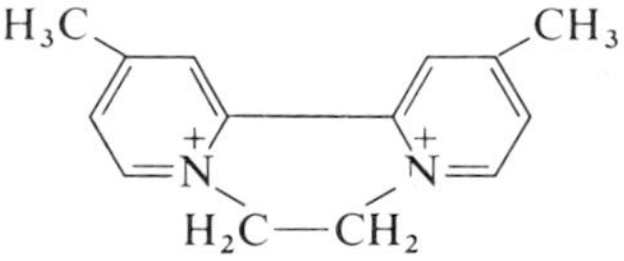 H₃C — CH₃ 2Br	−487	100	1.62		
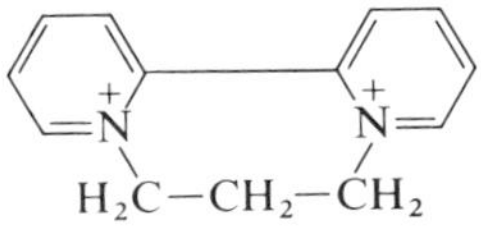 2Br	−548	500	0.79	300	0.48

[a]Threshold concentrations (T) and derived free radical concentrations (FR)(*11*).

seems probable that the dipyridylium herbicides are rapidly reduced to free radicals in the plant by electron transfer from water during photosynthesis. As these free radicals are reoxidized by molecular oxidation, they lead to the formation of peroxide radicals that are highly toxic to the plant tissues(*11*).

4. Biological Activity. The bipyridylium herbicides are very effective contact dessicants and have become widely used in the preharvest dessication of soybeans, sorghum, alfalfa, potato tops, cotton, sugar cane, and other crops; for aquatic weed control, and for postemergent nonselective weed control. Diquat is somewhat more toxic to broad-leaved weeds and paraquat is more toxic to grasses. These compounds are rapidly absorbed by plants but are not translocated in sufficient quantities to kill the roots of perennial weeds. Diquat and paraquat are very strong bases because of their quaternary ammonium structures and are rapidly absorbed and inactivated by many soils because of base exchange with cations of many clay minerals. Therefore, these compounds are not effective as pre-emergent herbicides.

5. Toxicology. The rat oral LD_{50} values are diquat 400 to 500 and paraquat 57 mg/kg. When properly used, these compounds appear to present no hazard to human or animal health. However, when ingested in large amounts, they have caused progressive and fatal pneumonia in man.

I. Miscellaneous Herbicides

Several groups of important herbicidal compounds that are not conveniently classified otherwise are discussed under this heading.

1. Toluidines. The first important herbicide of this group was *dipropalin* or N,N-dipropyl-2,6-dinitro-4-methylaniline, m.p. 42°C, b.p. 118°C at 0.1 mm, soluble in water to 0.03% at 27°C and readily soluble in common organic solvents. A newer product is the closely related *trifluralin*, or N,N-dipropyl-2,6-dinitro-4-trifluoromethylaniline, m.p. 46°C, soluble in water to 0.0024%, and soluble in most organic solvents.

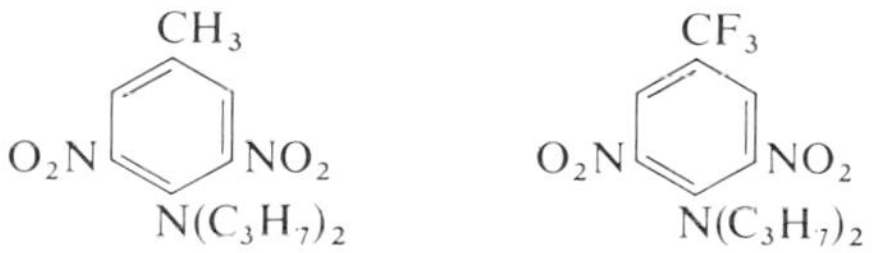

$$O_2N \quad \overset{CH_3}{\underset{N(C_3H_7)_2}{\bighexagon}} \quad NO_2 \qquad\qquad O_2N \quad \overset{CF_3}{\underset{N(C_3H_7)_2}{\bighexagon}} \quad NO_2$$

dipropalin   trifluralin

These two herbicides have shown unusual effectiveness for the selective control of seedling grasses in established turf grasses and as pre-emergence herbicides for annual weeds in cotton, soybeans, peanuts, sweet potatoes, and other crops. The rat oral LD_{50} values are dipropalin 3600 and trifluralin 10,000 mg/kg. A still newer product is BALAN which is the N-ethyl, N-butyl analog of trifluralin.

2. Endothal. *Endothal* is disodium 3,6-endoxohexahydrophthalate, m.p. 263°C, (the acid anhydride m.p. 116°C), which is soluble in water to 21 g/100 ml. It was developed in 1948, as a defoliant for cotton, soybeans, and legume seed crops; as a dessicant for small grains, and to kill potato tops. Endothal is also a contact-pre-emergence herbicide for beets, spinach, aquatic weeds, and turf. Of the three stereoisomers, the *endo–cis* configuration is most active(*88*).

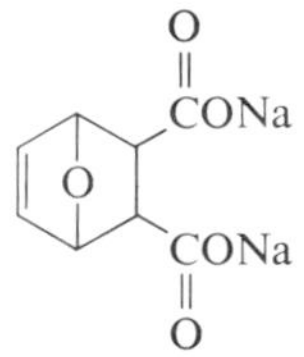

endothal

Little is known about the mode of action of endothal, which produces mitotic effects such as chromosome fragmentation and subsequent structural rearrangements similar to those following ionizing radiation (*14*). Endothal has a rat oral LD_{50} of 35 mg/kg.

3. Phthalamic Acids. A series of N-arylphthalamic acids was found to have interesting growth regulator properties by Hoffman and Smith(*62*). In this series of compounds activity was similar in esters, amides, N-aryl phthalimides, and in soluble salts. The N-1-naphthyl-phthalamic acid or *naptalam*, m.p. 185°C, water solubility 0.02% has proved to be a selective pre-emergence herbicide for soybeans, vegetable crops, nursery stock, and for crabgrass in turf.

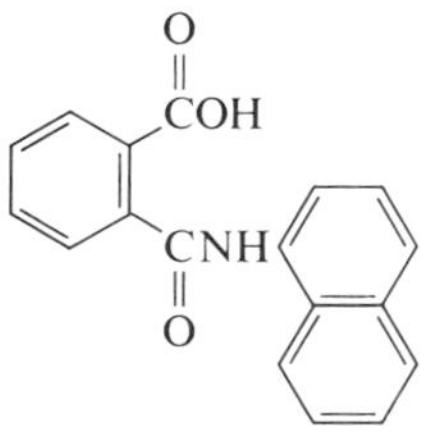

naptalam

The N-aryl group of the phthalamic acid was found to influence markedly the formative effect and the following minimum concentrations in ppm were determined(*62*): phenyl > 2000, *p*-tolyl 200, *o*- and *m*-tolyl 632, *o*- and *p*- nitrophenyl 2000, *m*- and *p*- chlorophenyl 632, *o*-chlorophenyl 20, and 1-naphthyl 0.1. Brian(*14*) suggested that naptalam interferes with auxin transport. Naptalam has a rat oral LD_{50} of 8200 mg/kg.

4. Maleic Hydrazide. Maleic hydrazide, MH or 1,2-dihydropyridazine-3,6-dione m.p. 296°C, is soluble to 0.4 g/100 ml in water and to about 0.1 g in xylene. It forms salts with bases. Its growth regulating properties were first described by Schoene and Hoffman(124). Maleic hydrazide is largely used as a plant growth inhibitor to prevent suckers on tobacco berries and cranberries, and to inhibit growth of trees, shrubbery, and grass.

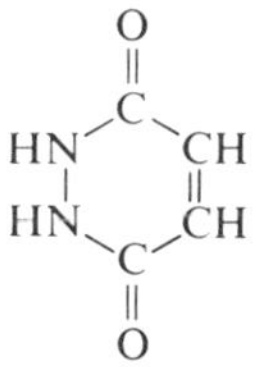

$$
\begin{array}{c}
O \\
\parallel \\
C \\
HN \diagup \quad \diagdown CH \\
\mid \qquad \parallel \\
HN \diagdown \quad \diagup CH \\
C \\
\parallel \\
O
\end{array}
$$

maleic hydrazide

Maleic hydrazide inhibits mitosis in the actively growing tissues of treated plants and also has profound effects on the rate of respiration. Hopkins et al.(63) showed that maleic acid, which reacts with —SH compounds such as cysteine and glutathione, competes with receptor sites of the —SH enzyme, succinic dehydrogenase. It seems reasonable that inhibition of such essential —SH enzymes in plant respiration may represent the site of action of maleic hydrazide in plants(14). Maleic hydrazide has a rat oral LD_{50} of 4000 mg/kg.

5. Amitrole. Amitrole, 3-amino-1,2,4-triazole, m.p. 153°C, has a water solubility 28 g in 100 ml at 23°C. This herbicide is very effective for direct foliage application to eradicate poison ivy, poison oak, scrub oaks, thistles; and is also used to suppress grass in roadsides and orchard floors. It also acts as a cotton defoliant.

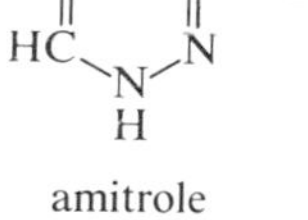

$$
\begin{array}{c}
N \text{———} CNH_2 \\
\parallel \qquad \parallel \\
HC \diagdown \quad \diagup N \\
N \\
\mid \\
H
\end{array}
$$

amitrole

Amitrole in low concentration stimulates plant growth and at high levels inhibits growth and interferes with photosynthesis by causing chlorosis. This effect may be related to interference with purine metabolism as adenine, quanine, hypoxanthine, and riboflavin are antagonists that reduce the growth inhibiting effect(138). Sund and Little(137) have shown that amitrole at concentrations as low as $10^{-5} M$ blocks the synthesis of riboflavin necessary for chloroplast development, perhaps by interfering with the enzymatic incorporation of adenine into riboflavin. The rat oral LD_{50} for amitrole is 1100 mg/kg.

6. Uracils. The herbicidal activity of the substituted uracils, which are related to the pyrimidine bases, was first reported by Bucha(*20*). The 3-butyl-6-methyluracil was selective to many annual weeds in peas and peanuts. *Bromacil* or 5-bromo-6-methyl-3-sec-butyl uracil, m.p. 158°C, is soluble in water to 815 ppm and soluble in most organic solvents. It has a rat oral LD_{50} of 520 mg/kg. *Isocil* is the closely related 5-bromo-6-methyl-3-isopropyl uracil (m.p. 158°C, solubility in water 2150 ppm, rat oral LD_{50} 3400 mg/kg). These compounds are nonselective pre-emergent herbicides most effective against broad-leaved weeds. The derivative 3-cyclohexyl-5,6-trimethylene uracil or duPont 634 has a much lower water solubility of 6 ppm and is a more persistent pre-emergence herbicide. It has a rat oral LD_{50} of > 11,000 mg/kg.

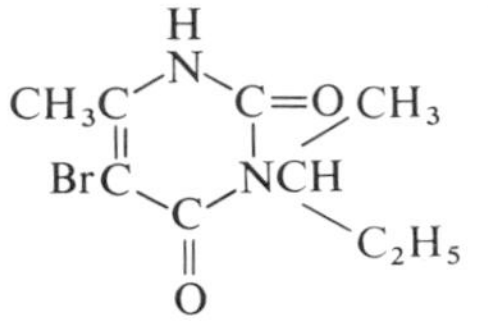

bromacil 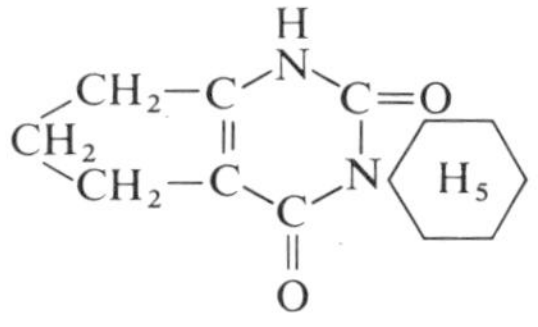   DuPont 634

The uracil herbicides appear to interfere with the photosynthetic processes in the plant, possibly through competition with the naturally occurring pyrimidines.

7. Acrolein. *Acrolein* or propenal, $CH_2{=}CHCHO$, b.p. 52.5°C, is a water-soluble herbicide used to control submerged water weeds in irrigation channels and ponds. Acrolein is a potent lachrymator and irritant and has an oral LD_{50} to the rat of 46 mg/kg.

8. Dinitrophenols. 4,6-Dinitro-*o*-cresol and 4,6-dinitro-*o*-sec-butylphenol (described under Insecticides, Sec. 1-4, G) are general herbicides that are very toxic to all growing plants and are used for postemergence weed control against annual weeds. The ammonium and triethanolamine salts of these compounds are water soluble and insoluble in oils and are used for selective postemergence weed control in small grains and legumes. These salts are also applied for pre-emergence weed control in potatoes, soybeans, peanuts, corn, peas, and beans. These compounds are also used as desiccants and for chemical thinning of deciduous fruits.

1-4 INSECTICIDES

A. Inorganic Arsenical Insecticides

1. Introduction and Chemistry. Various compounds of the element arsenic have been widely used as stomach poisons for insects, since the introduction of *paris green*, $Cu(C_2H_3O_2)_2 \cdot 3Cu(AsO_2)_2$, for the control of the Colorado potato beetle *Leptinotarsa decemlineata* about 1865. It contains 44.3% arsenic with 2 to 3% water solubility and is now used only as a mosquito larvicide. *Arsenious oxide* or arsenic trioxide, As_2O_3 contains 75.7% arsenic and is secured from flue dust after the roasting of various metallic ores. This oxide is water soluble to 2.04 g/100 g at 25°C and forms H_3AsO_3, a weak monobasic acid $K_1 = 6.0 \times 10^{-10}$ at 25°C. It reacts with strong bases to give a solution of arsenite ion $H_2AsO_3^-$, and is used as a poison bait and cattle dip. Three series of arsenite salts exist: orthoarsenites as Na_3AsO_3; pyroarsenites Na_4AsO_5; and metarsenites, $NaAsO_2$. The latter compound contains 57.7% arsenic which is water soluble and is used as a poison bait, a cattle dip, and as an herbicide (Sec. 1-3, A). Arsenic pentoxide As_2O_5 is water soluble to 150 g in 100 ml at 16°C and forms tribasic orthoarsenic acid $K_1 = 2.5 \times 10^{-4}$, $K_2 = 5.6 \times 10^{-8}$, $K_3 = 3 \times 10^{-13}$. Pyroarsenic acid $H_4As_2O_7$ and metarsenic acid $HAsO_2$ are also known, and all three readily form metallic salts (94).

The insecticidal activity of an arsenical is usually directly related to the percentage of metallic arsenic it contains, although other metals in combination such as lead or copper may also add to the toxicity. The phytotoxicity of an arsenical is directly related to the per cent of water-soluble arsenic that can enter the living leaf tissue and poison it. Thus the ideal arsenical is one having a high arsenic content, none of which is soluble in water but all of it readily soluble in the digestive fluids of the insect gut.

The arsenites (trivalent) are less stable and more toxic to insects and plants and have been used chiefly as the toxicants in baits. The arsenates (pentavalent), although less insecticidal, are more stable and safer to plants and are favored as general purpose stomach poisons.

Lead arsenate, $PbHAsO_4$, has 21.6% arsenic and 0.25% water-soluble arsenic. It was developed in 1892 for the control of the gypsy moth *Porthetria dispar* in forests of the eastern United States. It is the safest and most stable of all arsenicals and has been widely used for the control of the codling moth and many other chewing insects. Lead arsenate is a very fluffy powder that wets and suspends well. In water, especially that containing strong alkali, it undergoes the following reaction which produces plant burn:

$$5PbHAsO_4 + H_2O \longrightarrow Pb_4(PbOH)(AsO_4)_3 + 2H_3AsO_4$$

The addition of lime or zinc sulfate to the water as a precipitant for the arsenic acid serves as a "safener."

Basic lead arsenate, $Pb_4(PbOH)(AsO_4)_3$ with 14% arsenic, is preferred for use on delicate foliage in foggy or humid regions where the above reaction occurs.

Calcium arsenate was used as early as 1907 but caused severe plant damage until a "safe" preparation was produced in 1915. The undiluted dust was found to be effective against the cotton boll weevil *Anthonomus grandis* in 1920 and very large amounts have been used for control. Commercial calcium aresenates are mixtures of $Ca_3(AsO_4)_2$ (37.6% As) and $CaHAsO_4$ with an excess of lime and calcium carbonate. They tend to decompose as shown under lead arsenate above. A "safe" form of basic calcium arsenate, $[Ca_3(AsO_4)_2]_3 \cdot Ca(OH)_2$, has 25% metallic arsenic and 0.4 to 0.5% water-soluble arsenic.

2. Mode of Action. The arsenates produce regurgitation, torpor, and quiescence in insects. The epithelial cells of the mid-gut of poisoned insects are generally disintegrated with vacuolized cytoplasm and the chromatin of the nuclei clumped. These are obvious manifestations of a more fundamental biochemical lesion as arsenical poisoning slowly decreases the oxygen consumption of the insect. This is associated with a reaction among the trivalent arsenic and the sulfhydryl (—SH) groups of glutathione and various enzymes such as pyruvate oxidase.

3. Biological Activity. Despite the development of newer organic insecticides, substantial quantities of arsenicals are still utilized for control of the "resistant" cotton boll weevil, for cattle dips, for control of fruit pests, and as a mosquito larvicide. The arsenical insecticides are general biocides, and arsenic trioxide and sodium arsenite have also been used as rodenticides and herbicides (Sec. 1-3, A). The arsenicals have the disadvantages of being hazardous to man and domestic animals and are cumulative poisons when applied to the soil as soil sterilants or as "run-off" from dusts or sprays.

4. Toxicology. The arsenicals are generally highly toxic to mammals and the following rat oral LD_{50} values have been recorded: arsenic trioxide 138, calcium arsenate 40–100, acid lead arsenate 800. Symptoms of acute poisoning include abdominal pain and vomiting and profuse diarrhea, with exfoliative dermatitis and neuritis. Treatment involves emptying the stomach and administration of BAL or 2,3-dimercaptopropanol, a specific antidote that readily complexes and solubilizes trivalent arsenic(*55*).

B. Fluoride Insecticides

1. Introduction and Chemistry. Sodium fluoride was patented in England as an insecticide in 1896 and was used in the United States before 1900 for cockroach control and in 1915 for the control of poultry lice. Cryolite was introduced as an insecticide by Marcovitch in 1928 and its use became general during the 1930's as the disadvantages of arsenical residues on food crops became apparent. Fluoride compounds, which have had insecticidal usage, are salts of hydrofluoric acid HF, fluosilicic acid H_2SiF_6, and fluoaluminic acid H_3AlF_6. The insecticidal properties of these compounds are approximately related to the fluorine content, and the solubility in the digestive juices of the insect. The compounds with high water solubility such as sodium fluoride and sodium fluosilicate produce severe plant damage (94).

Cryolite or sodium fluoaluminate Na_3AlF_6 occurs naturally as a mineral of Greenland and is also produced synthetically. Cryolite contains 54.2% fluorine and is soluble in water to about 0.06%. The synthetic form is lighter and fluffier and is generally preferred. Because of its low water solubility and low mammalian toxicity, it has been used extensively as a stomach poison, applied as dust or spray to fruits and vegetables. Cryolite although relatively insoluble in water is soluble in dilute acids and alkalis, and spray residues have been removed from apples by washing in dilute hydrochloric acid. With lime, cryolite undergoes the following reaction:

$$Na_3AlF_6 + 3Ca(OH_2) \rightarrow Na_3AlO_3 + 3CaF_2 + 3H_2O$$

which results in plant injury.

Sodium fluoride, NaF, contains 45.2% fluorine and is soluble in water to 4.3%. *Sodium fluosilicate*, Na_2SiF_6, contains 60.6% fluorine and is soluble in water to 0.65%. Both of these compounds are highly toxic to foliage.

2. Mode of Action. In insects the fluoride compounds produce spasms, regurgitation, flaccid paralysis, and death. Fluoride inhibits the enzymes such as enolase that require magnesium as a prosthetic group, by precipitating a complex magnesium fluorophosphate and thus prevents phosphate transfer in oxidative metabolism.

3. Biological Activity. The fluoride insecticides are not spectacular in their insecticidal action and are little used at present. Cryolite has had extensive usage as a stomach poison when used on fruits and vegetables for the control of the codling moth, Mexican bean beetle, tomato worm, orange tortrix, and flea beetles. Sodium fluoride has been widely used for the control of cockroaches, chewing lice, and in poison baits for grasshoppers and silverfish. Sodium fluosilicate is an effective moth-proofing agent for woolen goods and has been used in poison baits for grasshoppers, crickets, and cutworms.

4. Toxicology. The rat oral LD_{50} values for the fluoride insecticides are: sodium fluoride 200, sodium fluosilicate 125, and cryolite 13,500 mg/kg. Cryolite is, therefore, one of the safest of all insecticides to mammals.

C. Nicotinoids

1. Introduction and Chemistry. Nicotine from tobacco was recommended for use as early as 1763 as a tea for the destruction of aphids. The alkaloid nicotine was discovered in 1828. *Nicotine, 1*-1-methyl-2-(3'-pyridyl)-pyrrolidine, b.p. 247°C, d. 1.009, is found in the leaves of *Nicotiana tobacum* and *N. rustica* in amounts ranging from 2 to 14% and is also found in *Duboisia hopwoodii* and in *Aesclepias syriaca*. It occurs as the principal alkaloid along with at least 12 other alkaloids of which *nornicotine*, 2-(3'-pyridyl)-pyrrolidine, b.p. 270°C, d. 1.07, and *anabasine, 1*-2-(3'-pyridyl)-piperidine b.p. 281°C, d. 1.048, are of insecticidal importance. Nornicotine occurs as both the *d*- and *l*- forms, the

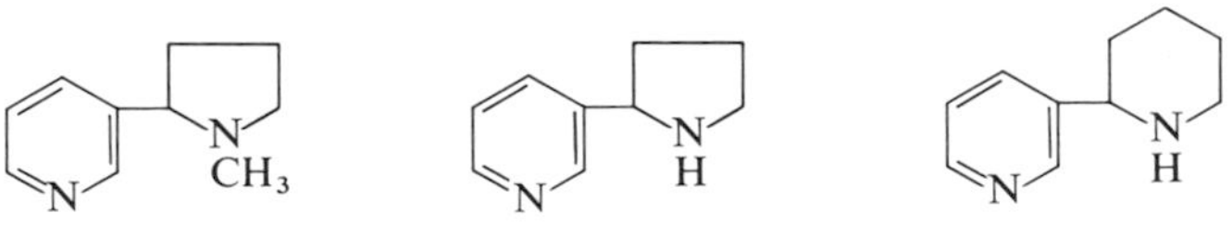

nicotine   nornicotine   anabasine

former in *D. hopwoodii* and the latter commonly predominating in *Nicotiana*. Anabasine is the chief alkaloid of *Anabasis aphylla*, where it occurs from 1 to 2% in the shoots and is also found to about 1% in *Nicotiana glauca*. These nicotinoids are appreciably volatile (nicotine v.p. 0.0425 mm at 25°C) and although colorless liquids when pure rapidly darken upon exposure to air. They are highly basic ($K_{b1} = 1 \times 10^{-6}$, $K_{b2} = 1 \times 10^{-11}$) and readily form salts with acids and many metals. Nicotine sulfate $(C_{10}H_{14}N_2)_2 \cdot H_2SO_4$ is widely used as an insecticide because it is more stable and less volatile.

2. Structure–Activity Relationships. It has been pointed out(*161*) that nicotinoids that have the highest insecticidal action have the highest pKa' and consequently exist largely in the ionized form at physiological pH. This relationship is shown in Table 1-14. This results in the anomaly that the compounds that are most highly ionized react most rapidly with the receptor protein, yet are less able to penetrate through the ionic barrier surrounding the insect nerve synapse. This has been demonstrated by the lack of toxicity of the completely ionized nicotine methiodide and nicotine dimethiodide upon injection into *Musca(161)*. The necessity for the nicotinoids to penetrate the insect nerve in the un-ionized state doubtlessly explains their relative lack of insecticidal action against

TABLE 1-14
RELATIONSHIP OF BASICITY TO RELATIVE TOXICITY OF NICOTINOIDS(*161*)

Nicotinoid		Basicity, pKa	% ionized, pH = 7	Relative toxicity to *Aphis, Musca, Oncopeltus*
l-Nicotine		7.9	88.8	1.0
l-Nornicotine		9.0	99.0	2.0
l-Anabasine		8.7	98.0	10.0
Dihydronicotyrine		7.4	71.5	1–3
Anabaseine		6.7	33.4	0.3–0.6
Myosmine		5.5	3.1	0.03–0.27
Nicotyrine		4.7	0.5	0.08–0.1

many insects whose nervous tissues are protected by a more efficient ion barrier than that of the highly susceptible aphids.

Therefore, the essential requirements for toxicity are (a) a pyridine ring, (b) a highly basic nitrogen at an optimum distance of about 4.2 Å from the pyridyl nitrogen, (c) an unsubstituted α-position of the pyridine ring. These requirements are met by several 3-pyridylmethyl N-dialkylamines, which Yamamoto(*161*) found to be toxic to the housefly.

3. Mode ot Action. Nicotine and anabasine affect the ganglion of the insect central nervous system, facilitating trans-synaptic conduction at low concentrations and blocking conduction at higher levels. The extent of ionization of the nicotinoids plays an important role in both their penetration through the ionic barrier of the nerve sheath to the site of action and in their interaction with the site of action which is believed to be the acetylcholine receptor protein. The similarity in dimensions between acetylcholine and the nicotinium ion is shown below(*6*).

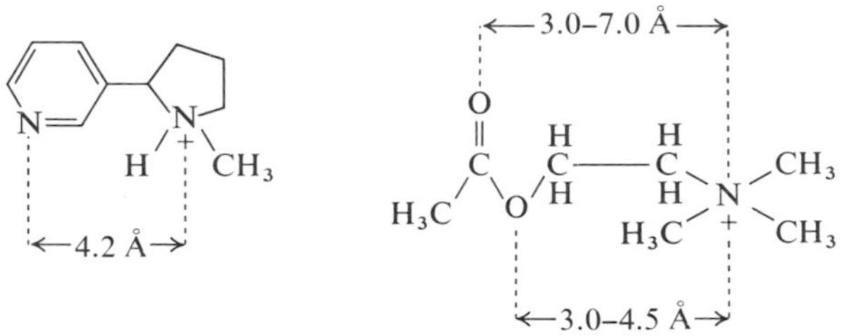

acetylcholine

4. Biological Activity. Nicotine, and to a lesser extent anabasine, are used as contact insecticides for aphids attacking fruits, vegetables, and ornamentals; and as fumigants in greenhouses and against poultry mites. Nicotine sprays commonly contain 0.05 to 0.06% nicotine and dusts 1 to 2% nicotine.

5. Toxicology. Nicotine has an oral LD_{50} to the rat of 30 mg/kg and a dermal LD_{50} to the rabbit of 50 mg/kg. Thus, it is very highly toxic to man and higher animals. However, as every smoker knows, it is very rapidly detoxified and eliminated from the animal body and appears to pose no particular chronic hazard. Its rapid deterioration in light and air favors its use on leafy vegetables within a few days of harvest. Nicotine sulfate is safer and more convenient to use, and the free alkaloid is liberated by the addition of soap, lime, or ammonium hydroxide to the spray solution.

D. Pyrethroids

1. Introduction and Chemistry. The insecticidal properties of pyrethrums from *Chrysanthemum cinerariaefolium* and *C. coccineum* have been known since about 1800. The preparation originated in the Transcaucasus region of Asia where it was used as a flea and louse powder. A century of chemical investigation has disclosed that the insecticidal properties are due to five esters (the *pyrethrins* I and II, and the *cinerins* I and II, and *jasmoline* II) that are present mostly in the achenes of the flowers, ranging from 0.7 to as much as 3% in selected strains.

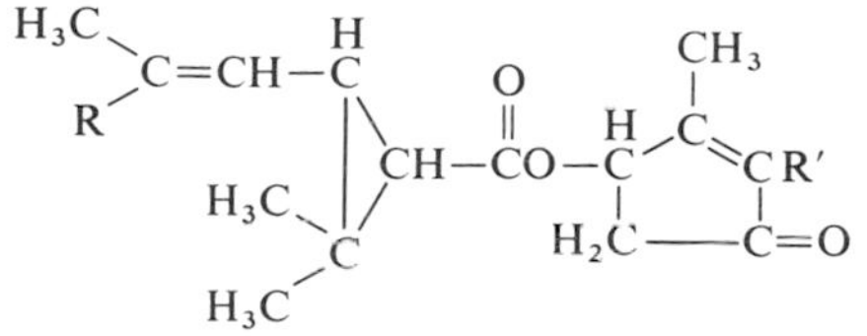

		R=	R'=
pyrethrin	I	CH_3	$-CH_2CH=CHCH=CH_2$
pyrethrin	II	$COOCH_3$	$-CH_2CH=CHCH=CH_2$
cinerin	I	CH_3	$-CH_2CH=CHCH_3$
cinerin	II	$-COOCH_3$	$-CH_2CH=CHCH_3$
jasmolin	II	$-COOCH_3$	$-CH_2CH=CHCH_2CH_3$
allethrin		CH_3	$-CH_2CH=CH_2$

These esters have asymmetric carbon atoms and double bonds in both alcohol and acid moieties. The naturally occurring forms are the *d*-trans-acid esters of the *d*-cis alcohols. The active principles are extracted from the ground flowers by petroleum ether, ethylene dichloride, or other organic solvent, and freed from wax by absorption on carbon to produce 90 to 100% pyrethrins. The pyrethroids are highly unstable to the action of light, air, moisture, and alkali, and residues deteriorate very rapidly after application.

Knowledge of the structure of the pyrethrins together with their high cost stimulated the synthesis of synthetic derivatives. *Allethrin* or *dl*-2-allyl-3-methylcyclopent-2-en-4-ol-1-onyl *dl-cis-trans*-chrysanthemate (see above) was synthesized by La Forge and co-workers in 1949 and is produced commercially by a complex 13-step synthesis. The commercial product is a clear brownish liquid d_{20} 1.005–1.015 containing 75–95% of eight individual optical and geometric isomers whose identity, abundance, and relative toxicity (7) are given in Table 1-15.

TABLE 1-15

COMPOSITION OF ALLETHRIN ISOMERS AND INSECTICIDAL ACTIVITY

	Abundance, %	Relative toxicity to *Musca domestica*
dl-Allylrethronyl *dl*-cis-trans-chrysanthemate		1.00
l-Allylrethronyl *d*-trans -chrysanthemate	12.4	0.58
d-Allylrethronyl *l*-trans -chrysanthemate	12.4	0.14
d-Allylrethronyl *d*-trans -chrysanthemate	22.8	3.37
l-Allylrethronyl *l*-trans -chrysanthemate	22.8	0.02
l-Allylrethronyl *d*-cis -chrysanthemate	8.0	0.33
d-Allylrethronyl *l*-cis -chrysanthemate	8.0	0.14
d-Allylrethronyl *d*-cis -chrysanthemate	6.8	1.77
l-Allylrethronyl *l*-cis -chrysanthemate	6.8	0.06

Other synthetic pyrethroids which have been produced commercially are also esters of chrysanthemic acid

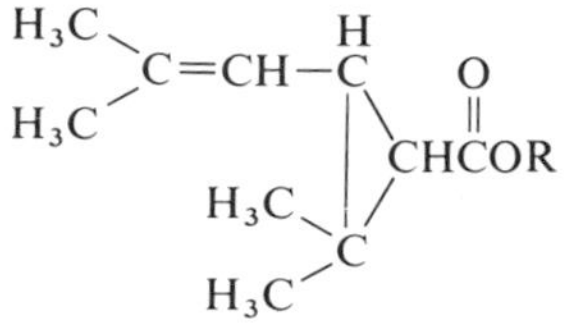

and include(7):   R =

cyclethrin

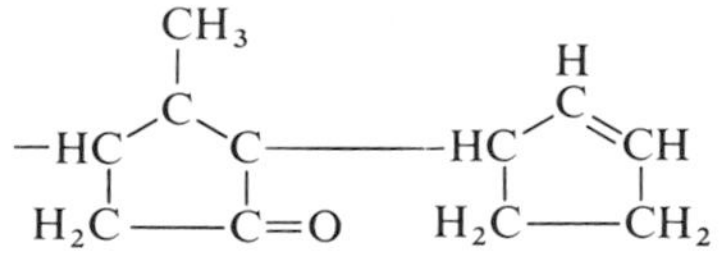

barthrin

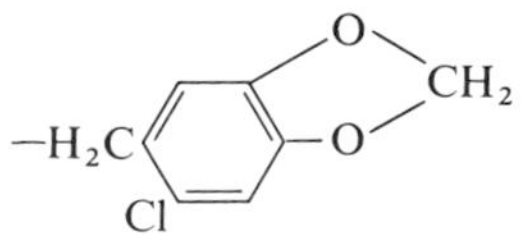

dimethrin

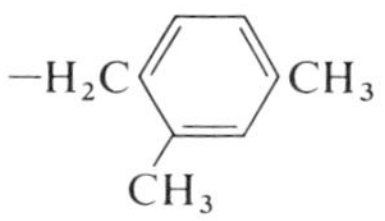

phthalthrin(68)

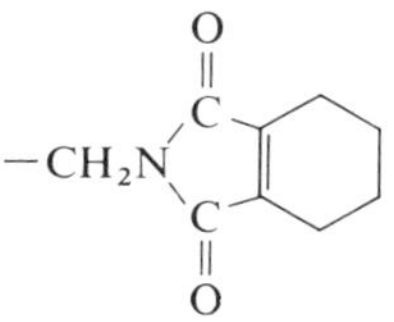

NRDC 104

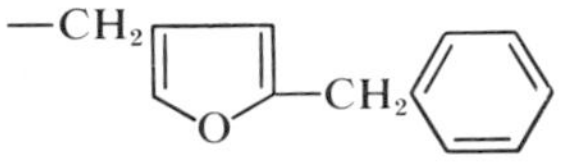

2. Structure–Activity Relationships. There are great variations in the activity of the pyrethroids, associated with optical and geometric isomerism as shown on page 74. The most active isomers are the *d*-trans acid esters of the *d*-cis-alcohols. This, together with the rigorous structural requirements of the isobutenyl and cyclopropane moieties of the chrysanthemic acid portion, suggest that biochemical activity is associated with a specific three-point contact between pyrethroid and biological receptor at (a) the isobutylene group and (b) the dimethyl cyclopropane ring of the acid, and (c) the unsaturated group of the keto alcohol(*95*).

3. Mode of Action. Despite a great deal of investigation, the specific biochemical lesion of pyrethroids intoxication in insects is unknown. The compounds readily penetrate the insect cuticle as shown by the topical LD_{50} to *Periplaneta* of 6.5 $\mu g/g$ vs 6 μg when injected. The pyrethrins when applied to the insect nerve axon produce a rhythmic spontaneous discharge at 0.01 to 0.1 ppm and a reversible blocking of conduction at 1 ppm. Paralyzed insects exhibit characteristic vacuolizations of the nerve tissue.

A characteristic of pyrethroid action on insects is the very rapid knockdown followed by substantial recovery, so that approximately a threefold increase in dosage is required to produce a per cent mortality equal to the per cent knockdown. This recovery from paralysis is the result of rapid enzymatic detoxication in the insect, apparently by "mixed function" oxidases. The detoxication enzyme(s) are inhibited by a number of compounds, especially the methylenedioxybenzene derivatives such as *piperonyl butoxide* or 3,4-methylenedioxypropyl benzyl butyldiethylene glycol ether; *propyl isome* or di-*n*-propyl-2-methyl-6,7-methylenedioxy 1,2,3,4-tetrahydronaphthalene-3,4-dicarboxylate; and *sulfoxide* or 1,2-methylenedioxy-4[2-(octyl sulfinyl)-propyl]-benzene. These synergists used at 10 parts to 1 part of pyrethroid may activate pyrethrins I about 30 times, allethrin 2 to 3 times, and cyclethrin 10 times. Another type of synergist is N-(2-ethylhexyl)-[2.2.1]-5-heptane-2,3-dicarboximide or MGK264. These pyrethrins synergists have considerable importance in "aerosol bombs" and household and cattle sprays and are also effective with carbamate and other insecticides(*95*).

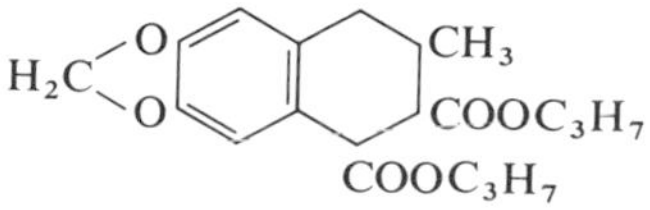

propyl isome 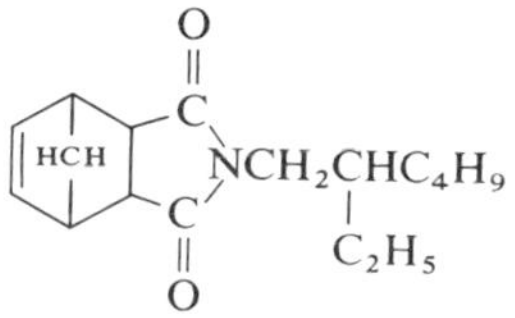   MGK 264

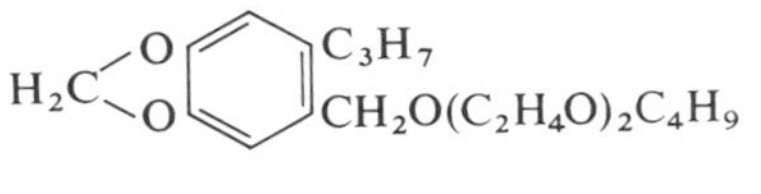

piperonyl butoxide 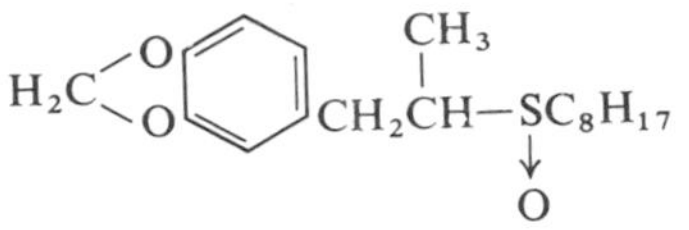   sulfoxide

Winteringham et al.(*158*) showed that female houseflies injected with 2 μg ^{14}C-pyrethrins, detoxified 50% in 24 hr, but when injected with 2 μg of pyrethrins plus 10 μg of piperonyl cyclonene synergist, no detoxication could be detected.

4. Biological Activity. Because of their very rapid knockdown, the pyrethroids have long been used in household, fly, and cattle sprays at about 0.03 to 0.1%, usually with 5 to 10 times this amount of synergist. "Aerosol bombs" commonly contain 0.04 to 0.25% pyrethroid plus synergist. Dimethrin is an effective mosquito larvicide especially for use in cisterns and in drinking water. The pyrethrins and piperonyl butoxide have been used in dust form as a grain protectant and on fruits and vegetables close to harvest.

5. Toxicology. The pyrethroids are of low mammalian toxicity because of their rapid detoxication following oral ingestion. The oral LD_{50} values to the rat are: pyrethrins 820, allethrin 920, cyclethrin 1400, and dimethrin 40,000 mg/kg. The synergists and combinations are also of low toxicity with oral LD_{50} values to the rat of: piperonyl butoxide 11,500; propyl isome 15,000; sulfoxide 2000; and MGK264 2800 mg/kg.

E. Rotenoids

1. Introduction and Chemistry. The earliest recorded use of rotenone as an insecticide was against leaf-eating caterpillars in 1848. However, plants containing these materials have been used as fish poisons for centuries. The principal commercial source of rotenoids are *Derris elliptica* and *D. malaccensis* from Malaya and *Lonchocarpus utilis* and *D. urucu* from South America. A number of toxicants are present in the roots and seeds, the most important of which is rotenone, m.p. 163°C, occurring in *Derris* roots to 4 to 9% and in *Lonchocarpus* to 8 to 11%(*95*).

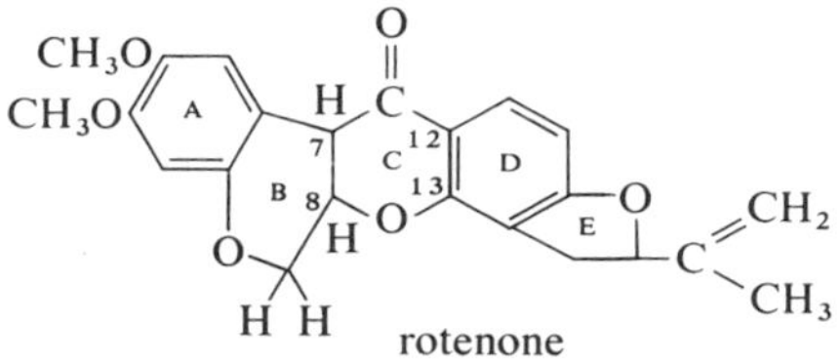

rotenone

Other extractives removed from the ground roots with carbon tetrachloride or ether and comprising 25 to 31% of the ground roots are *elliptone* in which a furan ring replaces ring B of rotenone, *sumatrol* which is 15-hydroxyrotenone, *malaccol* which is 15-hydroxyelliptone, *toxicarol* which has a hydroxy group at carbon 15 and a 2,2-dimethylpyrone ring in place of ring E of rotenone, and *deguelin* which has a hydrogen atom in place of the hydroxy group of toxicarol. Rotenone is, however, from 5 to 10 times as active insecticidally as the other extractives. Rotenone is readily converted to noninsecticidal derivatives by oxidation in the presence of light and air with the formation of first a hydroxy- derivative at C-7 and then dehydration to form dehydrorotenone with a C=C bond between rings B and C. These reactions render rotenone residues innocuous after 5 to 10 days in normal sunlight.

2. Structure–Activity Relationships. The other rotenoids plant constituents deguelin, toxicarol, sumatrol, and elliptone are only about 0.1 to 0.2 as active as rotenone in insecticidal activity. With the exception of hydrogenation of the isopropenyl side chain, which produces dihydrorotenone of nearly equal toxicity, all other chemical alterations greatly decrease the toxicity of the rotenone molecule. Dehydrorotenone and tetrahydrorotenone are nontoxic as are derivatives with major alterations in rings A, B, C, D, and E.

3. Mode of Action. Insects poisoned with rotenone exhibit a steady decline in oxygen consumption and the insecticide has been shown to have a specific action in interfering with the electron transport involved in the oxidation of DPNH to DPN by cytochrome b. Thus rotenone at 1×10^{-7} to $1 \times 10^{-6} M$ inhibits the mitochondrial oxidation by *Periplaneta americana* tissues of L-glutamate, α-ketoglutarate, malate, fumarate, oxalacetate, and citrate, all of which are catalyzed by DPN[37].

4. Biological Activity. Rotenone containing insecticides have been used as (a) dusts of ground roots, (b) dispersible powders, and (c) emulsive extracts for application to edible produce just prior to harvest, and for animal dusts to control ectoparasites and cattle grubs.

5. Toxicology. The oral LD_{50} of rotenone to the rat is 132 mg/kg. The rapid destruction of the compound by light and air renders residues innocuous within a matter of a few days.

F. Botanical Insecticides

In addition to nicotine, pyrethrins, and rotenone, other plant derivatives have shown insecticidal properties and two have attained commercial development[94].

1. Sabadilla. *Sabadilla* is in the ground seeds of *Schoenocaulon offi-cinale*, family Liliaceae of South and Central America. The seeds contain 2 to 4% of a mixture of *veratrine* alkaloids, with the following solubilities in gram per 100 milliliters: water 0.055, alcohol 35, chloroform 144, kerosene 0.1 to 0.2. The two important insecticidal alkaloids are *cevadine*, $C_{32}H_{49}O_9N$, m.p. 205°C, the angelic acid ester of cevine, and *veratridine*, $C_{36}H_{51}O_{11}N$, m.p. 160°C, the veratric acid ester of cevine (8). The seeds are activated by heating to 80°C, moistening with sodium carbonate solution or milling with hydrated lime. The veratrine alkaloids are rapidly destroyed by light but are highly poisonous to higher animals (veratridine has an intravenous LD_{50} to the mouse of 0.42 mg/kg) and are also very irritant to mucous membranes, but sabadilla has a rat oral LD_{50} of 5000 mg/kg.

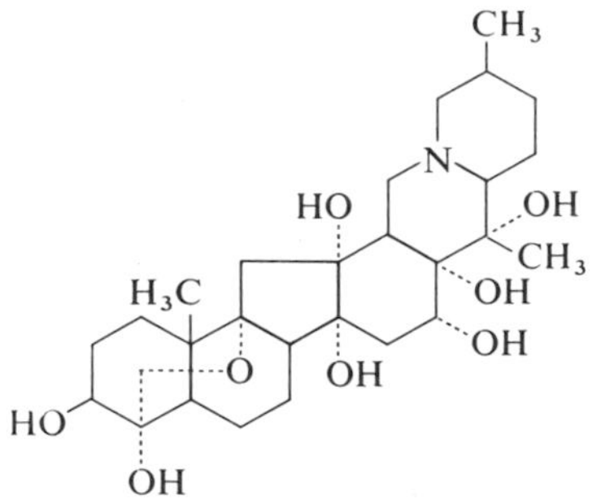

cevine

Sabadilla is used as a spray or dust to control thrips and various true bugs attacking vegetables. It may be applied safely to edible produce before harvest.

2. Ryania. *Ryania* is the ground roots and stem of *Ryania speciosa*, family Flacourtiaceae of South America. The wood contains 0.16 to 0.2% of insecticidal alkaloids of which *ryanodine* $C_{25}H_{35}NO_9$, m.p. 219°C, is the most important(*122*). Ryanodine is soluble in water, alcohol, acetone, and chloroform but not in petroleum oils. Ryania has a rat oral LD_{50} of 750 mg/kg. It is used to control the European corn borer and other lepidopterous larvae attacking fruits.

G. Dinitrophenols

1. Introduction and Chemistry. The dinitrophenols are compounds with a wide range of biocidal action and are useful as insecticides, acaricides, herbicides, and fungicides. Potassium-dinitro-*o*-cresylate, marketed in Germany in 1892, was the first synthetic organic insecticide. The compounds in use today are all derivatives of 4,6-dinitro-2-alkylphenols and their salts or esters:

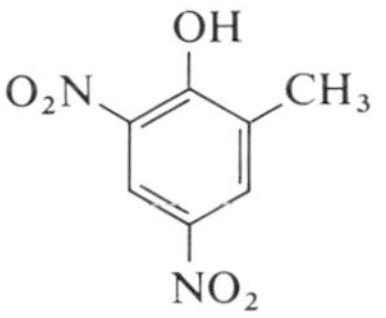

4,6-dinitro-cresol
DNOC

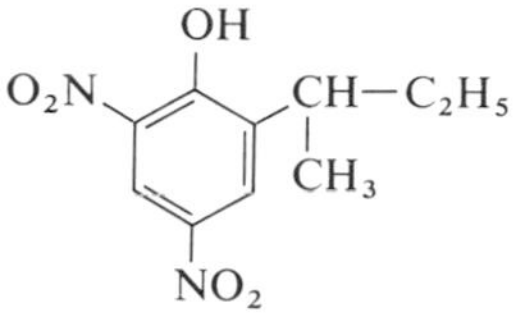

4,6-dinitro-*o*-sec-butylphenol
DNOSBP or *dinoseb*

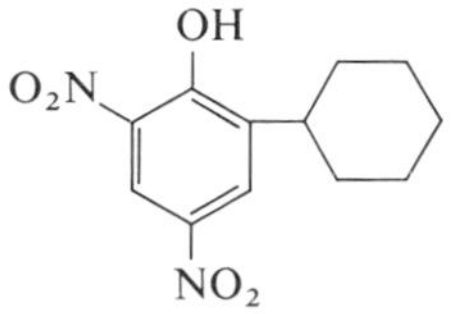

4,6-dinitro-*o*-cyclohexlphenol
DNOCHP or *dinex*

They are yellow, relatively odorless solids, slightly soluble in water, (DNOC, m.p. 85°C, to 0.014%, dinoseb, m.p. 42°C, to 0.0734%, and dinex, m.p. 106°C, to 0.0015%) and readily soluble in organic solvents. The compounds are acidic (DNOC $K_a = 2.5 \times 10^{-6}$) and readily form water soluble ammonium, sodium, potassium, and calcium salts(95). These water soluble salts are extremely toxic to plants and are used as herbicides (Sec. 1-3, I). For insecticidal and acaricidal use, acidity and the consequent phytotoxicity must be reduced. For this purpose the water insoluble triethanolamine and dicyclohexylamine salts have been employed as well as a variety of esters such as the 3,3-dimethylacrylate ester of dinoseb, *bipinacryl*, and the crotonate ester of 4,6-dinitro-2-caprylphenol, *dinocap* or KARATHANE, which is also an effective fungicide against powdery mildew (Sec. 1-2, B).

2. Structure–Activity Relationships. The structural requirements for toxicity are highly specific, the 4,6-dinitro-2-alkylphenols being the most toxic and activity depending on the length of the alkyl chain as shown in Table 1-16.

3. Mode of Action. The dinitrophenols are biologically active because of their ability to uncouple oxidative phosphorylation. As a result poisoned insects undergo pronounced increases in the rate of respiration, which may reach 3 to 10 times normal, and they die from metabolic exhaustion because of their inability to utilize the energy provided by respiration and glycosis for the conversion of ortho-phosphate to high-energy phosphate bonds as in ATP. Such inhibition of oxidative phosphorylation in muscle tissues of *Musca* and *Periplaneta* is produced in vitro by 10^{-6} M DNOC(3).

4. Biological Activity. The dinitrophenols are used as dormant ovicidal sprays for the control of mites, aphids, and scale insects. DNOCHP dicyclohexylamine salt has been used as a foliage spray to control mites

TABLE 1-16
TOXICITY OF DINITROPHENOLS TO 5TH INSTAR
SILK WORM *Bombyx mori* LARVA(66)

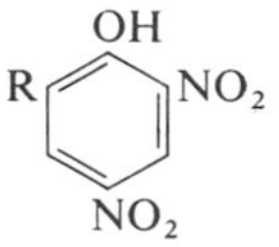

OH R⌬NO$_2$ NO$_2$	Oral LD$_{50}$, μg/g
R=CH$_3$	49
C$_2$H$_5$	29
C$_3$H$_7$	18
C$_4$H$_9$	9
C$_5$H$_{11}$	8
C$_6$H$_{11}$	4
C$_7$H$_{13}$	4
C$_8$H$_{15}$	10
cyclo-C$_5$H$_9$	9
cyclo-C$_6$H$_{11}$	7

on citrus and walnuts. DNOC in oil sprays is used to control grass-hoppers and as a dust for chinch bug barriers.

5. Toxicology. The dinitrophenols are highly toxic to mammals with the following oral LD$_{50}$ values to the rat: DNOC 30, DNOCHP 80, DNOSBP 37, bipinacryl 165, and dinocap 980 mg/kg. DNOC, for example, is a cumulative poison with a dangerous dose in man of about 2 g, and is excreted very slowly. These compounds are uncouplers of oxidative phosphorylation and increase the oxidative metabolism and heat production of the body. In the diet the tolerated concentrations in parts per million are DNOC 100, DNOSBP 100, and DNOCHP 500. Signs and symptoms of poisoning include nausea, gastric distress, restlessness, flushing, sweating, deep and rapid respiration, tachycardia, cyanosis, collapse, and coma. Treatment consists of ice bath to reduce fever, administration of oxygen, and infusion of large quantities of iso-tonic saline(55).

H. Organothiocyanates

1. Introduction and Chemistry. β-Butoxy-β'-thiocyanodiethyl ether (LETHANE 384) was the first widely used synthetic organic insecticide and was marketed in 1932 as a rapid knockdown agent for the control of household insects. A wide variety of organic compounds incorporating the thiocyano —SCN or isothiocyano CNS group have useful insecticidal properties. In addition to β-*butoxy*-β'-*thiocyanodiethyl ether* $C_4H_9OCH_2$-$CH_2OCH_2CH_2SCN$ b.p. 124°C at 0.25 mm, β-*thiocyanoethyl laurate* $C_{11}H_{23}COOCH_2CH_2SCN$, β,β'-*dithiocyanodiethyl ether* $NCSCH_2CH_2$-OCH_2CH_2SCN, *lauryl thiocyanate* $C_{12}H_{25}SCN$, and *isobornyl thiocyano acetate* (THANITE) have been marketed as insecticides(95)

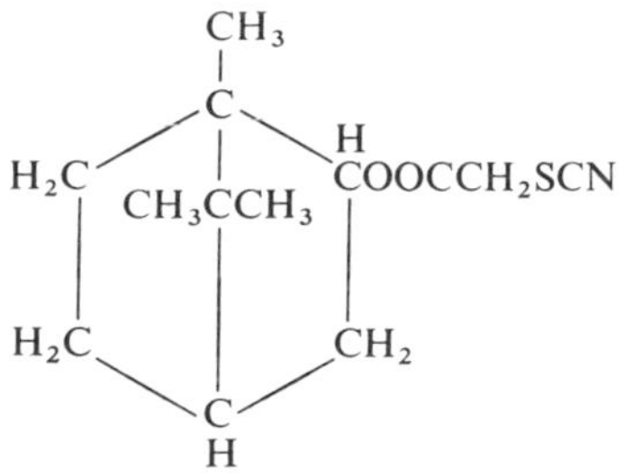

THANITE, isobornyl thiocyano acetate

The thiocyanates readily isomerize to the isothiocyanates upon heating.

2. Mode of Action. These thiocyanates have a rapid paralytic action to insects that is clearly attributable to the action of the thiocyano group. These compounds represent the classical concept of a conductophoric group, which produces an effective concentration of the molecule at the site of action and a toxophoric group (SCN), which produces the biochemical lesion. The alkyl thiocyanates provide an interesting example of this as given in Table 1-17.

The maximum activity of the C_{12} compound can be explained by the distribution coefficients (aqueous phase/nonaqueous phase) which fall in the same geometric progression as the relative toxicities. In an ascending series, the rate of decrease in aqueous solubility is greater than the rate of decrease in distribution coefficient. Hence, a point of maximum activity is reached beyond which effectiveness falls off very rapidly with increasing chain length(95).

TABLE 1-17
TOXICITY OF ALKYL THIOCYANATES

	Green chrysanthemum aphid LC_{50}, ppm
$C_6H_{13}SCN$	833
$C_8H_{17}SCN$	400
$C_{10}H_{21}SCN$	357
$C_{12}H_{25}SCN$	333
$C_{14}H_{29}SCN$	370
$C_{16}H_{33}SCN$	588

3. Biological Activity. The thiocyanates are used principally as household aerosols and sprays for the control of flies, mosquitoes, cockroaches, and as dairy and livestock sprays for fly control. Their rapid knockdown makes them particularly useful for the control of flying insects but they tend to have rather unpleasant odors.

4. Toxicology. The thiocyanates are of relatively low toxicity to mammals with oral LD_{50} values to the rat of 250 for β-butoxy-β'-thiocyanodiethyl-ether and 1600 mg/kg for THANITE. They are, however, somewhat volatile and should always be used with good ventilation.

I. DDT and Analogs

1. Introduction and Chemistry. Although *DDT* or dichlorodiphenyl trichloroethane was synthesized by Zeidler in 1874, its insecticidal properties were discovered by Müller only in 1939(*108*). The proper chemical name is 2,2-bis-(*p*-chlorophenyl)-1,1,1-trichloroethane, and this white powder has become the most widely used pesticide with an annual production in excess of 160,000,000 lb. Pure DDT, m.p. 108–109°C, v.p.$_{20}$ 1.5×10^{-7} mm Hg is virtually insoluble in water (0.00001%) but is soluble in many organic solvents — xylene 60%, carbon tetrachloride 47%, ethyl acetate 68%, and kerosene 8%. In the presence of strong alkali, DDT is dehydrochlorinated to the noninsecticidal 2,2-bis-(*p*-chlorophenyl)-1,1-dichloroethylene or DDE, m.p. 85°C; and this can be oxidized to *p,p'*-dichlorobenzophenone, reactions which account for much of the decomposition of DDT residues(*95*).

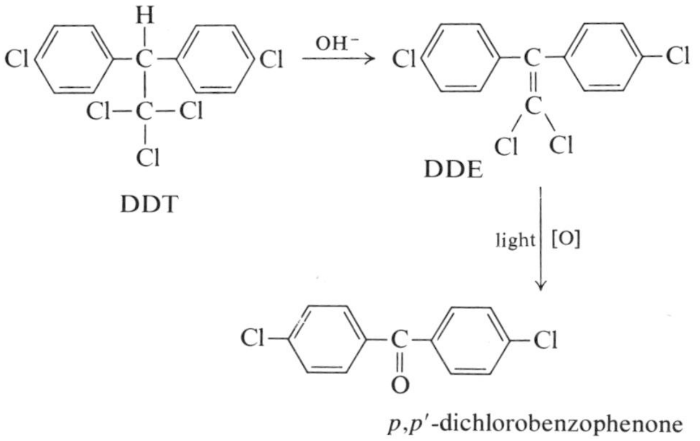

p,p'-dichlorobenzophenone

A considerable number of close relatives of DDT have insecticidal action and some have useful practical application as insecticides. *DFDT* or "Gix" 2,2-bis-(*p*-fluorophenyl)-1,1,1-trichloroethane, m.p. 45°C, was used as an insecticide by the German army in World War II. 2,2-

Bis-(p-methoxyphenyl)-1,1,1-trichloroethane or *methoxychlor*, m.p. 89°C, is an especially useful compound as it is not concentrated in animal fat. The 2,2-bis-(p-chlorophenyl)-1,1-dichloroethane (*TDE* or *DDD*), m.p. 112°C, has a slightly different spectrum of activity. The corresponding 2,2-bis-(p-ethylphenyl)-1,1-dichloroethane, m.p. 56°C, is PERTHANE.

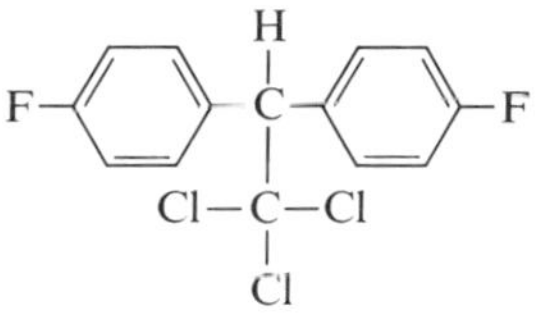

DFDT 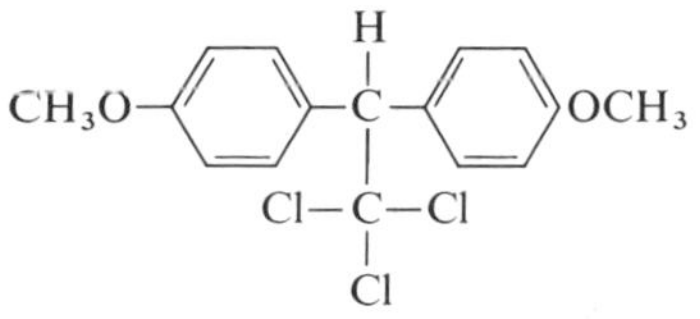   methoxychlor

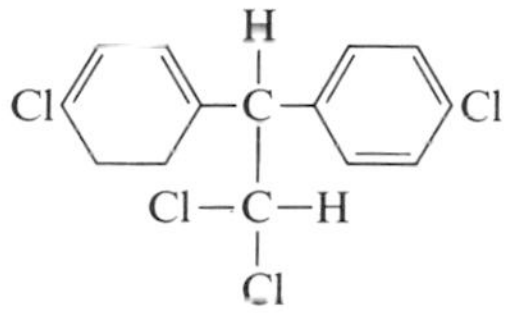

TDE 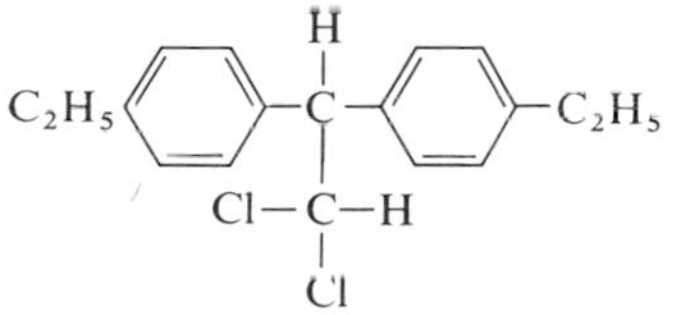   PERTHANE

Ortho-chloro-DDT or 2-(4'-chlorophenyl)-2-(2', 4'-dichlorophenyl)-1,1,1-trichloroethane m.p. 103°C, although of lesser activity than DDT, is not detoxified by DDT'ase and is, therefore, highly effective against DDT-resistant insects. *Deutero-DDT* in which the α-hydrogen is replaced by deuterium has proven more effective than DDT and is active against DDT-resistant *Aedes aegypti* mosquitoes for the same reason.

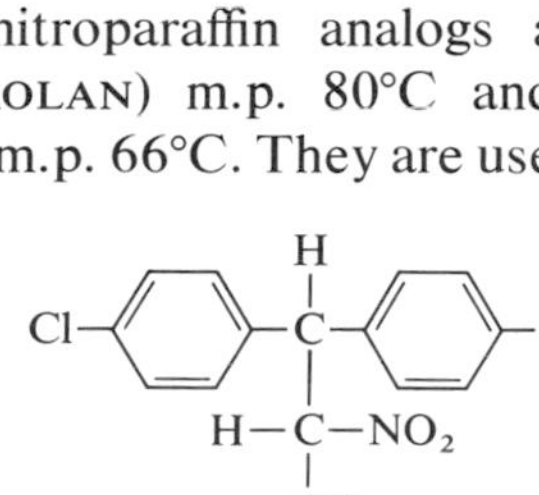

ortho-chloro-DDT 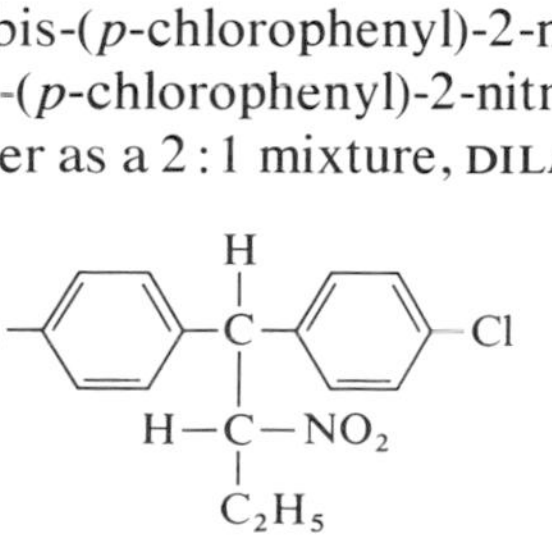   deutero-DDT

Two nitroparaffin analogs are 1,1-bis-(p-chlorophenyl)-2-nitro-propane (PROLAN) m.p. 80°C and 1,1-bis-(p-chlorophenyl)-2-nitrobutane, (BULAN) m.p. 66°C. They are used together as a 2:1 mixture, DILAN.

PROLAN    BULAN

All of these DDT analogs are unstable in strongly alkaline solutions, and share the general chemical and biochemical properties of DDT.

2. Structure–Activity Relationships. Although many hundreds of DDT-analogs have been evaluated as insecticides, only a few such as TDE, methoxychlor, and Gix have proved effective enough or cheap enough for commercial development. Some of the structure–activity relationships for the common DDT analogs are shown in Table 1-18. For maximum activity, the DDT-type molecule must be substituted in the p,p' position with halogen, or short-chain alkyl or alkoxy groups. The nature of the substituents for the ethane moiety is also critical and this portion of the molecule needs to have an optimum complementarity to a hypothetical protein receptor. This is expressed by activity in the following order: $CCl_3 > CHCl_2 > CH(NO_2)CH_3 > C(CH_3)_3$. The substituent on the α-carbon not only affects the optimum configuration of the molecule but also influences detoxication in the insect which can be reduced by α-D or α-F analogs.

3. Mode of Action. DDT has a highly specific insecticidal action as the oral LD_{50} to the rat is 200 mg/kg and the dermal LD_{50} is 3000 mg/kg. Its insecticidal activity is related to the ease with which it is absorbed by

TABLE 1-18
EFFECT OF STRUCTURE ON THE TOXICITY OF DDT ANALOGS TO INSECTS(98)

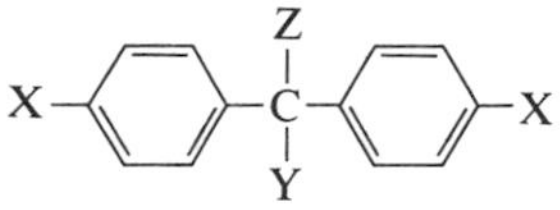

X	Y	Z	Topical LD_{50} fly *Musca domestica*, mg/kg	LD_{50} mosquito larva *Culex quinque fasciatus*, ppm
H	CCl_3	H	> 500	1.1
F	CCl_3	H (Gix)	5.0	0.074
Cl	CCl_3	H (DDT)	1.6	0.070
Br	CCl_3	H	1.9	0.018
CH_3	CCl_3	H	9.0	0.080
CH_3O	CCl_3	H (methoxychlor)	3.4	0.067
NO_2	CCl_3	H	> 500	> 10
Cl	$HCCl_2$	H(TDE)	6.5	0.038
Cl	$=CCl_2$	–(DDE)	> 500	> 10
Cl	CF_3	H	> 500	2.2
Cl	CBr_3	H	> 500	0.55
Cl	$C(CH_3)_3$	H	30	0.39
Cl	$CH(NO_2)CH_3$	H	8.5	0.064
Cl	CCl_3	Cl	> 500	> 10
Cl	CCl_3	F	125	0.092
Cl	CCl_3	D	2.3	0.016

the insect cuticle, e.g., to *Periplaneta americana* the topical LD_{50} is 10 mg/kg compared to the injected LD_{50} of 7. Despite years of intensive study, the critical biochemical lesion is unknown. DDT affects peripheral sensory organs to produce violent trains of afferent impulses that produce hyperactivity and then convulsion of the insect. The paralysis and death that ensue are thought to occur from metabolic exhaustion or from elaboration of a naturally occurring neurotoxin.

DDT in insects is metabolized by dehydrochlorination to form DDE or by α-hydroxylation to form 2,2-bis-(*p*-chlorophenyl)-2-hydroxy-1,1,1-trichlorethane (KELTHANE, Sec. 1-4, θ). In mammals, DDT is dehydrochlorinated to DDE and this is converted to water soluble bis-(*p*-chlorophenyl) acetic acid, DDA, which is excreted in urine.

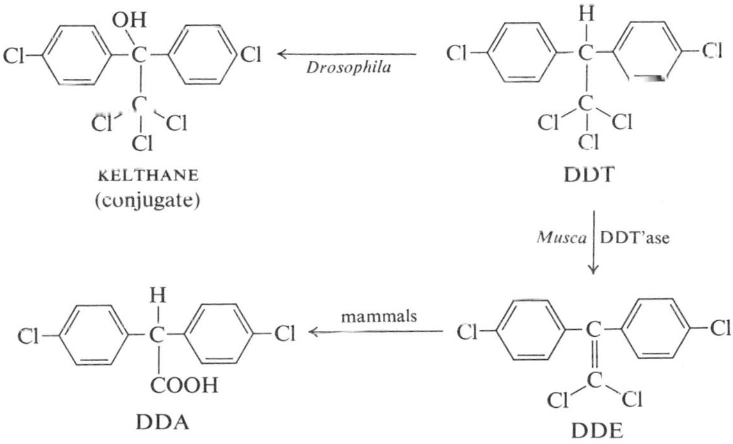

KELTHANE
(conjugate)

DDT

DDA

DDE

Insect resistance to DDT has become widespread in many species. This has been shown to be attributable to selection of naturally occurring genotypes that possess enhanced amounts of the enzyme DDT'ase, which has been isolated and purified(*81*) and which detoxifies DDT to the inactive DDE. Competitive inhibitors of DDT'ase such as 1,1-bis-(*p*-chlorophenyl) ethanol m.p. 69.5–70°C, *DMC* and N,N-dibutyl-*p*-chlorobenzenesulfonamide m.p. 33°C, *Antiresistant* have sometimes been used successfully as synergists

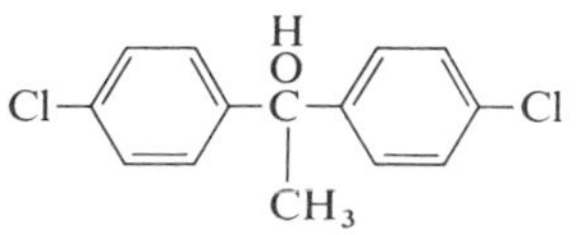

DMC

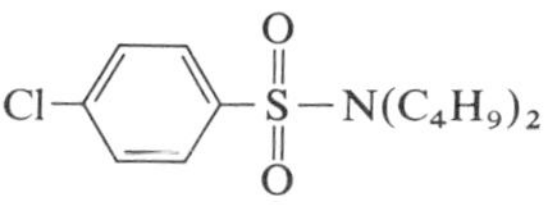

"Antiresistant"

with DDT to restore activity against resistant races of the housefly.

A peculiar and unfortunate problem encountered in the use of DDT is its affinity for lipid tissues in animals. When ingested in trace amounts, DDT is concentrated and stored in fatty tissues at a rate of 10 to 20 times that of ingestion and these storage deposits are only slowly converted to DDE and ultimately excreted as DDA.

4. Biological Activity. DDT is the most permanent and durable of all contact insecticides because of its insolubility in water, its very low vapor pressure, and its resistance to destruction by light and oxidation. Residues of DDT applied to indoor locations may remain effective for periods as long as a year. Thus, DDT applied at 2 g/m², has become the principal weapon in the residual spraying of dwellings for malaria eradication, and is widely used for the control of other mosquito, fly, louse, and flea vectors of human and animal disease.

DDT is also highly persistent on foliage and has been employed for the control of hundreds of species of insect pests of orchard, garden, field, and forest. It is generally applied as a dust or a water spray of suspension or emulsion of DDT at 0.1 to 5%.

5. Toxicology. DDT is relatively nontoxic to higher animals with an oral LD_{50} of 200 and a dermal LD_{50} of 3000 mg/kg to the rat. The comparable oral LD_{50} values for the other DDT analogs are: TDE 3400, methoxychlor 6000, PERTHANE 8170, PROLAN 4000, and BULAN 330 mg/kg (*55*). Hundreds of millions of pounds of DDT are used in human habitations throughout the world in the malaria eradication program without any evidence of adverse effects.

However, despite its evident safety, the use of DDT upon crops intended for animal forage or in forest areas, soils, etc., is objectionable because of its stability and lipoid solubility, which cause it to accumulate in animal tissues and to be magnified in concentration through food chains. Methoxychlor is largely free from these difficulties and is favored for many applications.

J. Lindane

The active constituent of benzene hexachloride is γ-1,2,3,4,5,6-hexachlorocyclohexane or *lindane*, which has the following structural formula:

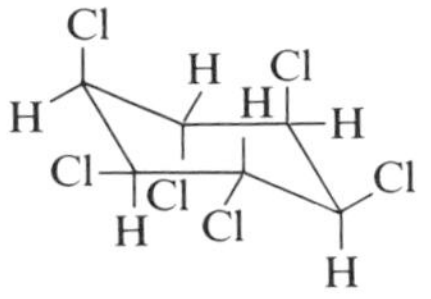

lindane

This compound was first prepared by Faraday in 1825, but its insecticidal properties were not discovered until about 1940 when British and French workers independently developed the material. Benzene hexachloride is prepared by the chlorination of benzene in the presence of sunlight. The crude product is a greyish or brownish amorphous solid with a very characteristic musty odor and begins to melt at 65°C. It consists of 10 to 18% of the active γ-isomer (configuration *aaaeee*), m.p. 112°C together with at least four other nearly inactive stereoisomers α-isomer (*aaeeee* and *aeeeea*), m.p. 157°C, 55 to 70%; β-isomer (*eeeeee*), m.p. 309°C, 5 to 14%; δ-isomer (*aeeeee*), m.p. 138°C, 6 to 8%; ε-isomer (*aeeaee*), m.p. 219°C, 3 to 4%, and a trace of η-isomer (*aeaaee*), m.p. 90°C. Also present are a heptachlorocyclohexane, up to 4%, and a trace of an octachlorocyclohexane, which are insecticidally inactive. The γ-isomer is by far the most toxic of the isomers to insects, being from 500 to 1000 times as active as the α-isomer and about 5000 to 10,000 times as active as the δ-isomer, while the β- and ε-isomers are nontoxic. The pure γ-isomer has a slight aromatic odor, d. 1.85, v.p. 9.4×10^{-6} mm at 20°C. It is very stable to the action of heat, light, and oxidation and can be burned without appreciable decomposition, but is readily decomposed by alkaline materials to form principally 1,2,4-trichlorobenzene and 3 moles of hydrogen chloride. The solubilities of the γ-isomer in grams per 100 g of solvent at 20°C are: acetone, 43; benzene, 29; carbon tetrachloride, 7; cyclohexanone, 37; ethyl acetate, 36; ethyl alcohol, 6; ethylene dichloride, 29; diesel oil, 4; deodorized kerosene, 2; dioxane, 31; mineral oil, 2; xylene, 25; and water, 0.001 (*95*).

The crude benzene hexachloride has been used extensively as a soil poison, and seed treatment, as a toxicant for grasshopper control, and against cotton insects. Great care should be exercised in applying benzene hexachloride to edible crops or to soil in which such crops are to be grown as severe off-flavors of the produce may result. This difficulty has been largely overcome by marketing the 99% pure γ-isomer, *lindane*, for use on food crops and as a household and public health insecticide. Lindane is of moderate toxicity to mammals with an oral LD_{50} of about 100 and a dermal LD_{50} of about 1000 mg/kg to the rat. Its mode of action is unknown but the specific toxicity of the γ-isomer suggests that like DDT it may interact with the pores of the lipoprotein structure of the insect nerve causing distortion and consequent excitation of nerve impulse transmission.

K. Cyclodienes

1. Introduction and Chemistry. The cyclodiene insecticides are highly chlorinated cyclic hydrocarbons with endomethylene bridged structures prepared by the Diels–Alder diene reaction. The development of these materials resulted from Hyman's discovery in 1945 of chlordene, the adduct of hexachlorocyclopentadiene and cyclopentadiene (70).

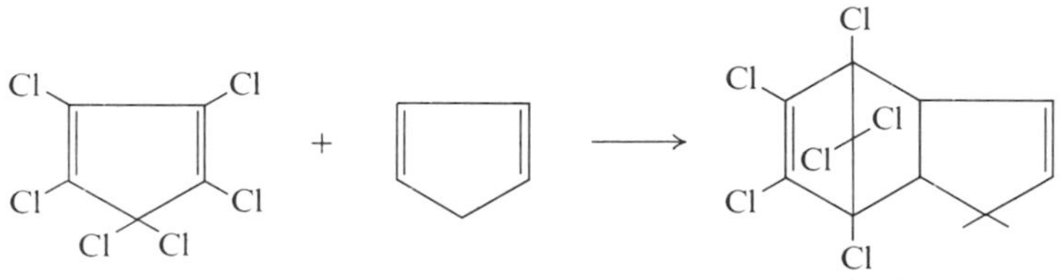

chlordene

a. CHLORDANE AND HEPTACHLOR. The chlorination of chlordene with SO_2Cl_2 adds two chlorines across the double bond of the five-membered ring to form two isomers of *chlordane* or 2,3,4,5,6,7,8,8-octochloro-2,3,-3a,4,7,7a-hexahydro-4,7-methanoindene: α-cis, m.p. 106.5°C, and β-trans, m.p. 104.5°C. The technical product is a brownish liquid, b.p. 175°C, at 2 mm Hg, d. 1.61, which contains about 60% of these isomers plus variable amounts of *heptachlor* or 1,4,5,6,7,8,8 heptachloro-3a,4,7,7a-tetrahydro 4,7-methanoindene m.p. 95°C, v.p. 3×10^{-4} mm at 25°C. Technical chlordane is dehydrohalogenated in the presence of alkali and is soluble in kerosene, xylene, esters, ketones, ethers, but is insoluble in water (95).

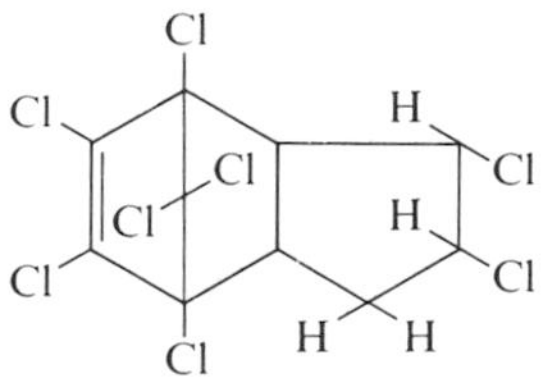

β-chlordane 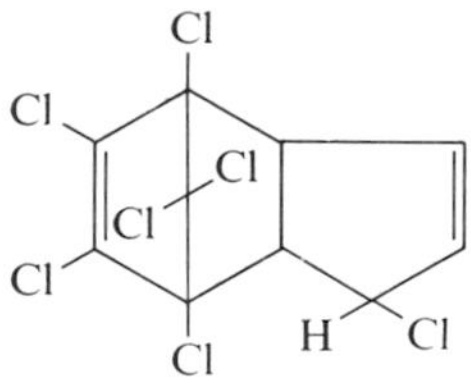   heptachlor

Heptachlor residues in and on plants and in animal tissues are slowly converted to *heptachlor epoxide*, m.p. 159°C, which is of about equivalent insect and animal toxicity. Hydrogenation of heptachlor gives β-dihydroheptachlor, m.p. 135°C, which retains high insecticidal activity but has much reduced mammalian toxicity (22).

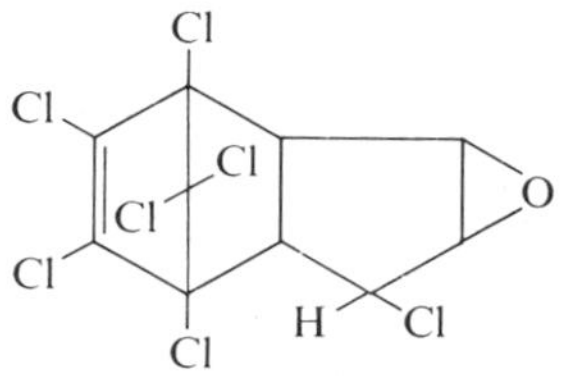

heptachlor epoxide 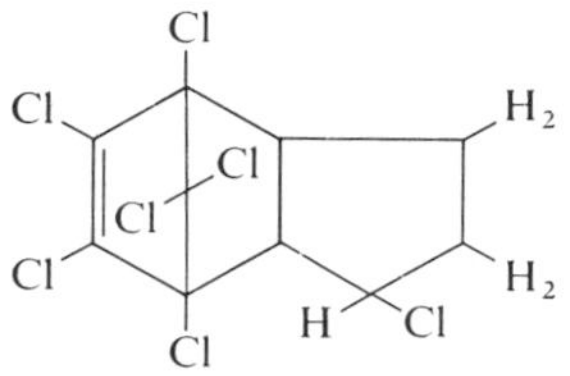   dihydroheptachlor

Isobenzan or TELODRIN or 1,3,4,5,6,7,8,8-octachloro-3a,4,7,7a-tetrahydro-4,7-methanophthalan, m.p. 122°C, v.p. 3×10^{-6} mm at 20°C, is closely related to heptachlor.

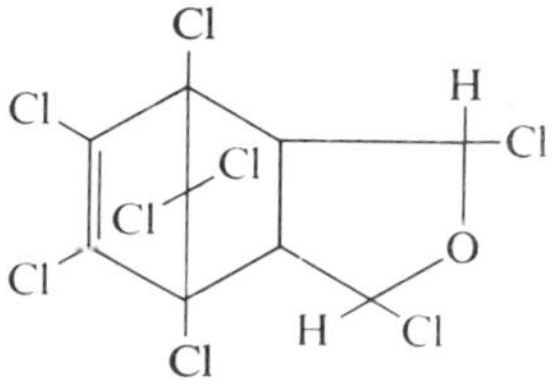

isobenzan

Endosulfan or THIODAN is the adduct of hexachlorocyclopentadiene and 1,4-dihydroxy-2-butene subsequently chlorinated with SO_2Cl_2 to produce 6,7,8,9,10,10-hexachloro-1,5,5a,6,9,9a-hexahydro-6,9-methano-2,4,3-benzodioxathiepin-3-oxide. The technical material is a brownish solid, m.p. 70–100°C, consisting of about 4 parts of α-*cis*-isomer, m.p. 108°C, and 1 part of β-*trans*-isomer, m.p. 206°C. The α-isomer is slowly converted to the more stable β- form at high temperatures and both isomers are slowly oxidized in air and in biological systems and are rapidly oxidized by peroxides or permanganate to endosulfan sulfate m.p. 181°C. In acidic media the isomers hydrolize to form endosulfan alcohol, m.p. 203°C, which has the endo-cis configuration.

Alodan or 1,2,3,4,7,7-hexachloro-bis-5,6-chloromethyl-bicyclo-[2.2.1]-2-heptene, m.p. 105°C, has a very low mammalian toxicity and is used to control animal ectoparasites. *Mirex* is the dimer of hexachloro-cyclopentadiene 1,2,3,4,5,5,6,7,8,9,10,10-dodecachloro-octahydro-1,3,4-methano-2H-cyclobuta-[c,d]-pentalene. Its 2-keto-derivative, m.p. 349°C dec., is KEPONE. These compounds are largely stomach poisons used in baits.

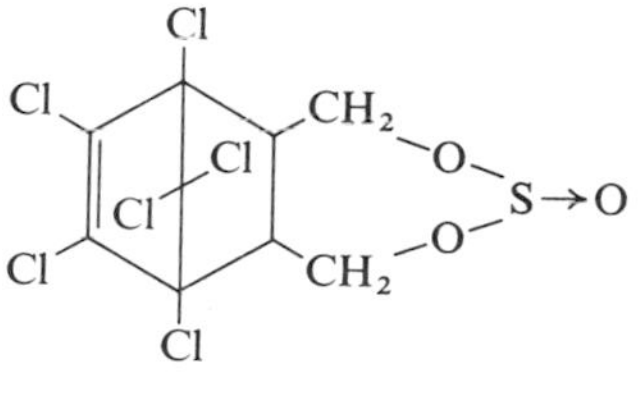

endosulfan 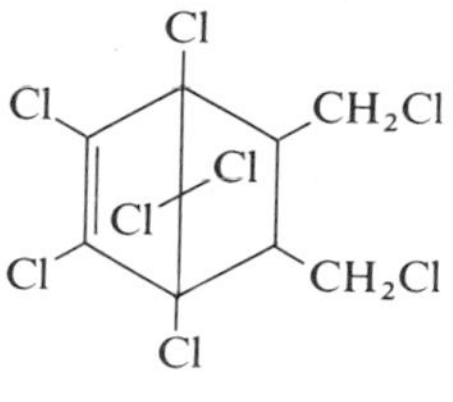   alodan 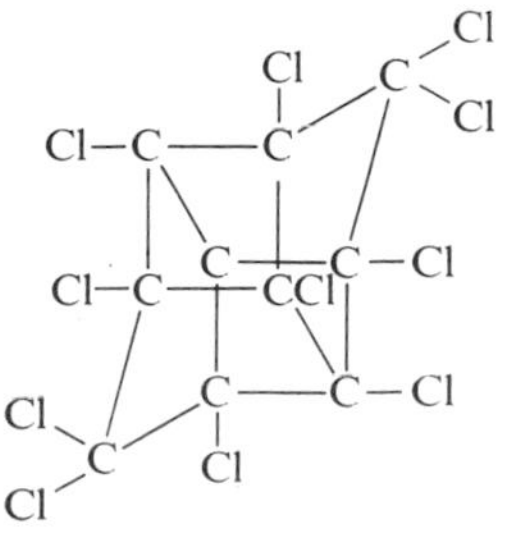   mirex

b. ALDRIN AND DIELDRIN. Many other variations of the Diels–
Alder reaction with hexachlorocyclopentadiene produce useful insecticides. The adduct with bicycloheptadiene is *aldrin* or 1,2,3,4,10,10-
hexachloro-1,4,4a,5,8,8a-hexahydro-1,4-*endo,exo*-5,8-dimethanonaph-
thalene, m.p. 104°C. Epoxidation of aldrin with peracetic or perbenzoic
acids forms the 6,7-epoxy-derivative *dieldrin*, 1,2,3,4,10,10-hexachloro-
6,7-epoxy-1,4,4a,5,6,7,8,8a-octahydro-1,4-*endo,exo*-5,8-dimethanonaph-
thalene, m.p. 176°C. Aldrin is slowly converted to dieldrin in plant and
animal tissues and in soil.

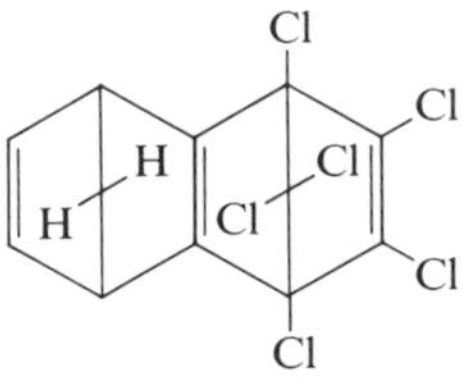

aldrin 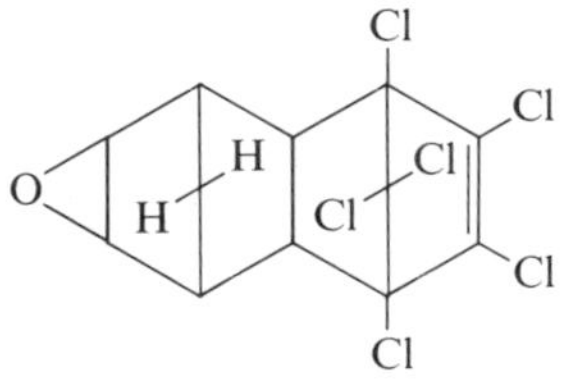   dieldrin

Endrin or 1,2,3,4,10,10-hexachloro-6,7-epoxy-1,4,4a,5,6,7,8,8a-
octahydro-1,4-*endo,endo*-5,8-dimethanonaphthalene, m.p. 245°C dec.,
is the endo,endo-isomer of dieldrin and differs in the spatial arrangement
of the two rings (*95*).

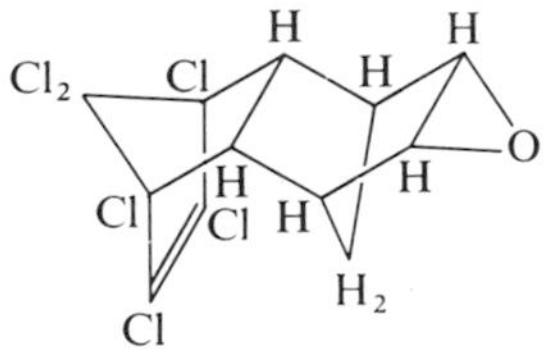

dieldrin 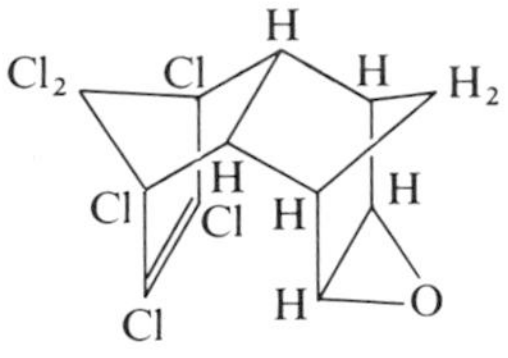   endrin

2. Structure–Activity Relationships. The relative toxicities to insect and
mammal for some of the important cyclodiene insecticides are given in
Table 1-19. The stereochemical configuration of the rings, e.g., endo,exo-
or endo,endo- may be varied by the method of synthesis and has a substantial effect upon the spectrum of insecticidal activity. The activities
of the various adducts of hexachlorocyclopentadiene vary enormously
with specific chemical structure and have been reviewed by Soloway
(*133*). He concludes that the highest activity is shown by those compounds that contain a polychlorinated center together with another
electronegative site such as a double bond or Cl, O, S, or N atom. The
positioning of these two electronegative centers has been shown by

TABLE 1-19

RELATIONS OF STRUCTURE AND TOXICITY FOR CYCLODIENE INSECTICIDES(*100*).

	Musca domestica topical LD_{50}, $\mu g/g$	*Culex pipiens quinque fasciatus* larval LC_{50}, ppm	Rat oral LD_{50}, mg/kg
Aldrin	3.0	0.0055	39–60
Alodan	10.0	0.076	1500
Dieldrin	0.9	0.0078	46
Endrin	3.0	0.014	7.5–7.8
α-Endosulfan	3.9	0.077	18–43
Heptachlor	2.2	0.0056	100–162
Heptachlor epoxide	1.0	0.0063	
Isobenzan	1.1	0.0022	4.8 5.5

molecular models to be similar in distance and in orientation for the various active compounds.

3. Mode of Action. The symptoms of poisoning in insects are hypersensitivity, hyperactivity with violent bursts of convulsions, and finally complete prostration with convulsive movements. The site of these disturbances lies in the ganglia of the central nervous system rather than in the peripheral nerves as with DDT(*43*). The precise biochemical lesion responsible for this toxic action is unknown. Perhaps the most promising theories related to the absorption and critical fit of the active compounds into the pores of nerve membranes to produce convulsant action.

The cyclodienes are readily absorbed by the insect cuticle and the topical LD_{50} values to *Periplaneta americana* for aldrin, 1.9; and for dieldrin 1.3 $\mu g/g$, are only slightly larger than the respective injected LD_{50} values of 1.5 and 1.1 $\mu g/g$. In insects as in other living systems, aldrin and heptachlor are readily converted to their epoxides, dieldrin and heptachlor epoxide.

4. Biological Activity. Chlordane is used for the control of cockroaches, ants, termites and other household pests, soil insects, and certain pests of vegetable and field crops. Heptachlor has much the same spectrum of effectiveness and is also used for grasshopper control. Aldrin is a broad spectrum insecticide used for the control of insect pests of fruits, vegetables, cotton, and as a soil insecticide and for seed treatment. Dieldrin is more stable and residual and is used in addition for the control of grasshoppers, insects of public health importance, termites, and for mothproofing. Endrin is especially used for control of lepidopterous larvae attacking cotton, field, and vegetable crops. THIODAN is a broad spectrum insecticide for pests of fruits, vegetables, and ornamentals.

The oxidative conversion of heptachlor and aldrin to their much more stable epoxides is an important reaction in both plant and animal tissues. On alfalfa the half-lives of the parent materials are less than 1 day but that of heptachlor epoxide is 8 days, and of dieldrin is 7 days. Heptachlor fed to animals is concentrated about 20 times and stored as heptachlor epoxide. Thus, these reactions may pose serious problems of environmental contamination.

5. Toxicology. The acute oral LD_{50} values to the rat for the important cyclodienes are given in Table 1-19. These compounds range from the extremely toxic TELODRIN LD_{50} 5 mg/kg to β-dihydroheptachlor LD_{50} > 4000 mg/kg. Most of the cyclodienes are readily absorbed by the mammalian skin and their dermal LD_{50} values are only about 2 times the oral LD_{50} values. The cyclodienes are primarily nervous system stimulants and cause reflex excitability, convulsions, brachycardia, and vasodepression. Phenobarbital and pentobarbital are effective antidotes and recovery is usually prompt with no after effects (55).

The chronic toxicity is more subtle. Expoxidation and storage in animal fat has already been mentioned, and this leads to undesirable biological magnification in food chain organisms.

L. Chlorinated Terpenes

A group of incompletely characterized insecticidal compounds has been produced by the chlorination of the naturally occurring terpenes. *Toxaphene* is prepared by the chlorination of the bicyclic terpene, camphene, to contain 67 to 69% chlorine, and it has the empirical formula $C_{10}H_{10}Cl_8$. The technical product is a yellowish, semicrystalline gum, m.p. 65 to 90°C, d.^{25} 1.64, and is a mixture of isomers. Toxaphene is unstable in the presence of alkali, upon prolonged exposure to sunlight, and at temperatures above 155°C, liberating hydrogen chloride and losing some of its insecticidal potency. It is very soluble in organic solvents, the solubilities in g/100 ml of solvent being benzene, carbon tetrachloride, ethylene dichloride, and xylene, > 450; kerosene, 280; mineral oil, 55 to 60; but it is insoluble in water. The oral LD_{50} to the rat is 69 mg/kg (95).

A product of similar composition, STROBANE, is prepared by chlorinating a mixture of camphene and pinene to contain 66% chlorine. The technical material is a straw-colored liquid, d. 1.60, v.p. 3×10^{-7} mm, oral LD_{50} to the rat, 200 mg/kg.

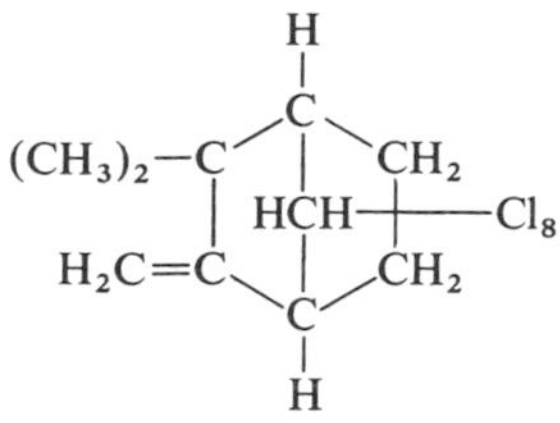

$$(CH_3)_2-C \overset{\overset{\displaystyle H}{|}\underset{\displaystyle C}{}}{\underset{\overset{\displaystyle C}{|}\underset{\displaystyle H}{}}{\underset{H_2C=C}{}}} \begin{matrix} CH_2 \\ HCH-|---Cl_8 \\ CH_2 \end{matrix}$$

toxaphene

These materials are broad spectrum insecticides especially useful for the control of insect pests of cotton, field crops, and animals.

M. Organophosphorus Insecticides

1. Introduction and Chemistry. The discovery of the biological activity of the organophosphorus esters originated with the chance observation of Lange and Kreuger in 1932(*78*) that casual exposure to the vapor of diethyl phosphorofluoridate produced strong cholinergic effects in humans. The practical development of these compounds as insecticides is attributable to the original and extensive work of Schrader and co-workers beginning in 1937(*126*). In seeking substitutes for nicotine Schrader devised BLADAN, which contained tetraethyl pyrophosphate as its active principal. A further search for a lime-stable product to control the Colorado potato beetle, *Leptinotarsa decemlineata*, led to the development of parathion. There followed in rapid succession schradan and dimeton, the first practical systemic insecticides; azinophosmethyl or GUTHION, a broad spectrum insecticide; trichlorfon or DIPTEREX, a stomach poison; coumaphos or CO-RAL, for control of animal parasites; and numerous others. These successes attracted the attention of many other investigators and there are presently more than 40 commercially successful organophosphorus insecticides, a number of which are marketed in multimillion pound quantities(*94*). It is estimated that more than 100,000 different organophosphorus compounds have been synthesized and evaluated as pesticides.

Tetraethyl pyrophosphate or *TEPP*, was first synthesized by Clermont in 1854, but its insecticial properties were discovered by Schrader in 1939. TEPP is produced in 20 to 40% yield from the reaction of phosphorus oxychloride or phosphorus pentoxide with triethyl phosphate together with inactive ethyl polyphosphates. Pure tetraethyl pyrophosphate can be separated from this mixture by vacuum distillation and is a water-white liquid, b.p. 104 to 110°C at 0.08 mm, d. 1.1845 at 25°C. The pure compound decomposes above 200°C to produce ethylene and meta-phosphoric acid, and various technical products undergo this decomposition above 140°C. TEPP is miscible in water, ethyl alcohol, ethyl acetate, carbon tetrachloride, benzene, zylene, and methylated naphthalenes but not in kerosene or mineral oils. Tetraethyl pyrophosphate hydrolyzes in water to produce two equivalents of nontoxic diethyl-*o*-phosphoric acid, the times for 50 and 99% hydrolysis at 25°C being 6.8 and 45 hr, respectively. This rapid destruction permits its use on edible produce immediately before harvest. Such solutions are strongly acid (pH of hydrolyzed 1% solution is about 1.5) and will corrode black iron, zinc, tin, and some porcelains. All the insecticidal products containing TEPP are hygroscopic and should be protected from moisture in order to preserve the insecticidal activity.

Sulfotepp or *tetraethyl dithionopyrophosphate* is a yellowish liquid with a garlic odor, b.p. 110 to 113°C at 0.2 mm, d. 1.196. It is only about 0.002% soluble in water, is soluble in organic solvents, and hydrolyzes only in alkaline solution. Sulfotepp is especially useful where a more stable product than TEPP is wanted, especially in aerosols or smokes for the control of greenhouse pests. The corresponding tetrapropyl analog also has useful insecticidal properties.

$$(C_2H_5O)_2\overset{\displaystyle O}{\overset{\|}{-}}P\overset{}{-}O\overset{\displaystyle O}{\overset{\|}{-}}P\overset{}{-}(OC_2H_5)_2 \qquad\qquad (C_2H_5O)_2\overset{\displaystyle S}{\overset{\|}{-}}P\overset{}{-}O\overset{\displaystyle S}{\overset{\|}{-}}P(OC_2H_5)_2$$

tetraethyl pyrophosphate   sulfotepp

Parathion or O,O-diethyl O-*p*-nitrophenyl phosphorothionate, also discovered by Schrader, has become the most widely used of all the organophosphorus insecticides. The technical material is a dark brown liquid, d. 1.265, which has an unpleasant garlic odor and is about 98% pure. The pure compound, m.p. 6°C and calculated b.p. 375°C, has a v.p. 0.00004 mm at 27°C. Parathion is very resistant to aqueous hydrolysis but is hydrolyzed by alkali to form inactive diethylphosphorothioic acid and *p*-nitrophenol. The times for 50% hydrolysis at 25°C are 120 days for a saturated aqueous solution, and 8 hr for a solution in limewater. At temperatures above 130°C, parathion slowly isomerizes to form O,S-diethyl O-*p*-nitrophenyl phosphorothioate, which is much less stable and less effective as an insecticide. Parathion is readily reduced to the nontoxic O,O-diethyl O-*p*-aminophenyl phosphorothionate and oxidized with difficulty to the highly toxic diethyl *p*-nitrophenyl phosphate. Parathion is completely miscible in benzene, xylene, phthalates, and glycols, but is almost insoluble in kerosene and mineral oils and is soluble in water to about 0.00002%.

$$\text{(C}_2\text{H}_5\text{O)}_2-\overset{\overset{\text{S}}{\|}}{\text{P}}\text{O}-\!\!\langle\bigcirc\rangle\!\!-\text{NO}_2 \qquad \text{(CH}_3\text{O)}_2-\overset{\overset{\text{S}}{\|}}{\text{P}}\text{O}-\!\!\langle\bigcirc\rangle\!\!-\text{NO}_2$$

parathion   methyl parathion

A limiting factor in the use of parathion has been its high toxicity to warm-blooded animals, and much effort has been devoted to finding safer derivatives. Certain closely related compounds have proved especially useful.

Methyl parathion is O,O-dimethyl O-*p*-nitrophenyl phosphorothionate, a white solid, m.p. 36°C, d. 1.358. This compound is about as widely used as parathion and is very similar in properties, but hydrolyzes and isomerizes more easily and is therefore less stable both in storage and as an insecticide residue. Methyl parathion is more effective than parathion against aphids and beetles and is somewhat less toxic to mammals.

Two chlorinated analogs of methyl parathion have greatly reduced acute mammalian toxicities and are useful as household insecticides and for certain agricultural pests. *Dicapthon* is O,O-dimethyl O-2-chloro 4-nitrophenyl phosphorothionate, a solid, m.p. 53°C. CHLORTHION is O,O-dimethyl O-3-chloro-4-nitrophenyl phosphorothionate, an oil, d.[20] 1.4330.

$$\text{(CH}_3\text{O)}_2-\overset{\overset{\text{S}}{\|}}{\text{P}}\text{O}-\!\!\langle\bigcirc\rangle\!\!-\text{NO}_2 \qquad \text{(CH}_3\text{O)}_2-\overset{\overset{\text{S}}{\|}}{\text{P}}\text{O}\!\!\langle\bigcirc\rangle\!\!-\text{NO}_2$$

Cl   Cl
dicapthon   CHLORTHION

Fenthion or BAYTEX is O,O-dimethyl O-3-methyl-4-methylthiophenyl phosphorothionate, a brownish liquid, b.p. 105°C at 0.01 mm, d.[25] 1.245, v.p. 2.15×10^{-6} mm at 20°C. Fenthion is a persistent, general-purpose insecticide especially useful as a residual spray for fly and mosquito control and as a mosquito larvicide. The closely related O,O-diethyl O-(*p*-methylsulfinylphenyl) phosphorothionate or DASANIT, b.p. 138–141°C at 0.01 mm, d. 1.202; is especially effective as a soil insecticide and nematocide. It has a rat oral LD_{50} of 5 mg/kg.

Ronnel is O,O-dimethyl O-2,4,5-trichlorophenyl phosphorothionate, a white crystalline solid, m.p. 35 to 37°C, which is soluble in organic solvents and to about 0.008% in water. It is a persistent, general-purpose insecticide and is also an animal systemic insecticide. *Bromophos*, the 4-bromo analog of ronnel is a persistent general purpose insecticide of low acute toxicity to mammals.

$$(CH_3O)_2\overset{\displaystyle S}{\overset{\|}{P}}O-\underset{\underset{\displaystyle CH_3}{}}{\bigcirc}-SCH_3$$

fenthion

$$(CH_3O)_2\overset{\displaystyle S}{\overset{\|}{P}}O-\underset{\underset{\displaystyle Cl}{}}{\overset{\overset{\displaystyle Cl}{}}{\bigcirc}}-Cl$$

ronnel

$$(CH_3O)_2-\overset{\displaystyle S}{\overset{\|}{P}}O-\underset{\underset{\displaystyle CH_3}{}}{\bigcirc}-NO_2$$

fenitrothion

$$(C_2H_5O)_2-\overset{\displaystyle S}{\overset{\|}{P}}O-\bigcirc-\overset{\displaystyle O}{\overset{\uparrow}{S}}CH_3$$

DASANIT

Fenitrothion or SUMITHION is O,O-dimethyl O-3-methyl-4-nitrophenyl phosphorothionate, a brownish liquid, b.p. 140–145°C at 0.1 mm, d.[25] 1.3227, soluble in alcohols, ethers, ketones, and aromatic hydrocarbons, and to 0.002% in water. It is a persistent general purpose insecticide for agricultural and public health uses with greatly increased safety to mammals.

Diazinon is O,O-diethyl O-2-isopropyl-4-methylpyrimidyl-6 phosphorothionate, a brownish liquid, b.p. 83–84°C, d. 1.11, which is soluble in most organic solvents but only to about 0.004% in water. Diazinon is a persistent general purpose insecticide used for the control of soil insects, pests of vegetables, fruits, and field crops, and as a residual spray for dairy and livestock pests.

Azinphosmethyl or GUTHION is O,O-dimethyl S-4-oxo-1,2,3-benzo-triazin-3-(4-H)-yl-methyl phosphorodithioate, a white solid, m.p. 73–74°C, d.[20] 1.44. It is soluble in organic solvents but only to about 0.003% in water. The O,O-diethyl analog, *azinphosethyl*, m.p. 53°C is also a commercial insecticide. Both compounds are persistent broad spectrum insecticides effective against pests of fruits, vegetables, cotton, and ornamentals.

$$(CH_3)_2CH\underset{\underset{\displaystyle N}{}}{\overset{\overset{\displaystyle CH_3}{\displaystyle N}}{\bigcirc}}\overset{\displaystyle S}{\underset{\|}{O\overset{\|}{P}(OC_2H_5)_2}}$$

diazinon

$$\bigcirc\underset{\underset{\displaystyle N}{\displaystyle N}}{\overset{\overset{\displaystyle O}{\overset{\|}{C}}}{}}NCH_2S\overset{\displaystyle S}{\overset{\|}{P}}(OCH_3)_2$$

azinphosmethyl

Malathion is O,O-dimethyl S-(1,2-dicarboxyethyl) phosphorodithioate, a brownish liquid, b.p. 156 to 157°C at 0.7 mm. The technical material is 95 to 98% pure, with an unpleasant odor, d.[25] 1.23. It is soluble in most organic solvents, slightly soluble in mineral oils, and soluble in water to 0.0145%. Malathion is easily hydrolyzed above pH 7.0 and below pH 5.0 and is incompatible with alkaline materials. It is a very safe, persistent, general purpose insecticide, especially suited for household, home garden, vegetable, and fruit insect control, and for the control of mosquitoes, flies, and lice of public health importance.

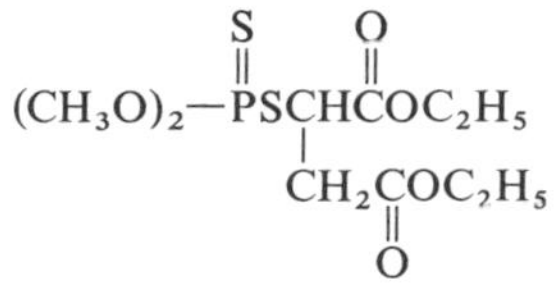

$$
(CH_3O)_2{-}\overset{\displaystyle S}{\overset{\|}{P}}SCH\overset{\displaystyle O}{\overset{\|}{C}}OC_2H_5
$$
$$
\underset{\underset{O}{\|}}{CH_2C}OC_2H_5
$$

malathion

Trichlorfon or DIPTEREX is O,O-dimethyl(1-hydroxy-2,2,2-trichloro-ethyl-)phosphonate, $(CH_3O)P(=O)CH(OH)CCl_3$, a white solid, m.p. 83–84°C, d. 1.73. It is soluble in water to 15% and in aromatic hydrocarbons, alcohols, and ketones. Its butyl ester, *butonate* or dimethyl 2,2,2-trichloro-1-(butyryloxy)-ethylphosphonate, an oily liquid that is used as a household insecticide. Trichlorfon is a stomach poison used in dry sugar bait for the control of flies and on foliage for chewing insects. Above pH 6, trichlorfon is rapidly converted to *DDVP* or *Dichlorvos*, dimethyl dichlorovinyl phosphate $(CH_3O)_2P(=O)OCH=CCl_2$, a reaction that accounts for its toxic action. DDVP is a colorless liquid, b.p. 120°C at 14 mm, d. 1.415, with a half-life in water at pH 7.0 of about 8 hr. The compound is very effective in baits and aerosols for the rapid knockdown of flies, mosquitoes, moths, etc., and has fumigant action against household pests. *Naled* or DIBROM is brominated DDVP or dimethyl 1,2-dibromo-2,2-dichloroethyl phosphate, $(CH_3O)_2P(=O)OCHBrCBrCl_2$, m.p. 26°C. It is insoluble in water and soluble in aromatic solvents. DIBROM is destructively hydrolyzed in water in about 2 days and provides a short-lived insecticide for use on plants, in the household, and for fly and mosquito control.

Other more recently developed persistent, general-purpose insecticides used for the control of various insect and mite pests of fruits, vegetables, forage crops, and animals include:

Dioxathion or DELNAV or 2,3-*p*-dioxanedithiol S,S-bis-(O,O-diethyl-phosphonodithioate), a brown liquid, d. 1.257, insoluble in water and soluble in organic solvents.

EPN or O-ethyl O-*p*-nitrophenyl phenylphosphonothionate, m.p. 36°C. The technical material is a brown liquid, d. 1.27, v.p. 0.03 mm at 100°C, insoluble in water, and soluble in organic solvents.

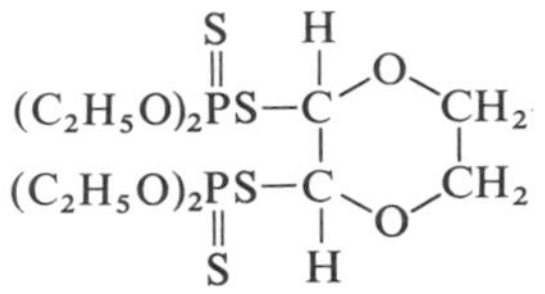

dioxathion 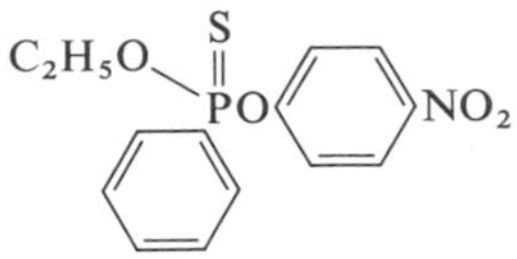   EPN

IMIDAN or O,O-dimethyl S-phthalimidomethyl phosphorodithioate, m.p. 71–72°C.

Carbophenothion or TRITHION is O,O-diethyl S-p-chlorophenylthio-methyl phosphorodithioate, a light amber liquid, d. 1.265, insoluble in water, and soluble in kerosene, xylene, alcohols, ketones, and esters. The O,O-dimethyl ester, methyl TRITHION is also a commercial insecticide.

Ethion or O,O,O′,O′-tetraethyl S,S′-methylene-bis-phosphorodithioate, $(C_2H_5O)_2P(\!\!=\!\!S)\!\!-\!\!SCH_2S\!\!-\!\!P(\!\!=\!\!S)(OC_2H_5)_2$, is a liquid, m.p. -15 to $-12°C$, d. 1.22, insoluble in water, and soluble in most organic solvents. Ethion is an acaricide and insecticide used on fruits and vegetables.

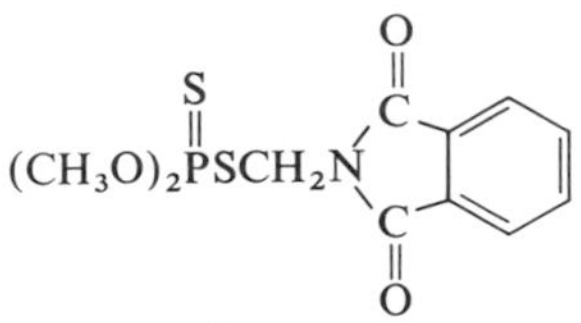

IMIDAN    TRITHION

a. SYSTEMIC INSECTICIDES FOR PLANTS. When applied to seeds, roots, or leaves of plants, systemics are absorbed and translocated to the various plant parts in amounts lethal to insects feeding thereon. This method of plant protection has the decided advantages of (a) minimizing to some extent the inequalities of spray coverage, (b) increasing the length of residual control by protection of the spray residue from attrition by weathering, (c) protecting new plant growth formed subsequent to application, and (d) having less damaging effects on beneficial predatory and pollinating insects(120). Systemic compounds may be applied as direct sprays to the foliage, by drenching the soil as in the irrigation water, by direct application or injection of concentrates to the trunk or stem, by side-dressing about the roots with granular or encapsulated material, or by treatment of seeds before planting(116). Because of the possible contamination of edible portions of the plant, the use of systemics on food crops must be predicated on a thorough knowledge of the nature and magnitude of toxic residues.

Schradan or octamethyl pyrophosphoramide $[(CH_3)_2N]P(=O)$—O—$P(=O)[N(CH_3)_2]_2$ was among the first practical systemics. The technical material is a brown liquid, b.p. 154°C at 2 mm, d.[25] 1.13, which contains about 40% of the octamethyl pyrophosphoramide and an equivalent amount of decamethyl triphosphoramide,

$$[(CH_3)_2N]_2\overset{\overset{\displaystyle O}{\|}}{P}—O—\overset{\overset{\displaystyle O}{\|}}{\underset{\underset{\displaystyle N(CH_3)_2}{|}}{P}}—O—\overset{\overset{\displaystyle O}{\|}}{P}[N(CH_3)_2]_2$$

which is of about the same activity. The product is water-miscible and also soluble in most organic solvents, but only slightly soluble in petroleum hydrocarbons. Schradan is remarkably stable to alkaline hydrolysis but hydrolyzes much more rapidly in acids to form dimethyl amine and phosphoric acid. It possesses only a weak contact action to most insects, although it is fairly toxic to Hemiptera and Homoptera and has been shown to be slowly converted in plants and rapidly in animals to the monoamide oxide, which is the actual toxicant. When absorbed into plants, schradan provides long-term protection from aphids and mites.

Dimefox is bis-(dimethylamino) phosphoryl fluoride, $[(CH_3)_2N]_2$-$P(=O)F$, which is a colorless liquid, b.p. 67°C at 4 mm, d. 1.12, v.p. 0.4 mm at 30°. This compound is water soluble and has been used to control the mealybug vector of the swollen shoot disease of cacao by implanting soluble capsules about the roots. Because of its volatility, dimefox has a very short residual activity and is not suited for foliage application.

Demeton or SYSTOX is a mixture of 2 parts of O,O-diethyl O-2-(ethylthio)-ethyl phosphorothionate, $(C_2H_5O)_2P(=S)OCH_2CH_2SC_2H_5$, b.p. 94°C at 4 mm, d. 1.119, water-solubility 0.0066%, and 1 part of O,O-diethyl S-2-(ethylthio)-ethyl phosphorothiolate, $(C_2H_5O)_2P(=O)SCH_2$-$CH_2SC_2H_5$, b.p. 110°C at 0.4 mm, d.[20] 1.132, water solubility 0.2%. The technical material is a yellowish liquid, d. 1.83, soluble in organic solvents, and unstable in alkaline solution. Demeton provides a long-lasting systemic insecticide rapidly absorbed by roots, stems, or foliage. Internally in plant tissues, both isomers are rapidly oxidized to the sulfoxide and sulfone derivatives as shown with phorate.

Methyl demeton or METASYSTOX is O,O-dimethyl S-2-(ethylthio)-ethyl phosphorothiolate, $(CH_3O)_2P(=O)SCH_2CH_2SC_2H_5$. It has very similar properties and behavior to demeton. *Oxydemeton methyl* or META SYSTOX sulfoxide, $(CH_3O)_2P(=O)SCH_2CH_2S(=O)C_2H_5$, is more water soluble, and is a very active systemic, especially suited for granular applications to the soil.

Dimethoate is O,O-dimethyl S-(N-methylcarbamoyl)-methyl phosphorodithioate, $(CH_3O)_2P(\!=\!S)SCH_2C(\!=\!O)NHCH_3$, a white crystalline solid, m.p. 51°C. It is soluble in organic solvents and soluble in water to about 7%. Because of its water solubility it has proved especially useful as a persistent systemic for fruit fly larvae and for side-dressing of soil about plants.

Mevinphos or PHOSDRIN is dimethyl 2-carbomethoxy-1-methylvinyl phosphate, $(CH_3O)P(\!=\!O)OC(CH_3) = CHC(\!=\!O)OCH_3$. The technical material is a colorless liquid, b.p. 106 to 107.5°C at 1 mm, d.[20] 1.25, v.p. 0.0029 mm at 21°C. It is miscible in water, xylene, and acetone, but is less than 5% soluble in kerosene. The technical product is composed of about 2 parts of trans- and 1 part of the cis-isomer, the former of which is about ten times as active as the latter. PHOSDRIN is especially useful for the treatment of edible produce close to harvest time since it is rapidly dissipated by volatilization and enzymatic decomposition in the plant.

Phosphamidon is dimethyl 2-chloro-2-diethylcarbamoyl-1-methylvinyl phosphate, $(CH_3O)_2P(\!=\!O)OC(CH_3) = CClC(\!=\!O)N(C_2H_5)_2$, a colorless liquid, b.p. 160°C at 1.5 mm, that is miscible in water and organic solvents. It is rapidly absorbed by plant surfaces and quickly decomposed in the plant to provide a short-lived systemic.

Bidrin is 3-(dimethoxyphosphinyloxy)-N,N-dimethylcrotonamide, $(CH_3O)_2P(\!=\!O)OC(CH_3)\!=\!CHC(\!=\!O)N(CH_3)_2$, a brown liquid, b.p. 115–120°C at 0.001 mm, d.[23] 1.19, soluble in water, ketones, and alcohols. Bidrin is especially useful for the systemic treatment by trunk injection of trees for the control of pests attacking bark and foliage.

b. SEED AND SOIL TREATMENTS. Several systemic compounds of low water solubility are especially suited for the protection of young plants from attack by mites, thrips, aphids, and leafhoppers by application (a) as a 50% charcoal powder to the seeds of cotton, alfalfa, or sugar beets before planting, (b) by granular application at time of transplanting, or (c) as a granular side-dressing applied at planting (*116*).

Phorate or THIMET is O,O-diethyl S-2-(ethylthio)methyl phosphorodithioate, a yellowish liquid, b.p. 118–120°C at 0.8 mm, d.[25] 1.167. It is soluble in organic solvents and to 0.0085% in water. In plant tissue, phorate is rapidly oxidized to the sulfoxide and sulfone metabolites that are responsible for the systemic toxicity of the compound. They are more water soluble and less stable than the parent material and are decomposed in the plant to nontoxic hydrolysis products.

Disulfoton or DI-SYSTON is O,O-diethyl S-2-(ethylthio)-ethyl phosphorodithioate and is thus the dithioanalog of demeton. This compound $(C_2H_5O)_2P(\!=\!S)SCH_2CH_2SC_2H_5$, is a pale-yellow liquid, b.p. 62°C at

0.01 mm, d.[20] 1.144. It is soluble in organic solvents and to 0.0066% in water. Disulfoton is oxidized in plant tissue in the same manner as phorate, the final active metabolites being the same as the sulfoxide and sulfone metabolites of demeton thiol-isomer. *Dithiomethyldemeton* O,O-dimethyl S-2-(ethylthio)-ethyl phosphorodithioate, $(CH_3O)_2P(=S)SCH_2$-CH_2-SC_2H_5, has similar properties.

c. SYSTEMIC INSECTICIDES FOR ANIMALS. When fed or topically applied to domestic animals, systemics have been found to move through the body tissues in quantities lethal to such internal parasites as cattle grubs, screw-worm larvae, and helminths and over a shorter period to external parasites such as horn and stable flies, mites, lice, and ticks, without injury to the host animal. The insecticides are slowly destroyed by enzymatic action in the animal body, so that after a safe period of 60 days, the animal can be used for milk production or slaughtered for meat. (*94*).

Ronnel or TROLENE, O,O-dimethyl O-2,4,5-trichlorophenyl phosphoro-thionate, is effective when fed at 100 mg/kg of animal weight.

CO-RAL is O,O-diethyl O-(3-chloro-4-methyl-2-oxo-2H-1-benzopyran-7-yl) phosphorothionate, a tan crystalline solid, m.p. 90 to 92°C, v.p. 10^{-7} mm at 20°C. It is insoluble in water and soluble in organic solvents, and is applied as a 0.25 to 0.5% spray.

RUELENE is O-methyl O-(4-*tert*-butyl-2-chlorophenyl) methylphos-phoramidate, m.p. 61°C, soluble in most organic solvents and insoluble in water. It is effective either as an 0.75% spray or when fed at 20 to 25 mg/kg of animal weight.

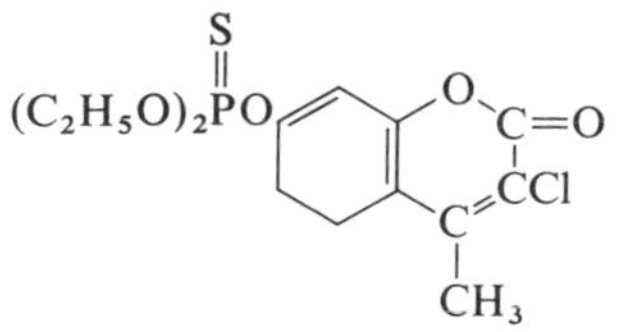

CO-RAL 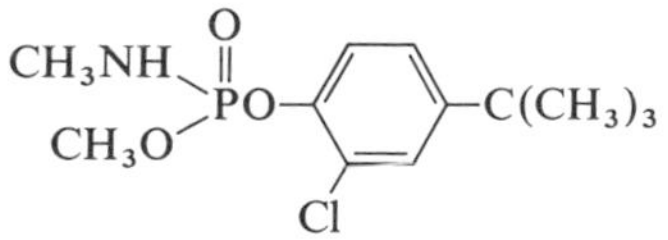   RUELENE

CYTHIOATE or O,O-dimethyl O-*p*-sulfamoylphenyl phosphorothioate, $(CH_3O)_2P(S)OC_6H_4SO_2NH_2$-*p* and *famphur* or O,O-dimethyl O-*p*-dimethylsulfamoylphenyl phosphorothioate, $(CH_3O)_2P(S)OC_6H_4SO_2N$-$(CH_3)_2$-*p* are new animal systemic insecticides.

CIODRIN or O,O-dimethyl O-[1-methyl-2-(1-phenylcarbethoxy)-vinyl] phosphate, a straw-colored liquid, b.p. 135°C at 0.03 mm, is an effective livestock spray for the control of biting flies and other ecto-parasites.

$$\underset{\text{CIODRIN}}{(CH_3O)_2\overset{\displaystyle O}{\overset{\displaystyle \|}{P}}\overset{\displaystyle CH_3}{\overset{\displaystyle |}{OC}}=CH\overset{\displaystyle O}{\overset{\displaystyle \|}{C}}\underset{\underset{\displaystyle CH_3}{\displaystyle |}}{OCH}—}$$

2. Structure–Activity Relationships. The enormous structural diversity of the organophosphorus insecticides is apparent from the discussion in Section 1-4, M, 1. However, all of these compounds can be represented by the classical hypothetical phosphorylating agent as originally proposed by Schrader(*126*):

$$\underset{R'}{\overset{R}{\diagdown}}\overset{\displaystyle O(S)}{\overset{\displaystyle \|}{P}}—X$$

where R and R′ are short chain alkyl, alkoxy, alkylthio, or amide groups, and X is a labile leaving group or a group that can be metabolized in vivo to a labile entity.

The mode of action of these phosphorus esters as inhibitors of the enzyme cholinesterase is discussed in Sec. 1-4, M, 3. Their toxicity depends upon the reactivity of the compound and this is determined by (a) the magnitude of the electrophilic character of the phosphorus atom, (b) the strength of the bond P—X, and (c) the steric nature of the substituents(*38*). The electrophilic nature of the central P atom is a function of the relative electronegativity values of the atoms bonded to P (e.g., P 2.1, O 3.5, S 2.5, N 3.0, and C 2.5). Thus in the phosphate esters (P=O), the P is much more electrophilic than in the phosphorothionate esters (P=S). The latter are generally so stable as to be unreactive with cholinesterase and are activated in the animal body by oxidation to the corresponding P=O compound as discussed in Sec. 1-4, M, 3.

The necessity for the toxic phosphorus compound to undergo a bimolecular reaction with cholinesterase (Sec. 1-4, M, 3) implies that the bond P—X is broken during the inhibition process. Therefore, the toxicity of the compounds is strongly influenced by groups that increase the lability of this bond. This is well illustrated by the data for the para-substituted phenyl diethyl phosphates shown in Table 1-20. Here both reactivity with cholinesterase and toxicity to the housefly are directly correlated with the electron-withdrawing capacity of the substituent as measured by Hammett's σ(*39*). The most toxic compounds are those which incorporate a strongly electron-withdrawing substituent such as p-NO_2 or a group such as CH_3S— that can be oxidized in vivo to the strongly electron-withdrawing CH_3SO— and CH_3SO_2—, because these

TABLE 1-20

RELATIONSHIP BETWEEN ELECTRON WITHDRAWING POWER (σ) OF SUB-
STITUENT AND CHOLINESTERASE INHIBITION AND INSECT TOXICITY OF
SUBSTITUTED PHENYL DIETHYL PHOSPHATES(39)

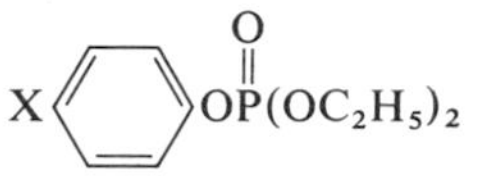

$$X \hspace{-0.5em}\left\langle \bigcirc \right\rangle\hspace{-0.5em} O\overset{O}{\overset{\|}{P}}(OC_2H_5)_2$$

X =	σ	K_e fly ChE	Topical LD$_{50}$ fly Musca domestica, $\mu g/g$
CH$_3$O	−0.135	2.9×10^1	> 5000
CH$_3$	−0.170	2.0×10^1	> 5000
CH$_3$S	−0.047	2.3×10^3	2.0
H	0.0	2.6×10^1	1500
Cl	0.227	2.3×10^3	150
CH$_3$SO	0.730	2.4×10^5	1.5
CN	0.891	2.6×10^5	3.5
CH$_3$SO$_2$	1.049	1.9×10^6	2.5
NO$_2$	1.267	2.7×10^7	1.0

substituents enhance the electrophilic nature of the P atom by inductive
and mesomeric effects:

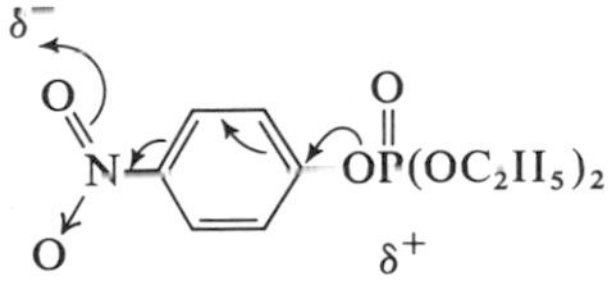

$$\overset{\delta^-}{O} \overset{O}{\underset{O}{\overset{\|}{N}}} \hspace{-0.5em}\left\langle \bigcirc \right\rangle\hspace{-0.5em} O\overset{O}{\overset{\|}{P}}(OC_2H_5)_2 \quad \delta^+$$

The alkyl and alkoxy substituents of phosphate or phosphonate esters
also affect the electrophilic nature of the P-atom through steric and
inductive effects as shown in Table 1-21 where the reactivity with
cholinesterase and toxicity to the housefly are directly correlated with
the quantitative measure of these effects, Taft's polar substituent constant σ^*. Thus toxicity decreases and stability increases with increasing
chain length and with chain branching(40).

The development of malathion in 1950 was an important milestone
in the emergence of selective insecticides. As illustrated in Table 1-22,
this material is from $\frac{1}{2}$ to $\frac{1}{20}$ as toxic to insects as parathion but is only
about $\frac{1}{200}$ as toxic to mammals. Its world-wide usage in multi-million
pound quantities in the home, garden, field, orchard, woodland, on
animals, and in public health programs has demonstrated an extraordinary degree of safety coupled with pest control effectiveness. The
biochemical basis for the selectivity of malathion lies in its rapid detoxi-

TABLE 1-21

RELATIONSHIP BETWEEN STERIC AND INDUCTIVE EFFECTS (σ^*) OF
ALKYL SUBSTITUENT AND CHOLINESTERASE INHIBITION AND INSECT
TOXICITY OF p-NITROPHENYL ETHYL ALKYL PHOSPHONATES(40)

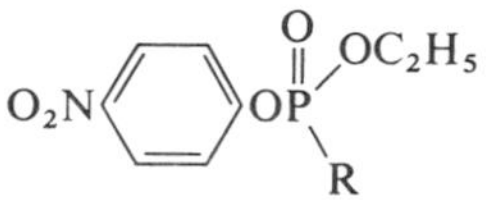

R =	σ^*	K_e fly ChE	Topical LD$_{50}$ fly *Musca domestica*, $\mu g/g$
CH_3	0.000	7.6×10^6	1.0
C_2H_5	−0.100	1.5×10^7	1.2
C_3H_7	−0.115	1.0×10^7	2.0
C_4H_9	−0.130	4.0×10^6	2.4
C_5H_{11}	−0.145	8.9×10^5	12.0
C_6H_{13}	−0.160	6.9×10^5	62
$(CH_3)_2CH$	−0.190	1.1×10^5	28
$(CH_3)_3C$	−0.300	6.7	> 5000

cation in the mammalian liver, but not in the insect, through the attack of carboxyesterase enzymes upon the aliphatic ester moieties of the molecule.

Extraordinary selectivity has also been accomplished with the para-thion-type compounds as shown in Table 1-22. The incorporation of —Cl or —CH$_3$ groups in the meta-position of the aryl ring greatly reduces mammalian toxicity without materially decreasing the insecticidal effectiveness. The compounds CHLORTHION, dicapthon, ronnel, fenthion, and fenitrothion are in extensive use in household, public health, and agricultural pest control. Selectivity in this group of insecticides results from the net differences in rates of activation, of detoxication, and in cholinesterase target sensitivity between insects and mammals. As the science of pest control becomes increasingly sophisticated, such selective compounds are playing increasingly vital roles not only in elimination of the hazards of insecticide application but also in the development of *integrated control programs* whereby insecticides are used to supplement or augment the activities of beneficial parasites and predators in reducing populations of injurious insects below the levels producing economic damage. (See page 106 for Table 1-22.)

3. Mode of Action. The organophosphorus compounds despite their diverse structures all owe their biological activities to the capacity of the central P atom to phosphorylate the esteratic site of the enzyme cholinesterase (ChE), which is an essential constituent of the nervous system not only of the Insecta but also of all higher animals(56, 109). The phosphorylated enzyme is irreversibly inhibited and is therefore no longer able to carry out its normal function of the rapid removal and destruction of the neurohormone (ACh) from the nerve synapse. As a result, ACh accumulates and disrupts the normal functioning of the nervous system, giving rise to the typical cholinergic symptoms associated in insects with O—P poisoning, i.e., hyperactivity, tremors, convulsions, paralysis and death. In higher animals these cholinergic effects are translated into muscarinic effects such as nausea, salivation, lachrymation, and myosis; nicotinic effects such as muscular fasciculations, and central effects such as giddiness, tremulousness, coma, and convulsions.

The reaction between esterase and phosphorus inhibitor (B) is bimolecular of the well-known S_N2 type and represents the attack of a nucleophilic group (possibly the imidazole ring of a histidine residue at the active site) upon the electrophilic phosphorus atom, and mimics the normal three-step reaction between enzyme and substrate (A):

(A)

$$CH_3\overset{\overset{\displaystyle O}{\|}}{C}OCH_2CH_2\overset{+}{N}(CH_3)_3 + EH \underset{k_{-1}}{\overset{k_1}{\rightleftharpoons}} CH_3\overset{\overset{\displaystyle O}{\|}}{C}OCH_2CH_2\overset{+}{N}(CH_3)_3 \cdot EH \quad (1)$$

$$CH_3\overset{\overset{\displaystyle O}{\|}}{C}OCH_2CH_2\overset{+}{N}(CH_3)_3 \cdot EH \xrightarrow{k_2} CH_3\overset{\overset{\displaystyle O}{\|}}{C}\cdot E + HOCH_2CH_2\overset{+}{N}(CH_3)_3 \quad (2)$$

$$CH_3\overset{\overset{\displaystyle O}{\|}}{C}\cdot E + H_2O \xrightarrow{k_3} EH + CH_3\overset{\overset{\displaystyle O}{\|}}{C}OH$$

$$(3)$$

(B)

$$(RO)_2\overset{\overset{\displaystyle O}{\|}}{P}X + EH \underset{k_{-1}}{\overset{k_1}{\rightleftharpoons}} (RO)_2\overset{\overset{\displaystyle O}{\|}}{P}X \cdot EH$$

$$(1)$$

$$(RO)_2\overset{\overset{\displaystyle O}{\|}}{P}X \cdot EH \xrightarrow{k_2} (RO)_2\overset{\overset{\displaystyle O}{\|}}{P}\cdot E + HX$$

$$(2)$$

$$(RO)_2\overset{\overset{\displaystyle O}{\|}}{P}\cdot E + H_2O \xrightarrow{k_3} EH + (RO)_2\overset{\overset{\displaystyle O}{\|}}{P}OH$$

$$(3)$$

In the normal process (A), step 3 occurs very rapidly and step 1 is the rate-determining step, while in the inhibition process (B) step 3 occurs very slowly, generally over a matter of days, so that it is rate determining. Thus it has been demonstrated with ChE that (a) insecticides such as tetraethyl pyrophosphate and mevinphos engage in first-order reactions

TABLE 1-22
SELECTIVE TOXICITY OF PARATHION, MALATHION, AND ANALOGS (96)

		Topical LD_{50} fly *Musca domestica*, $\mu g/g$	LC_{50} mosquito larva *Culex quinquefasciatus*, ppm	Oral LD_{50} rat, mg/kg
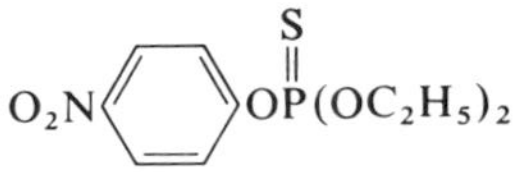 O_2N—⟨⟩—$\overset{S}{\underset{\parallel}{O\!P}}(OC_2H_5)_2$	Parathion	0.9	0.0032	6
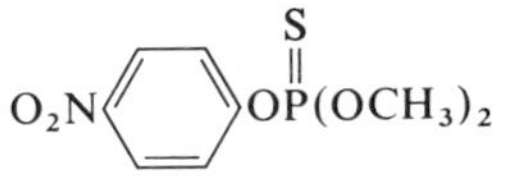 O_2N—⟨⟩—$\overset{S}{\underset{\parallel}{O\!P}}(OCH_3)_2$	Methyl parathion	1.3	0.018	15
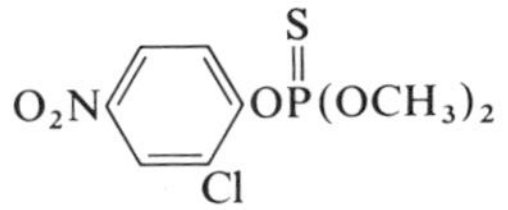 O_2N—⟨⟩(Cl)—$\overset{S}{\underset{\parallel}{O\!P}}(OCH_3)_2$	Dicapthon	1.6	0.027	400
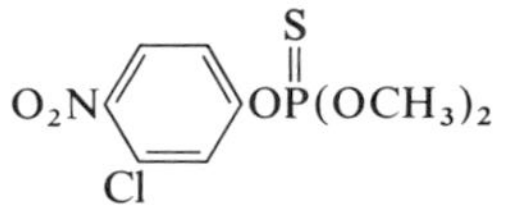 O_2N—⟨⟩(Cl)—$\overset{S}{\underset{\parallel}{O\!P}}(OCH_3)_2$	CHLORTHION	11.5	0.026	1500
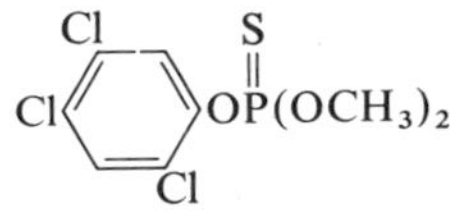 Cl—⟨⟩(Cl)—$\overset{S}{\underset{\parallel}{O\!P}}(OCH_3)_2$	Ronnel	2.7	0.030	1740

Structure	Name			
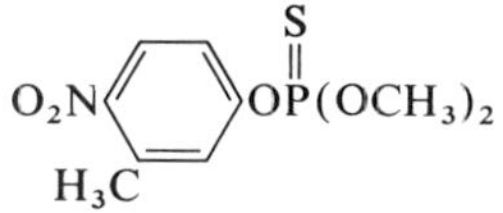 O_2N–⬡(–H_3C)–$OP(OCH_3)_2$ (S=)	Fenitrothion	2.6	0.0058	500
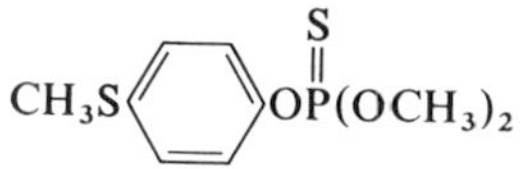 CH_3S–⬡–$OP(OCH_3)_2$ (S=)		2.0	0.0054	10
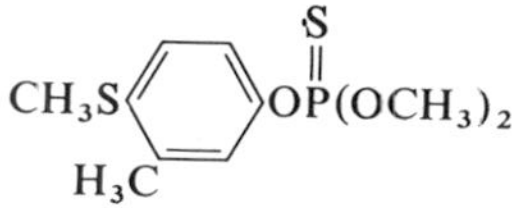 CH_3S–⬡(–H_3C)–$OP(OCH_3)_2$ (S=)	Fenthion	2.3	0.0045	500
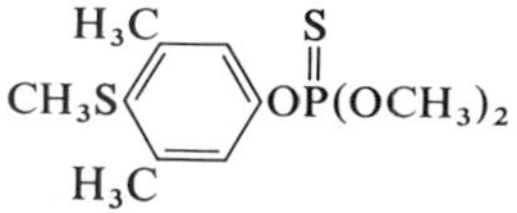 H_3C, CH_3S–⬡(–H_3C)–$OP(OCH_3)_2$ (S=)		4.3	0.11	3000
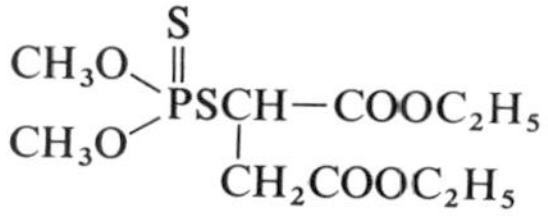 CH_3O, CH_3O–$PSCH$–$COOC_2H_5$ / $CH_2COOC_2H_5$ (S=)	Malathion	27.0	0.081	1500

with the enzyme, (b) the inhibited enzyme is a relatively stable phosphorylated compound containing one mole of phosphorus per mole
of enzyme, and (c) as a result of the reaction an equimolar quantity of
alcoholic or acidic product HX is liberated.

As has been discussed in Sec. 1-4, M,2, the phosphorothionate esters
(P=S) have substantially less electrophilic and reactive phosphorus
groups than the corresponding phosphates (P=O), and are generally
so stable as to be unable to combine with cholinesterase. The phosphorothionates owe their biological activity to in vivo oxidation by mixed
function oxidases which takes place in the insect gut and fat body tissues
and in the mammalian liver. A typical example is the oxidation of parathion to paraoxon:

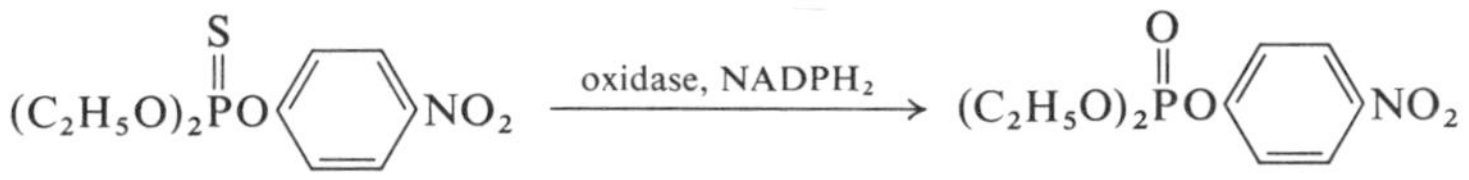

$$(C_2H_5O)_2\overset{\overset{\text{S}}{\|}}{P}O\!\!\left\langle\bigcirc\right\rangle\!\!NO_2 \xrightarrow{\text{oxidase, NADPH}_2} (C_2H_5O)_2\overset{\overset{\text{O}}{\|}}{P}O\!\!\left\langle\bigcirc\right\rangle\!\!NO_2$$

parathion   paraoxon

Another type of in vivo oxidation is of importance in the lethal synthesis through which electron-donating groups such as CH_3S— are
converted to the strongly electron withdrawing groups CH_3SO— and
CH_3SO_2— (note the equivalence in toxicity between these three groups
in Table 1-20). This type of reaction with its built-in delay principle is
involved in the mode of action of certain systemic insecticides such as
phorate and disulfoton where both P=S and C_2H_5S— oxidation occur
(95):

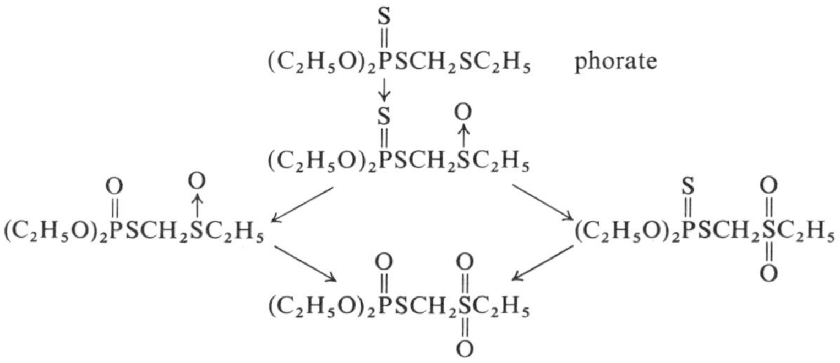

$$(C_2H_5O)_2\overset{\overset{\text{S}}{\|}}{P}SCH_2SC_2H_5 \quad \text{phorate}$$

$$\downarrow$$

$$(C_2H_5O)_2\overset{\overset{\text{S}}{\|}}{P}SCH_2\overset{\overset{\text{O}}{\uparrow}}{S}C_2H_5$$

$$(C_2H_5O)_2\overset{\overset{\text{O}}{\|}}{P}SCH_2\overset{\overset{\text{O}}{\uparrow}}{S}C_2H_5 \qquad\qquad (C_2H_5O)_2\overset{\overset{\text{S}}{\|}}{P}SCH_2\underset{\underset{\text{O}}{\|}}{\overset{\overset{\text{O}}{\|}}{S}}C_2H_5$$

$$(C_2H_5O)_2\overset{\overset{\text{O}}{\|}}{P}SCH_2\underset{\underset{\text{O}}{\|}}{\overset{\overset{\text{O}}{\|}}{S}}C_2H_5$$

In this type of activation, which occurs in both animal and plant tissues,
the original insecticide is relatively stable and can be translocated
through plant tissues without destructive hydrolysis until the oxidation

has occurred which then makes the insecticide both highly toxic and relatively unstable so that it is hydrolyzed rapidly to nontoxic products.

4. Biological Activity. It is apparent from the diversity of compounds and their pest control usages as outlined in Sec. 1-9, M, 1 that the organophosphorus insecticides represent the most versatile group of pesticidal compounds yet developed. There are available compounds with very short residual action such as tetraethylpyrophosphate and mevinphos or with prolonged residual action such as diazinon and azinphos. There are broad spectrum insecticides such as parathion and malathion and materials with very specific action such as schradan. The unique properties of demeton and dimethoate have resulted in successful plant systemic insecticides and this property has been refined further into seed and soil treatments with phorate or THIMET and disulfoton or DI-SYSTON, which protect newly developed seedlings from insect attack. RUELENE can be fed to cattle to kill grubs, *Hypoderma spp*, living in the animals' bodies, while others such as trichlorfon have pronounced activity as stomach poisons but almost no contact action. By taking advantage of differences in the processes of detoxication in Insecta and Mammalia, selective compounds such as malathion and fenitrothion or SUMITHION incorporate a high degree of insecticidal action along with a high degree of safety to the human user and his domestic animals.

5. Toxicology. Inasmuch as the target enzyme, cholinesterase, plays a vital role not only in Insecta but also in higher animals, the organophosphorus insecticides may be expected to be acutely toxic to many forms of animal life. As is pointed out in Sec. 1-4, M, 3, parathion and certain related compounds are generally toxic to nearly all types of animals. However, by taking advantage of selective activation, detoxication, and target sensitivity, it is possible to design organophosphorus insecticides such as malathion which are of a very low degree of toxicity to higher animals.

Hayes(55) gives an excellent discussion of the toxicology of organophosphorus insecticides. These compounds are absorbed into the body by respiratory, gastrointestinal, and cutaneous pathways. Signs and symptoms of poisoning relate almost entirely to the inhibition of cholinesterase and are discussed in Sec. 3. The relative acute oral LD_{50} values to the male rat in mg/kg are representative of the hazard associated with the use of specific compounds: TEPP 1.0, phorate 2.3, phosdrin 6.1, demeton 6.2, disulfoton 6.8, schradan 9.1, azinphosmethyl 13, parathion 13, methyl parathion 14, phosphamidon 23.5, carbophenothion 30, EPN 36, CO-RAL 41, dioxathion 43, ethion 65, dichlorvos 80, diazinon 108, dimethoate 215, fenthion 215, dicapthon 400, trichlorfon 630,

CHLORTHION 880, ronnel 1250, malathion 1375, and bromophos 3750 (55). Other values in the literature may differ somewhat from these (Table 1-22) but the trends are the same.

In general, because of rapid hydrolysis and excretion in the urine, the organophosphorus compounds are not accumulated and stored in the animal body. However, with the more toxic compounds, the cholinesterase of blood and nerve tissues may be progressively inhibited so that the animal can be precipitated into acute illness by absorption of less than a normal toxic dosage. Therefore, great care should be exercised in handling the concentrate materials of the highly toxic compounds. The use of rubber gloves, goggles, a respirator, and other protective clothing is advisable. Any material spilled on the skin should be immediately removed with soap and water. When spraying and dusting with these compounds, contaminated clothing should be changed frequently. Periodic estimation of blood cholinesterase levels is of value in the early detection of over-exposure. In the cases of acute poisoning by parathion or other organophosphorus insecticides, the victim should be given prompt medical attention, with the administration of therapeutic doses of atropine, which is a specific antidote. See Hayes(55) for recommendations of specific therapeutic measures.

N. Carbamate Insecticides

1. Introduction and Chemistry. The N-methyl and N,N-dimethyl-carbamic esters of a variety of phenols and heterocyclic enols possess useful insecticidal properties and are also effective molluscacides. These carbamates, which are aromatic carbamic esters, should not be confused with the carbamate herbicides (Sec. 1-3, F-6) which are carbanilate aliphatic esters, or the carbamate fungicides (Section 1-2, D).

Heterocyclic N,N-dimethylcarbamates compounds were discovered by Gysin(51). The two compounds of greatest importance are 1-iso-propyl-3-methylpyrazolyl-(5) N,N-dimethylcarbamate, *isolan*, b.p.$_{0.3}$ 105–107°C, oral LD$_{50}$ rat 54 mg/kg and 2-(N,N-dimethylcarbamyl)-3-methylpyrazolyl-(5) N,N-dimethylcarbamate, *dimetilan*, m.p. 68–71°C, oral LD$_{50}$ rat 65 mg/kg. *Pyrolan*, m.p. 50°C, is the 1-phenyl analog of isolan and has similar properties but is less active.

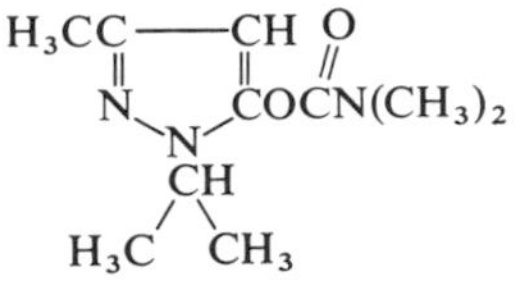

isolan

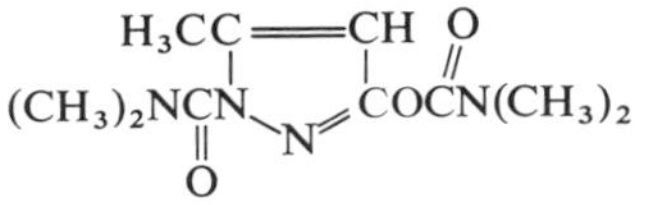

dimetilan

Aromatic N-methylcarbamates are derivatives of phenyl N-methyl-carbamate with a great variety of chloride, alkyl, alkylthio, alkoxy, and dialkylamino side chains. They are related to the alkaloid physostigmine from *Physostigma venenosum* and the first effective insecticide of this type, *m-tert*-butylphenyl *N*-methylcarbamate was described by Kolbezen, Metcalf, and Fukuto(76). A number of these compounds are shown in Table 1-23. These compounds are generally crystalline materials of

TABLE 1-23
RELATIVE TOXICITY OF SUBSTITUTED PHENYL N-METHYL CARBAMATES(99)

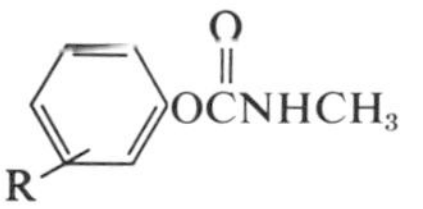

$R = $	I_{50}M fly ChE	Topical LD_{50} fly *Musca domestica,* mg/kg alone	1:5 P.B.[a]	LC_{50} mosquito larva *Culex quinquefasciatus,* ppm
H	2×10^{-4}	500	38	>10
o-Cl	5.0×10^{-6}	75	24	>10
m-Cl	5.0×10^{-5}	500	36	>10
o-CH_3	1.4×10^{-4}	500	85	10
m-CH_3	1.4×10^{-5}	50	27	3.3
o-CH_3O	3.7×10^{-5}	92.5	18	>10
m-CH_3O	2.2×10^{-5}	90.0	14.5	10
o-CH_3S	9.0×10^{-7}	48.5	14.0	3.9
m-CH_3S	7.0×10^{-6}	8.5	6.5	1.5
o-$(CH_3)_2CH$	6.0×10^{-6}	95	24	0.56
m-$(CH_3)_2CH$	3.4×10^{-7}	90	9	0.03
p-$(CH_3)_2CH$	7.0×10^{-3}	>500	500	>10
o-$(CH_3)_2CHO$	6.9×10^{-7}	25.5	7	0.3
D-2-$C_2H_5CH(CH_3)$	6.0×10^{-6}	515	41.0	2.12
L-2-$C_2H_5CH(CH_3)$	1.0×10^{-6}	171	23.5	0.26
3,5-diCH_3-4-$(CH_3)_2N$	3.3×10^{-6}	60	13.5	0.49
3,5-diCH_3-4-CH_3S	1.2×10^{-6}	24	12.5	0.23

[a]With 5 parts piperonyl butoxide synergist.

low water solubility but soluble in organic solvents. They are notably unstable in alkaline solution and when exposed to the enzymes of plant and animal tissues, but may persist for weeks or even months on inert surfaces.

Carbaryl or SEVIN, the most widely used, is 1-naphthyl N-methyl-carbamate. It has a m.p. 142°C, water solubility 0.004%, and is soluble to 10% in xylene and 20% in methyl isobutyl ketone. ZECTRAN is 4-dimethylamino-3,5-xylenyl N-methylcarbamate, m.p. 85°C, water solubility 0.01%. The closely related 4-dimethylamino-3-tolyl N-methyl-carbamate, m.p. 93°C is *aminocarb* or MATACIL.

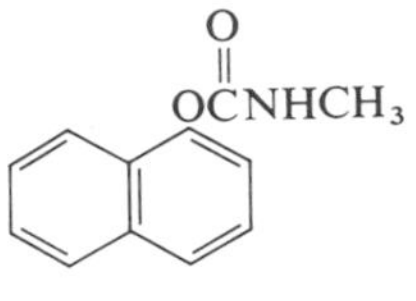

$$\underset{\text{carbaryl}}{}$$

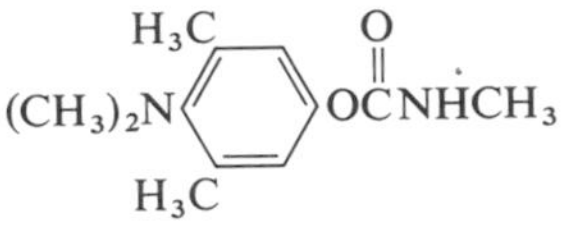

O
‖
OCNHCH$_3$

carbaryl

H$_3$C
(CH$_3$)$_2$N — O — OCNHCH$_3$
H$_3$C

ZECTRAN

O
‖
OCNHCH$_3$
O   CH$_3$
CH
CH$_3$

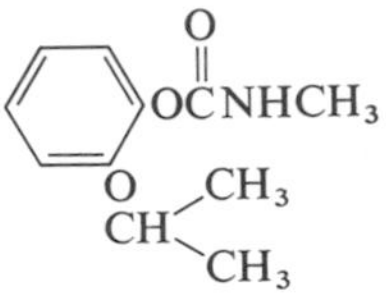

BAYGON

BAYGON or 2-isopropoxyphenyl N-methylcarbamate, m.p. 91°C, is a broad spectrum insecticide soluble in water to about 0.1% and MESUROL is 4-methylthio-3,5-xylenyl N-methylcarbamate m.p. 121°C.

Two newer compounds with outstanding systemic action are TEMIK or 2-methyl-2-(methylthio)-propionaldehyde O-(methylcarbamoyl)-oxime, m.p. 100°C, water soluble to 0.6%, and 2,2-dimethylbenzofuran-7-yl N-methylcarbamate, m.p. 147°C (carbofuran).

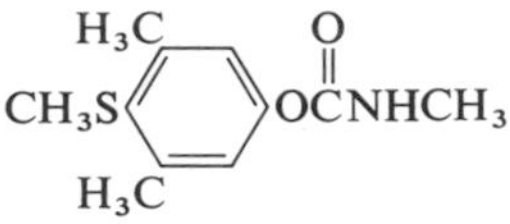

H$_3$C
CH$_3$S — O — OCNHCH$_3$
H$_3$C

MESUROL

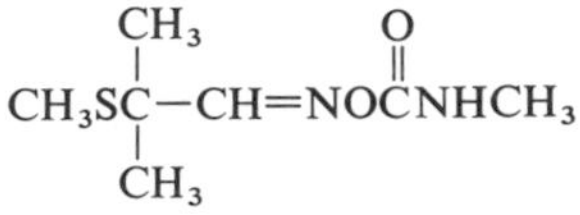

CH$_3$ O
‖   ‖
CH$_3$SC—CH=NOCNHCH$_3$
CH$_3$

TEMIK

O
‖
OCNHCH$_3$
H$_3$C   O
H$_3$C

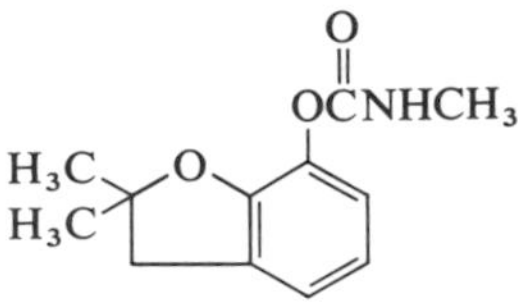

carbofuran

2. Structure–Activity Relationships. The key feature for activity in the insecticidal carbamates is structural resemblance to acetyl choline, so that the carbamate has a high affinity for the enzyme cholinesterase (see Sec. 3). The key features are a bulky side chain capable of interacting with the anionic site of cholinesterase and located about 5 Å from the C=O group. In the substituted phenyl N-methylcarbamate, this is demonstrated by the increase in I_{50} values for cholinesterase and in insecticidal activity as the substituent increases in Van der Waal's radius from H to CH$_3$ to (CH$_3$)$_2$CH, which increases the anticholinesterase activity about 1000 times. (Table 1-23). The critical nature of the spacing between side chain and C=O group is shown by the relative activities of the isopropylphenyl N-methylcarbamates, $p = 1$, $o = 800$, and $m = 5000$; and by L-(2-sec-butyl)-phenyl N-methylcarbamate which is 6 times as active as the D-isomer [Table 1-23,(99)].

3. Mode of Action. The insecticidal carbamates are highly cholinergic, and poisoned insects and other animals exhibit violent convulsions and other neuromuscular disturbances. The compounds are strong inhibitors of cholinesterase (I_{50} values 10^{-8} to 10^{-6} M in vitro), and may also have a direct effect on acetyl choline receptors, because of their pronounced structural resemblance to acetyl choline:

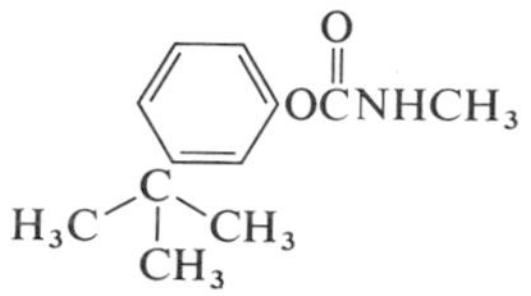

m-*tert*-butyl-phenyl
N-methylcarbamate

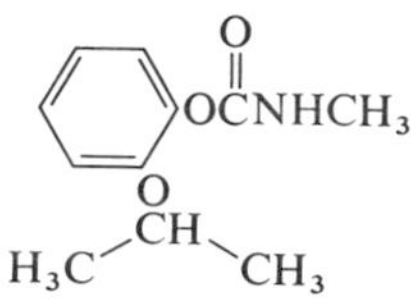

BAYGON

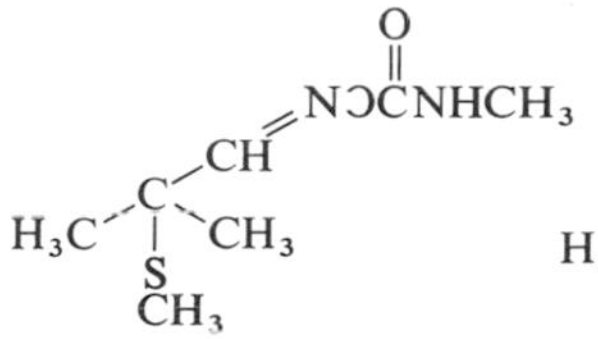

TEMIK

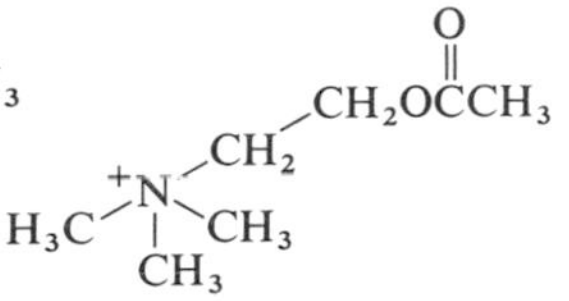

acetyl choline

The importance of structural complementarity to cholinesterase is discussed under "Structure–Activity Relationships."

The mechanism for carbamate reaction with cholinesterase is analogous to the normal 3-step hydrolysis of acetyl choline (Sec. 1-4, M)

$$CH_3NH\overset{O}{\overset{\|}{C}}OR + EH \underset{}{\overset{k_1}{\rightleftharpoons}} CH_3NH\overset{O}{\overset{\|}{C}}OR \cdot EH \qquad (1)$$

$$CH_3NH\overset{O}{\overset{\|}{C}}OR \cdot EH \overset{k_2}{\longrightarrow} CH_3NH\overset{O}{\overset{\|}{C}} \cdot E + ROH \qquad (2)$$

$$CH_3NH\overset{O}{\overset{\|}{C}} \cdot E + H_2O \overset{k_3}{\longrightarrow} EH + CH_3NH\overset{O}{\overset{\|}{C}}OH \qquad (3)$$

The carbamates penetrate rapidly into insects, from 40 to 80% of a topical dosage of 1 μg per house fly, penetrating within 4 to 8 hr. In some insect species detoxication is very rapid and proceeds by two general processes (a) hydroxylation of aryl ring or N-CH$_3$ group, and (b) hydrolysis of ester link. In mammals the phenols are rapidly eliminated as conjugates in the urine and the carbamates are not accumulated in the body or excreted in milk.

An interesting feature of the insecticidal action of these carbamates is the pronounced synergism shown by the addition of methylenedioxy compounds such as piperonyl butoxide, sesoxane, and propyl isome

(Sec. 1-3, D). These prevent the in vivo detoxication of the carbamates by inhibiting the "mixed function oxidase" detoxication system.

4. Biological Activity. The insecticidal carbamates are often highly specific in their action. Isolan and dimetelan have been used as baits with sugar or as impregnated bands and disks for the control of house flies and fruit flies (Trypetidae). Isolan is also an effective aphicide and has strong systemic action. Carbaryl is a very widely used general purpose insecticide registered for use on more than 100 crops and is especially useful on cotton, forage, fruits, and vegetables. ZECTRAN and MATACIL are broad spectrum compounds particularly effective against lepidopterous larvae and on ornamental pests including snails and slugs. BAYGON is highly active to flies, mosquitoes, cockroaches, ants, and other household pests. It is also a very effective residual treatment of dwellings for control of adult mosquitoes in malaria eradication. MESUROL is active against pests of fruits. TEMIK used as a 5% granular preparation for seed and soil treatments, is a very effective systemic insecticide.

5. Toxicology. The acute toxicities of the carbamates range from high to very low as shown by the following oral LD_{50} values to the rat: carbaryl 540, MESUROL 135, BAYGON 100, ZECTRAN 60, MATACIL 50, isolan 54, dimetilan 65, carbofuran 5, and TEMIK 1.0 mg/kg. However, the dermal toxicity values are generally very much lower, i.e., carbaryl > 4000 and BAYGON > 1000 mg/kg. These compounds are rapidly metabolized and excreted in the urine. They are reversible cholinesterase inhibitors and recovery from toxic dosage, which produces characteristic cholinergic symptoms, is unusually rapid. Atropine is of value as an antidote (55).

O. Acaricides

Most of the insecticidal chemicals discussed, with the exception of the dinitrophenol and organophosphorous insecticides, are not of practical value in controlling the mites and ticks of the order Acarina. However, a number of acaricidal chemicals have come into widespread use which have almost specific toxicity to the mites but are inactive against insects. In general, these acaricides are highly stable compounds with comparatively prolonged residual action and low mammalian toxicity. Certain of the compounds described below are effective only as ovicides, killing the eggs and sometimes the newly emerged nymphs, while others are active against all the stages of the mites. These acaricides exhibit a considerable degree of specificity for various species of Acarina and have been found most useful for the phytophagous Tetranychidae and Erio-phyidae(94, 95).

DMC or DIMITE is di-(*p*-chlorophenyl) methyl carbinol or 1,1-bis-(*p*-chlorophenyl)-ethanol, a white solid, m.p. 69.5 to 70°C. It is insoluble in water and has the following solubilities in gram per 100 ml of solvent at 25 to 30°C: petroleum ether, 4.3; ethanol, 125; and toluene, 110. The compound is readily dehydrated upon heating or in the presence of strong acids and forms the inactive 1,1-bis-(*p*-chlorophenyl)-ethylene. Technical DMC contains small amounts of the *o-p'* and *o-o'* isomers and traces of isomeric dichlorobenzophenones. DMC is active against all stages of mites. It has an oral LD_{50} to the rat of about 200 mg/kg.

Chlorobenzilate is ethyl *p,p'*-dichlorobenzilate, a yellowish viscous oil, b.p. 141 to 142°C at 0.06 mm. The technical material, d. 1.2816, contains about 90% of the active compound, is insoluble in water, and soluble to more than 40% in deodorized kerosene, benzene, and methyl alcohol. Chlorbenzilate is hydrolyzed in alkali and in strong acids to the inactive *p-p'*-dichlorobenzilic acid and ethanol. The compound is active against all stages of mites and has an oral LD_{50} to the rat of about 1000 mg/kg. The corresponding isopropyl ester, *chloropropylate*, has similar properties and uses with an oral LD_{50} to the rat of 34,600 mg/kg.

Dicofol or KELTHANE is 1,1-bis-(*p*-chlorophenyl)-2,2,2-trichloro-ethanol, a white crystalline solid, m.p. 78.5 to 79.5°C. This compound is insoluble in water and soluble in organic solvents, and in the presence of alkali forms the inactive *p-p'*-dichlorobenzophenone and chlorotorm. Dicofol is a long-lasting acaricide and is active against all stages of mites. It has an oral LD_{50} to the rat of about 1000 mg/kg.

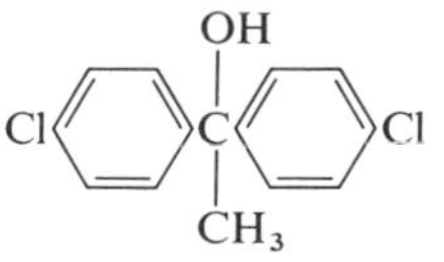

DMC 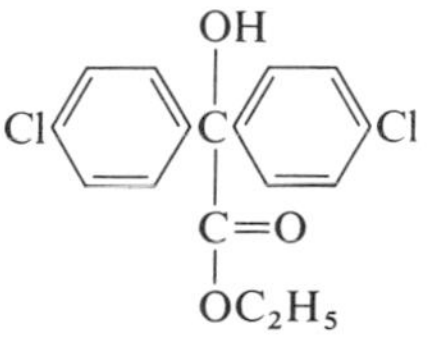   chlorobenzilate 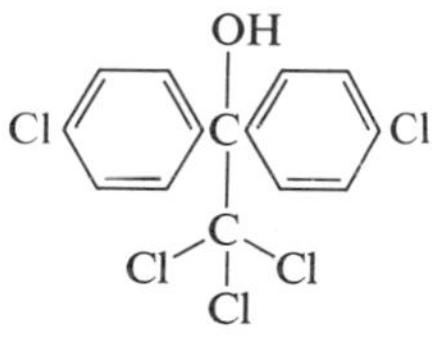   dicofol

Tetradifon or TEDION is 2,4,5,4'-tetrachlorodiphenyl sulfone, a crys-talline solid, m.p. 148°C. It is insoluble in water and has the following solubilities in gram per 100 g of solvent at 18°C: petroleum ether, 0.4; ethyl acetate, 7.1; carbon tetrachloride, 1.6; methyl ethyl ketone, 10.5; xylene, 11.5. Tetradifon is stable to the action of acids and alkalies, light and temperature, and has a very prolonged residual action. It is active against all stages of mites and has an oral LD_{50} to the rat of > 14,700 mg/kg. The corresponding sulfide, *tetrasul*, or 2,4,5,4'-tetrachlorodiphenyl sulfide, m.p. 88°C is also a practical acaricide with a similarly low acute toxicity.

SULPHENONE is *p*-chlorophenyl phenyl sulfone, a white solid, m.p. 98°C. The technical material consists of about 80% of this compound, with small amounts of *o*- and *m*-isomers, bis-(*p*-chlorophenyl) sulfone, and diphenyl sulfone. SULPHENONE is insoluble in water and has the following solubilities in gram per 100 g of solvent at 20°C: hexane, 0.4; xylene, 18.2; carbon tetrachloride, 4.9; and acetone, 74.4. SUL-PHENONE is effective against all stages of mites. It has an oral LD_{50} to the rat of > 2000 mg/kg.

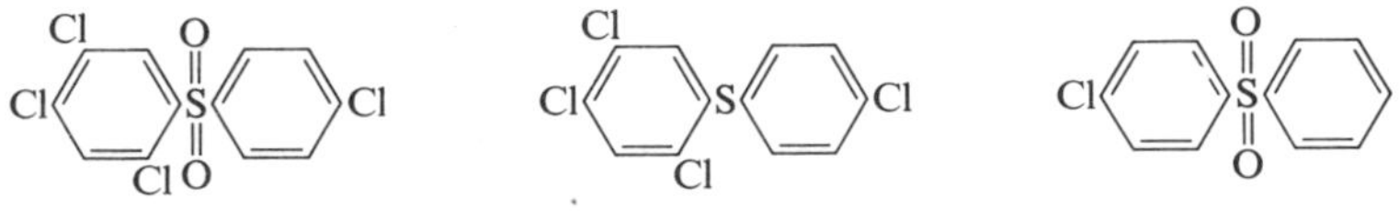

tetradifon   tetrasul   SULPHENONE

Chlorfenson, *ovex* or OVOTRAN is *p*-chlorophenyl *p*-chlorobenzene sulfonate, a white solid, m.p. 86.5°C. It is insoluble in water and has the following solubilities in gram per 100 g of solvent at 25°C: kerosene, 2; carbon tetrachloride, 41; cyclohexanone, 110; ethyl alcohol, 1; ethylene dichloride, 110; and xylene, 78. Ovex is effective only as an ovicide. It has an oral LD_{50} to the rat of 2000 mg/kg.

Two analogs of ovex have similar solubilities, ovicidal properties, and like ovex are hydrolyzed in alkali to form the phenol and benzene sulfonate salt. GENITE is 2,4-dichlorophenyl benzene sulfonate, m.p. 45 to 47°C, v.p. 2.7×10^{-4} mm at 30°C. The technical product is about 97% pure and has an oral LD_{50} to the rat of 1400 mg/kg. *Fenson* is *p*-chlorophenyl benzene sulfonate, m.p. 61 to 62°C, d. 1.33.

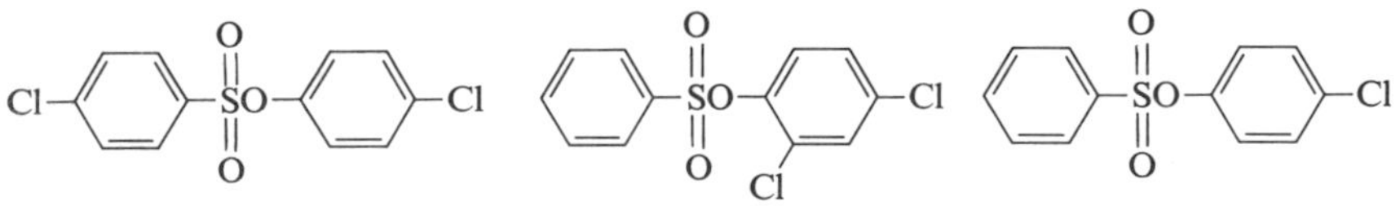

ovex   GENITE   fenson

Chlorbenside or MITOX, is *p*-chlorobenzyl *p*-chlorophenyl sulfide, a white crystalline solid, m.p. 74°C, v.p. 2.6×10^{-6} mm at 20°C. It is insoluble in water and has the following solubilities in grams per 100 g of solvent at 20°C: kerosene, 5 to 7.5; methyl ethyl ketone, 137; xylene,

93. The technical product contains about 90% *p-p'* isomer, 5% *o-p'* isomer, and 2.5% *m-p'* isomer. Chlorbenside is unaffected by reduction and by acid and alkaline hydrolysis but it is readily oxidized to *p*-chlorobenzyl *p*-chlorophenyl sulfoxide, m.p. 125°C, and more slowly to *p*-chlorobenzyl *p*-chlorophenyl sulfone, m.p. 150°C. These reactions occur on the leaf surface, and the oxidation products are acaricidal, but do not penetrate locally into the leaf tissue as does chlorbenside. Chlorbenside is active only against eggs and immature mites. The oral LD_{50} to the rat is > 10,000 mg/kg.

ARAMITE is 2-(*p-tert*-butylphenoxy)-isopropyl 2'-chloroethyl sulfite, a brownish oil, b.p. 175°C at 1 mm d. 1.148. The technical material contains at least 90% active ingredient and is insoluble in water and soluble in aromatic solvents. ARAMITE hydrolyzes in alkali to form 1-*p-tert*-butylphenoxypropan-2-ol, ethylene oxide, and inorganic sulfite and under strong sunlight liberates SO_2. It has an oral LD_{50} to the rat of 3900 mg/kg. ARAMITE is registered only for mite control on ornamentals and other nonfood crops. A closely related acaricide is OMITE, or 2-(*p-tert*-butyl-phenoxy)-cyclohexyl 2-propynyl sulfite, which has similar usage and an oral LD_{50} to the rat of 2500 mg/kg.

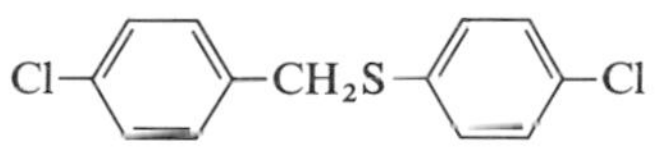

chlorbenside

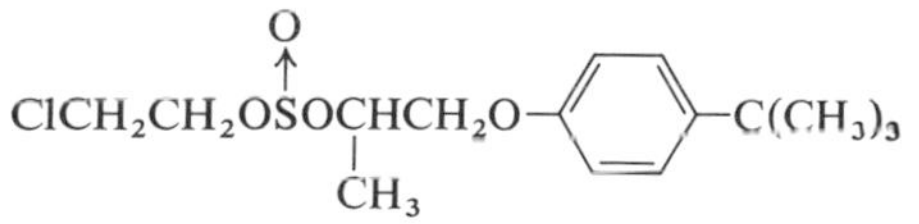

ARAMITE

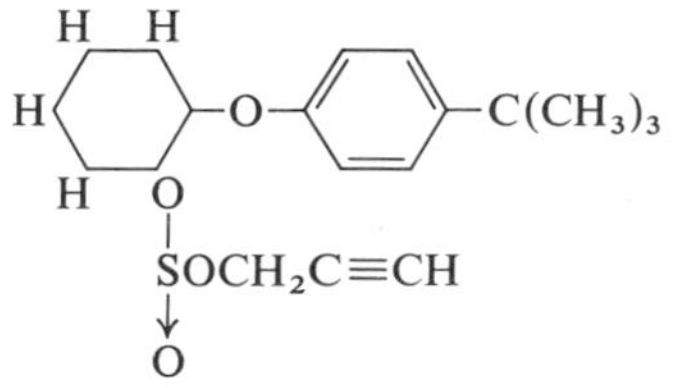

OMITE

Oxythioquinox or MORESTAN is 6-methyl-2,3-quinoxalinedithiol cyclic carbonate, a yellow crystalline solid, m.p. 172°C. It is insoluble in water and soluble in organic solvents and has an oral LD_{50} to the rat of 2500 mg/kg.

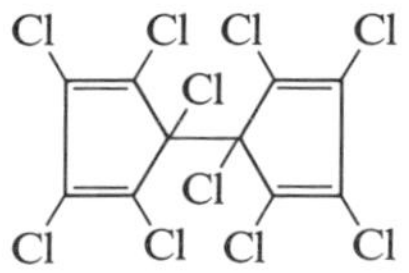

PENTAC

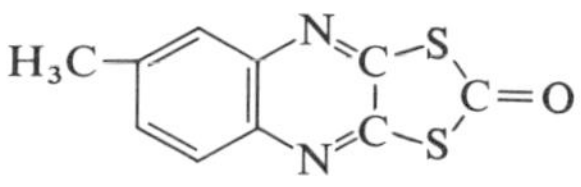

oxythioquinox

Several other acaricides have more selective action or specific use. NEOTRAN is bis-(p-chlorophenoxy)methane, m.p. 70 to 72°C, which has an acute oral LD_{50} to the rat of 5800 mg/kg. It is effective against all stages of the citrus and European red mites.

$$Cl-\langle\!\!\!\rangle-OCH_2O-\langle\!\!\!\rangle-Cl$$

NEOTRAN

Azobenzene, $C_6H_5N\equiv NC_6H_5$, is an orange dyestuff, m.p. 68°C, b.p. 297°C, d. 1.203, that is effective as an ovicide for spider mites in greenhouses as a spray or when volatilized by burning 45% pyrotechnic mixture or from a 70% paste applied to steam pipes. Eight grams are used per 1000 ft³. Azobenzene is insoluble in water and soluble in most organic solvents.

The fungicide *zineb* or zinc ethylene-bis-dithiocarbamate (Sec. 1-3, D) has proved to be especially effective for the control of the citrus rust mite.

P. Mothproofing Compounds

The clothes moths *Tinea pellionella*, *Tineola bisselliella*, and *Tricho-phaga tapetzella*; and the carpet beetles, *Anthrenus* spp., cause as much as $500,000,000 damage annually to clothing, rugs, and upholstery. There is, therefore, great demand for insecticidal protection of these articles. Several common insecticides are widely used for household mothproofing. *Sodium fluosilicate* (Sec. 1-3, B) used at 0.5 to 0.7% in water solution with 0.25 to 0.5% wetting agent is fixed in wool and is an effective mothproofing agent. *DDT* at 0.25 to 0.75% and *dieldrin* at 0.05% of the dry weight of cloth have been widely used to treat fabrics either by spraying in a volatile solvent, by application in dry cleaning bath, or by application of an aqueous emulsion. Such treatments are invisible, do not affect the properties of the fabric, and will remain effective against fabric pests for months to years if not repeatedly washed or dry cleaned(*94*).

Paradichlorobenzene and *naphthalene* (Sec. 1-9, S) are used as space fumigants to destroy fabric pests in confined storage spaces.

Thousands of organic chemicals have been evaluated as colorless dye-stuffs for application to fabrics during the dyeing operation. Two commercial compounds that are stated to provide protection from fabric pests during the lifetime of the fabric, when applied at 1 to 3% of the weight of cloth   are   5-chloro-α,α-bis-(3,5-dichloro-2-hydroxyphenyl)-*o*-toluene sulfonic acid or EULAN, and N-3,4-dichlorophenyl N'-2-(4-chlorophenoxy-2-sulfonic acid)-5-chlorophenyl urea or MITIN FF. Both compounds are applied to fabrics as the water-soluble sodium salts and are fixed to wool fibers by a chemical combination with basic groups of the proteins.

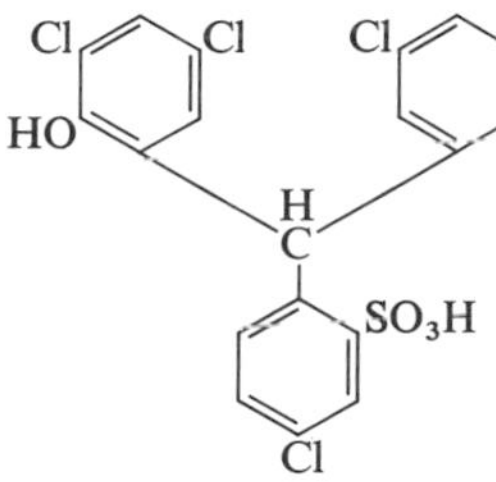

EULAN 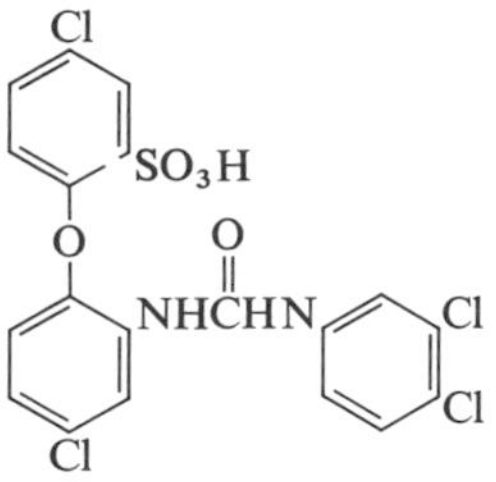   MITIN FF

Q. Microbial Insecticides

Several microbial products have been used as insecticides. They are distinguished by a very high degree of specificity and appear to be harmless to higher animals. Spores of *Bacillus popillae* diluted to 100×10^6 per gram of powder are applied to the soil to control the larvae of the Japanese beetle, *Popillia japonica*.

Bacillus thuringiensis contains a crystalline, proteinaceous toxin that has been used for the control of the cabbage looper, *Trichoplusia ni,* and the alfalfa caterpillar *Colias philodice eurytheme*. The products contain from 70 to 150×10^9 spores per gram of powder.

R. Petroleum Oils

Petroleum oils play a dual role as pesticides. First as carriers for many types of toxicants in solution or in emulsive formulations and second as the actual toxicant in acaricides, insecticides, herbicides, and fungicides. Consideration of the first use lies outside the scope of this chapter, but brief mention is made of the use of oils as pesticides.

1. Acaricides and Insecticides. Spray oils were introduced as insecticides as early as 1763, but it was not until the development of kerosene emulsions about 1870 that such materials could be applied safely to foliage, and in 1904 the first commercial oil emulsions were marketed (94). Insecticidal spray oils are used (a) for dormant sprays of scale insects, mites, insect eggs; (b) as foliage sprays for aphids, mealybugs, and scale insects; (c) as mosquito larvicides; and (d) as parasiticides for lice, fleas, and mites attacking domestic animals. Important characteristics in spray oils are the *volatility* or distillation range, the *viscosity* or resistance to flow, and the *purity* or degree of refinement as measured by the unsulfonated residue. In general, the higher the boiling point the more toxic the oil to the insect; the practical problem is to use the heaviest oil that can be used safely on the plant. Petroleum oils used as insecticides consist largely of aliphatic hydrocarbons but may be paraffinic or iso-paraffinic. Trends in the development of spray oils have been toward more highly refined narrow-cut fractions with low viscosities, which are safer on plants. The properties of typical foliage spray oils are(*94*):

Grade	Distillation at 636°F,%	Unsulfonated residue, %	Viscosity Saybolt sec, 100°F
Light medium	52–61	92	60–75
Medium	40–49	92	70–85
Heavy	10–25	94	90–105

Three types of oil formations are in use (1) oil emulsions containing 80 to 90% oil emulsified in a small amount of water, (2) emulsive oils containing only oil and emulsifier, and (3) tank mix oils which are to be added to the spray tank with water and emulsifier.

Oils kill insects and mites by covering them with a continuous film which causes death by suffocation.

2. Herbicides. Oils for killing plants are generally of lower boiling range with a gravity range of 25° to 30° A.P.I. and contain 50% or more of aromatic constituents(*29*). There are several theories of the mode of action of phytocidal action of oils. Van Overbeek and Blondeau(*147*) believe that oils become incorporated in the plasma membrane and cause a breakdown in the protoplasm. Thus, within any configuration series of oil molecules of increasing size, the small molecules are more toxic than the larger. Solubilization is very low with the high boiling foliage spray oils. Riehl and Wedding(*119*) have studied the effects of oils on photosynthesis and suggest that toxic oils inhibit this process by interference with gaseous exchange.

S. Fumigants

Fumigants are chemicals that are toxic to insects, nematodes, weeds, or microorganisms when present in a confined space in the gaseous phase. Therefore, fumigants must have a substantial vapor pressure so that a lethal concentration can exist in the gaseous state at normal environmental temperatures. This physical requirement greatly limits the number of compounds that may be usefully employed in fumigation. For general fumigation of structures where exposures are limited to a few hours, compounds that boil at or about room temperature, such as hydrogen cyanide, methyl bromide, and ethylene oxide, are most useful. In soil fumigation, however, exposures are much longer and the slower release of gas from substances such as dichloropropene, ethylene dibromide, or dichlorobromopropane, which boil as high as 200°C, is effective. Substances of relatively high vapor pressure, such as naphthalene and para-dichlorobenzene, that sublime readily at room temperatures are used in tight enclosures to fumigate for clothes moths and dermestid beetles. A number of contact insecticides, such as nicotine, dichlorvos or DDVP, mevinphos, lindane, and azobenzene, may kill insects by vapor action under certain circumstances, and their fumigant action may be enhanced through atomization, volatilization by heating, or through burning in pyrotechnic mixtures (94).

Perhaps the earliest fumigant was carbon bisulfide, CS_2, which was used in France in 1858 to control pests of stored grain. *Hydrogen cyanide*, HCN, was employed as early as 1886 to fumigate citrus trees infested with cottony cushion scale, *Icerya purchasi*, and was used to fumigate nursery stock in 1890, greenhouse pests in 1894, households in 1898, and mills in 1899. Soil fumigation techniques began with the use of carbon bisulfide against the grape phylloxera, *Phylloxera vitifoliae*, in France in 1872 and the development of soil fumigants as Nematocides is discussed in Sec. 1-6.

Present day fumigants are listed in Table 1-24 according to their properties and general areas of usefulness. The low-boiling fumigants, such as hydrogen cyanide, methyl bromide, ethylene oxide, sulfur dioxide, and sulfuryl fluoride, are available as compressed or liquified gases in cylinders. The soil fumigants (Sec. 1-6) and most of the grain and commodity fumigants are applied as liquids and allowed to volatilize. Special techniques are sometimes used to generate fumigant materials. Hydrogen cyanide can be released by the exposure of calcium cyanide to the moisture in the atmosphere:

$$Ca(CN)_2 + 2H_2O \longrightarrow Ca(OH)_2 + 2HCN$$

or by reaction of sodium cyanide with sulfuric acid

$$NaCN + H_2SO_4 \longrightarrow NaHSO_4 + HCN$$

TABLE 1-24
Fumigants, Their Uses, and Properties (94)

Compound	Formula	b.p., °C	Specific gravity		v.p., mm Hg, 20°C	Solubility, g/100 ml., H_2O, 20°C	Flammability in air vol. %	Safe limit, ppm
			liquid d_4^{20}	gas (air = 1)				
A. General commercial fumigants								
Acrylonitrile	$CH_2{=}CHCN$	78	0.797	1.8	83		3	20
Carbon bisulfide	CS_2	43.6	1.263	2.6	314	0.22	1	20
Carbon tetrachloride	CCl_4	76	1.595	5.3	89	0.08	F	10
1,1-Dichloro-1-nitroethane	$CH_3CCl_2NO_2$	124	1.415		$16.9^{29°}$	0.25		10
Ethylene dichloride	$ClCH_2CH_2Cl$	83.5	1.257	3.4	78	0.87	6	50
Ethylene oxide	$(CH_2)_2O$	10.7	0.887°	1.5	1095		3	50
Hydrogen cyanide	HCN	26	0.688	0.9	630		6	10
Methyl bromide	CH_3Br	4.5	$1.732^{0°}$	3.3	1420	$1.34^{25°}$	13.5	20
Methylene chloride	$ClCH_2Cl$	40	1.325			2	NF	500
Sulfur dioxide	SO_2	−10	1.434	1.433	2453	$8.5^{25°}$	NF	5
B. Soil fumigants								
Chloropicrin	Cl_3CNO_2	112	1.651	5.7	20	0.19	NF	0.1
1,2-Dibromo-3-chloropropane	$CH_2BrCHBrCH_2Cl$	199	2.08		0.58	0.12		1.0
β,β'-Dichlorodiethyl ether	$ClCH_2CH_2OCH_2CH_2Cl$	178	1.222	4.9	0.73	1.1	F	15
1,2-Dichloropropane	$CH_2ClCHClCH_3$	95.4	1.159		210	0.27		75
trans-1,3-Dichloropropene	$ClCH{=}CHCH_2Cl$	111	1.224		18.5	0.28	F	1
Ethylene chlorobromide	$ClCH_2CH_2Br$	107	1.689		$40.0^{30°}$	$0.7^{30°}$	F	
Ethylene dibromide	$BrCH_2CH_2Br$	131.6	2.172	6.5	7.7	0.34	NF	25
Methyl isothiocyanate	CH_3NCS	119	$1.069^{37°}$				NF	
Propargyl bromide	$CH{\equiv}CCH_2Br$	88	1.520					0.1

C. Commodity fumigants

Name	Formula	bp (°C)	d	vapor density	vapor pressure	solubility	flammability	TLV
Ethyl formate	$HCOOC_2H_5$	54	0.917			10		100
Methyl formate	$HCOOCH_3$	32	0.974			30		
Phosphine	PH_3	−87.4	$0.746^{-90°}$	1.2	$624^{25°}$		2	0.3
Trichloroacetonitrile	Cl_3CCN	85	$1.44^{25°}$		$73^{25°}$	insol.	NF	
Trichlorethylene	$ClCH{=}CCl_2$	86.7	$1.47^{15°}$			1		100
Propylene oxide	$CH_2CH(O)CH_3$	35	$0.859°$					

D. Structural fumigant

Name	Formula	bp (°C)	d	vapor density	vapor pressure	solubility	flammability	TLV
Sulfuryl fluoride	SO_2F_2	−55.2	$1.342^{25°}$	3.5	12010	$0.075^{25°}$	NF	5

E. Fabric, household, greenhouse

Name	Formula	bp (°C)	d	vapor density	vapor pressure	solubility	flammability	TLV
Dimethyldichlorovinyl phosphate	$(CH_3O)_2P(O)\text{-}OCH{=}CCl_2$	120/14 mm	1.415					
Naphthalene	$C_{10}H_8$	(m.p. 80) 218		4.4	$0.08^{25°}$	0.003	F	10
Paradichlorobenzene	$C_6H_4Cl_2$	(m.p. 53) 173		5.1	$1^{25°}$	$0.008^{25°}$	NF	75
Nicotine	$C_{10}H_{14}N_2$	247	1.009		$0.0425^{25°}$			

Phosphine is generated in situ as a grain fumigant by reaction of pellets of aluminum phosphide and ammonium carbamate to moisture:

$$AlP + 2NH_4OC(O)NH_2 + 3H_2O \longrightarrow PH_3 + Al(OH)_3 + 4NH_3 + 2CO_2$$

T. Wood Preservatives

Wooden articles particularly those in the soil and in water are often exposed to the attacks of a wide variety of organisms ranging from bacteria and more than 2000 species of Basidiomycetes fungi to termites and powder post beetles, *Lyctus* spp. The articles most frequently protected by treatment with wood preservatives are railroad ties, utility poles, construction timbers, and piles.

Inorganic salts such as *zinc chloride*, $ZnCl_2$, and *chromated zinc* chloride are used as water-soluble impregnants that can be applied to newly cut trees and dispersed throughout the wood by translocation.

Metal organic salts such as *copper naphthenate*, the copper salts of various naphthenic acids; and copper 8-hydroxyquinolinate (Sec. 1-2, I) are used as fungicides for field boxes, bins, and other harvesting and handling equipment.

The most widely used preservatives against termites and fungi are *pentachlorophenol*, C_6Cl_5OH, m.p. 190°C, soluble in water to 0.0014% and in *o*-dichlorobenzene to 8.5% and diesel oil to 3.1%; and *sodium pentachlorophenate*, which is soluble in water to 26%. These materials are highly irritant to eyes, nose, and skin and should be handled with care. The rat oral LD_{50} for pentachlorophenol is 180 mg/kg. The chlorinated benzenes: *o-dichlorobenzene*, $C_6H_4Cl_2$, b.p. 179°C, d.[15] 1.315 and a mixture of 1,2,3- and 1,2,4-*trichlorobenzene* b.p. 205 to 250°C, d. 1.460 to 1.477, are used as soil poisons for termites.

Creosote, which is a complex mixture of crude coal tar distillates, is a cheap and widely used wood preservative for crude timbers.

1-5 MOLLUSCACIDES

Snails and slugs are important pests of vegetable and flower gardens and citrus. Aquatic snails of several genera including *Bulinus*, *Oncomelonia*, and *Australorbis*, cause immense injury as the alternate hosts of the *Schistosoma* parasites that cause human schistosomiasis. Specific pesticides that control these snails and slugs are called molluscacides.

Metaldehyde, $[OCH-CH_3]_4$, m.p. 246°C, is a specific attractant and toxicant for garden snails and slugs and is used at 1.5% with moist bran bait. It is insoluble in water and soluble in benzene, chloroform, and alcohol. Calcium arsenate at 6 to 7% (Sec. 1-4, A) is often added to improve the toxicity of the bait. Sprays of *tartar emetic* or potassium antimonyl tartarate, $OSbOOC(CHOH)_2COOK$, with sugar are also used to control these snails.

The carbamate insecticide *isolan* or 1-isopropyl-3-methylpyrazolyl-(5) N,N-dimethylcarbamate (Sec. 1-4, N) has been found to be highly effective in baits and sprays against snails(*113*). Another insecticidal carbamate ZECTRAN or 4-dimethylamino-3,5-xylenyl *N*-methylcarbamate (Sec. 1-4, N) has also proved to be highly effective against garden snails and slugs used in baits of grain, apple, molasses, and amyl acetate(*42*).

The organophosphorus insecticide azinphosmethyl or GUTHION (Sec. 1-4, M) sprayed on citrus at 0.4% was highly effective in controlling the European brown snail, *Helix aspersa* but not the common garden slug(*114*).

Aquatic snails are controlled by a variety of toxicants. *Copper sulfate* (Sec. 1-2, A), *ziram* (Sec. 1-2, D) and copper dimethyldithiocarbamate or *cuprobam* are effective at 3 to 5 ppm. The dinitrophenols such as *dinex* or DNOCHP (Sec. 1-4, G) have also been used at concentrations of 3 to 5 ppm.

Trialkyltins especially tri-*n*-propyl tin hydroxide and tri-*n*-butyl tin acetate (Sec. 1-2, H) are extremely effective against aquatic snails and their eggs(*35*) at concentrations of 0.1 to 0.4 ppm. They persist in the aquatic environment for periods of up to 9 months and have the disadvantage of general toxicity to fish, amphibia, and other aquatic organisms.

BAYLUSCIDE or 5-chlorosalicylic 2'-chloro-4'-nitroanilide, m.p. 199°C, is a new molluscacide almost insoluble in water but which forms a water soluble ethanolamine salt. It is effective against most aquatic snails at 0.3 to 1.0 ppm and persists for several months(*34a*). It has the advantage of considerable selectivity and has a rat oral LD_{50} of > 5000 mg/kg.

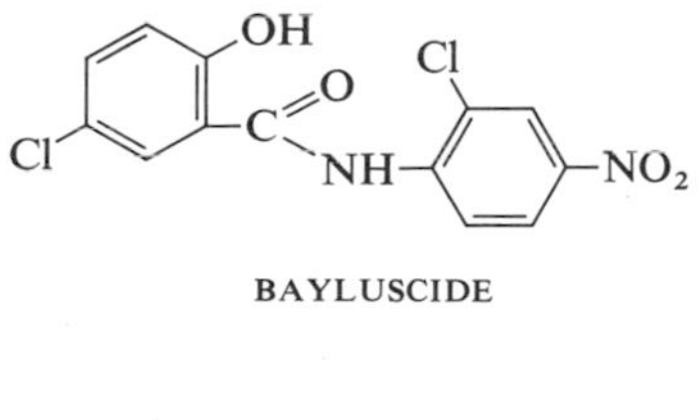

BAYLUSCIDE

1-6 NEMATOCIDES

A. Introduction and Chemistry

The plant parasitic nematodes attack the roots of most crop plants where they cause vast and often unappreciated damage. Chemical

control of these "eelworms" has been most successful through the
use of soil fumigants which have sufficient volatility to penetrate throughout the upper levels of the soil. Many of the chemicals in use are also
effective fumigants for insects and fungal spores, and their properties
are listed in detail in Table 1-24.

The first soil fumigant was *carbon bisulfide*, CS_2, which was used,
with indifferent results, by Kühn in Germany in 1871 against the sugar-
beet nematode. *Chloropicrin*, O_2NCCl_3, was first evaluated as a nematocide by Mathews in England in 1919 and was shown by extensive tests
in California in 1927–1928 and in Hawaii in 1934 to be a practical
nematocide when injected 6 to 8 in. deep at 150 to 170 lb/acre(*44*).
Large quantities of "war surplus" chloropicrin were routinely used for
soil fumigation in pineapple fields until stocks were exhausted(*143*).

The nematocidal effectiveness of *D-D mixture* (which consists of
about two parts of *1,3-dichloropropene* $CHCl{=}CHCH_2Cl$ and one part
of *1,2-dichloropropane*, $CH_2ClCHClCH_3$) at 250 lb/acre, was made
by Carter(*24*) and the development of this material began the era of
commercial soil fumigation. *Dichloropropene* alone was subsequently
introduced as a commercial fumigant.

The next year saw the beginning of the development of *1,2-dibromo-
ethane*, $BrCH_2CH_2Br$, or ethylene dibromide as a soil fumigant for
nematodes and wireworms. The most recent of these halogenated aliphatic soil fumigants to reach commercial development is *1,2-dibromo-3-
chloropropane*, $BrCH_2BrCHCH_2Cl$. This compound differs from the
other halogenated nematocides, which are used only in preplanting
treatments, in that it can be applied safely to certain perennial plants,
especially orchard trees and ornamentals, to control nematodes attacking
the roots(*90*).

Methyl bromide, CH_3Br, which is considerably more volatile than the
compounds discussed above (Table 1-24), has highly specialized use as
a nematocide in greenhouse benches and seed beds. It is effective only
when the gas is confined in containers or under plastic sheeting. A
closely related but higher boiling nematocide is *3-bromopropyne*, or
$CH{\equiv}CCH_2Br$.

In addition to these halogenated aliphatic nematocides, other types
of compounds have more recently been developed as soil fumigant
nematocides. *Metham* or VAPAM or sodium N-methyl dithiocarbamate,
$Ch_3NHC(S)$-SNa, is a water-soluble solid that is used as a nematocide
in seed beds, nursery crops, and turf and ornamentals and is also a soil
fungicide (Sec. 1-2, D). Dazomet or MYLONE or 3,5-dimethyl-1,3,5-2H-
tetrahydrothiadiazine-2-thion, m.p. 99.5°C, has similar uses. *Tetrachlorothiophene*, m.p. 29°C, is a promising new nematocide.

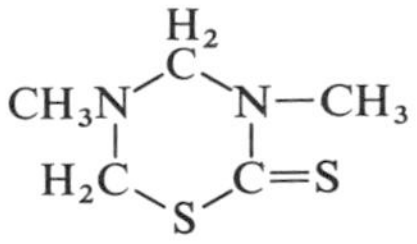

dazomet   tetrachlorothiophene

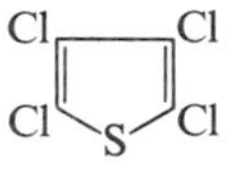

Both metham and dazomet readily form methylisothiocyanate, $CH_3N=$$C=S$, upon reaction in the soil, and this latter material is also available as a nematocide in 20% solution in chlorinated hydrocarbons as VORLEX (105).

$$CH_3NH-\overset{\overset{\displaystyle S}{\|}}{C}S^- \rightarrow CH_3N=C=S + HS^-$$

metham

$$\rightarrow CH_3N=C=S + CH_3NH_2 + HCHO + H_2S$$

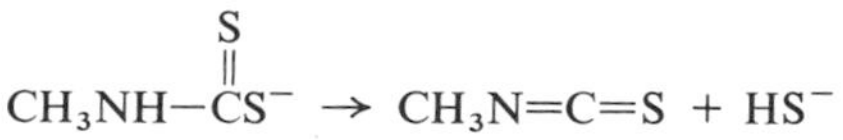

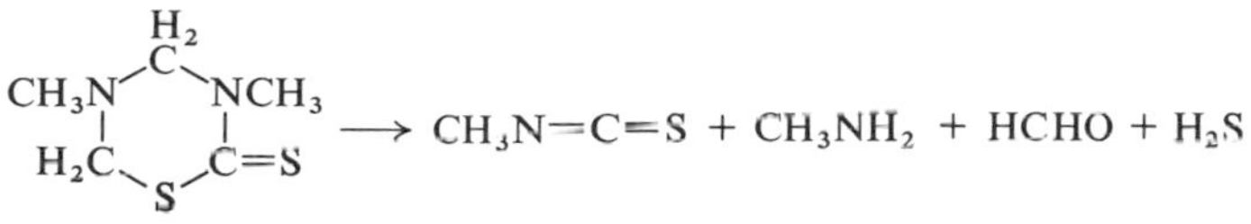

dazomet

The most recent developments in nematocides are the organophosphorus esters related to those used as insecticides (Sec. 1-4, M). These include O,O-diethyl O-(2,4-dichlorophenyl) phosphorothioate or *VC 13*, a yellowish liquid used in greenhouse and plant bed soils and in ornamentals, and O,O-diethyl O-(2-pyrazinyl) phosphorothioate or ZINOPHOS which is reported to have systemic nematocidal properties and is also an effective soil insecticide. O,O-diethyl O-(p-methylsulfinylphenyl) phosphorothionate or DASANIT (Sec. 1-4, M) is another effective anticholinesterase nematocide and soil insecticide.

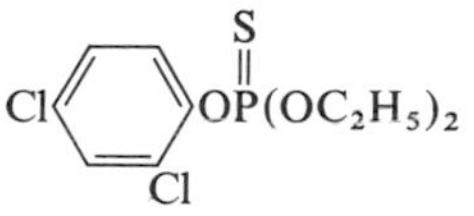

VC-13   ZINOPHOS

B. Structure–Activity Relationships

As is explained in Sec. C, the toxicity of the organohalide nematocides is a function of their reactivity with an essential nucleophilic center in the nematode, e.g., the —SH groups of an essential enzyme. Thus, the toxicity of these nematocides, as represented by the generalized formula

RX, is the nature of the leaving group X and the groups attached to the carbon atom in R. When the group R remains constant, the order of nematocidal activity with variation in X is generally I > Br > Cl(*106*). Thus, with the citrus nematode *Tylenchulus semipenetrans*, the relative order of effectiveness was $CH_3(CH_2)_2I > CH_3(CH_2)_3Br > CH_3(CH_2)_4$-Cl > CH_2=$CHCH_2I > CH_2$=$CHCH_2Br > CH_2$=$CHCH_2Cl$.   This order of activity corresponds to the relative displacement rate of the halogens in the biomolecular S_N2 reaction (Sec. C), which is Cl-2 × 10⁻², Br-1, and I-3(*105*).

With regard to the nature of the alkyl groups in the RX-type compound, the toxicity of alkyl halides decreases with increasing chain length of R in the order $CH_3 > C_2H_5 > C_3H_7$, again in the same order as the order of reactivity in S_N2 reaction (Sec. C), which for alkyl bromides is: CH_3 1, C_2H_5 1.3 × 10⁻², C_3H_7 8.5 × 10⁻³, $(CH_3)_3C$ 3.9 × 10⁻⁵(*105*).

The 2,3- unsaturated halides are from 50 to 100 times as reactive as the corresponding saturated halides in the S_N2 reaction and are always more toxic as nematocides(*105*). Some of the relationships of the activities of these unsaturated compounds as compared to their S_N2 reactivity is shown in Table 1-25.

TABLE 1-25

COMPARISON OF TOXICITY AND REACTIVITY OF UNSATURA-
TED HALIDES TO THE CITRUS NEMATODE *Tylenchulus
semipenetrans(104, 105)*

Compound	LD_{50}, M	Relative S_N2
CH_2=$CHCH_2Cl$	1.5 × 10⁻³	1.00
CH_2=$C(Cl)CH_2Cl$	2.6 × 10⁻⁴	0.72
CH_2=$CH(CH_3)CH_2Cl$	2.8 × 10⁻⁴	1.58
H≡CCH_2Cl	2.0 × 10⁻⁴	1.78
trans-$ClCH$=$CHCH_2Cl$	7.7 × 10⁻⁵	2.9
cis-$ClCH$=$CHCH_2Cl$	3.3 × 10⁻⁵	8.58
CH_2=$CHCH_2Br$	7.5 × 10⁻⁵	506
HC≡CCH_2Br	4.5 × 10⁻⁶	909

C. Mode of Action

The commercial nematocides are preponderantly saturated or unsaturated halides, e.g., methyl bromide, ethylene dibromide, chloropicrin, 1,2-dibromo-3-chloropropane, and 1,3-dichloropropene. These compounds are believed to act as described by Moje(*105*) through a chemical combination with some essential nucleophilic center, i.e., —SH, NH_2, or —OH groups, in a vital enzyme system in the nematode.

Moje considered that these nematocides (R : X) undergo a bimolecular displacement reaction of the S_N2 type with the essential protein (E:):

$$E: + R:X \longrightarrow [E:R:X] \longrightarrow E:R+X:$$

In this type of reaction, the most important factors affecting the toxicity of the nematocide RX are the nature of the leaving group X and the steric and electronic properties of the groups attached to the carbon atom in R as discussed under Sec. B above.

In addition to this chemical reactivity, narcosis may also play a role in toxicity. This is physical mechanism controlled by the lipid/water distribution of the compound and is a function of the chemical structure of the nematocide. There appears to be a continuous intergradation of these two mechanisms with the highly reactive halides acting largely by chemical and the stable halides largely by physical processes.

A third process of importance in the action of these soil fumigants is solvolysis or the reaction with an extraneous nucleophile, such as water in the soil, which is present in great excess. Therefore, the reaction appears to be unimolecular (S_N1) and dependent only upon the concentration of the fumigant (R : X)·

$$R:X + :OH \longrightarrow R:OH + X:$$

This reaction effectively prevents the fumigant from reaching the site of action in the nematode. Therefore, the most effective nematocides would appear to be compounds having low reactivity in S_N1 reactions and high reactivity in S_N2 reactions (105).

The newer sulfur-containing nematocides such as metham and dazomet owe their activity to their reaction with soil elements to produce methylisothiocyanate (Sec. A). This latter compound is also used as a soil fumigant. Methylisothiocyanate is highly reactive with nucleophilic centers such as —SH groups in vital enzymes of the nematode and thus appears to kill these organisms through the same S_N2 type reaction as discussed for the halogenated aliphatic nematocides above.

The organophosphorus ester nematocides such as VC-13 and ZINO-PHOS become highly active anticholinesterases after activation to the corresponding P=O derivatives either by soil microorganisms or in the nematode. They presumably kill the nematodes through inactivation of the neurohormone cholinesterase (Sec. 1-4, M).

D. Biological Activity

The chemical control of nematodes has developed slowly because of the dearth of information regarding the biology of the nematodes themselves and because of their inaccessibility in the soil. An acre foot of soil weighs approximately 4,500,000 lb and high dosages of fumigant from 50 to 400 lb/acre, are required to produce lethal concentrations. Therefore, soil fumigation has generally proved most practicable for crops that yield high returns such as pineapple, tobacco, citrus, vegetables, turf, ornamentals, seed beds, and in greenhouses.

The performance of the various fumigants is dependent upon the vapor pressure, diffusion coefficient, and distribution of the fumigant through air, water, and solid phases of the soil. The temperature and moisture content of the soil and its content of organic matter are very important. The most effective results are secured with light sandy loams; clay loams and mucks require two to three times more fumigant, and poor results are secured in peats even with very high applications. The pattern and depth of injection of the fumigant are also important in securing favorable results. The effects of all of these factors are discussed in detail by Goring(48).

The complexity of the soil fauna and flora also has an important bearing on the success of the soil fumigation, which is often conducted to control not only nematodes but also insects, fungi, and weed seeds. These vary in their susceptibility to the individual fumigants. For example, Goring (48) lists the relative numbers of units of fumigant required to control nematodes (and fungi) as dibromochloropropane 1 (150), ethylene dibromide 2 (>200), dichloropropene 8 (100), chloropicrin 12 (25), methyl bromide 15 (40), and carbon bisulfide > 100 (>200). Intricate biological interrelations may produce indirect control of fungi by controlling nematodes, and certain fungi may prey on nematodes.

E. Toxicity

Prolonged exposure to the vapors of any of the soil fumigants is inadvisable. The following safe limits for concentrations in air (in ppm) have been established: chloropicrin 0.1, dichloropropene 1, dibromochloropropane 1, methyl bromide 20, carbon bisulfide 20, and ethylene dibromide 25. The rat oral LD_{50} values are: D-D mixture 140, dibromochloropropane 173, ethylene dibromide 117, dazomet 650, metham 820, tetrachlorothiophene 93, and ZINOPHOS 12.

Chloropicrin is a very active lachrymator and is so irritating that dangerous concentrations cannot be tolerated. Methyl bromide is a dangerous cumulative poison with delayed symptoms of central nervous system intoxication that may appear as long as several months after exposure. Dichloropropene is irritant to the eyes and mucous membranes and acts as a vesicant on the skin.

1-7 RODENTICIDES

A. Introduction and Chemistry

Rodenticides are substances used to control rodent pests such as rats, mice, gophers, and ground squirrels. Historically the plant toxins such as strychnine and red squill have been used as rodenticides for many years together with inorganic compounds of arsenic, barium, thallium, and phosphorus. More recently synthetic organic rodenticides have proved more effective and specific and have replaced the older materials for most uses.

Red squill is obtained from the flesh of the inner bulb of *Urginea* (=*Scilla*) *maritima*, family Liliaceae(*110*). This plant contains a number of cardiac glycosides such as the scillarins, scillin, scillitin, scillain, scillipicrin, and sinistrin(*93*). A substance highly toxic to rats, however, is *scilliroside*, yellow crystals, m.p. 168°C, soluble in alcohols, dioxane, ethylene glycol, and acetic acid, and almost insoluble in water. Scilliroside has a rat oral LD_{50} of 0.43 ♀ and 0.7 ♂ mg/kg and reaches a concentration of 0.037% in red squill but is not found in the white squill(*135*). The structure of scilliroside has been determined by Stoll and Renz (*136*) as:

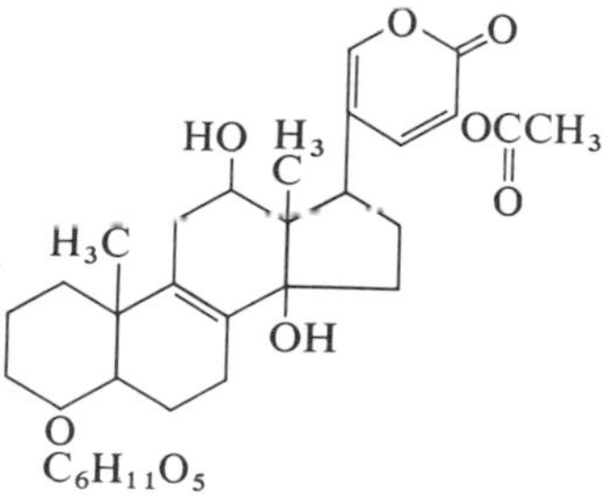

scilliroside

The action of scilliroside is that of a powerful cardiac glycoside and produces convulsions and respiratory failures. It is particularly effective against rats and mice because they are unable to regurgitate it while its powerful emetic action protects other animals.

Strychnine is obtained from the seeds of *Strychnos nux-vomica* and other species, family Loganiaceae. Strychnine is a complex dibasic alkaloid, m.p. 268–290°C, K_1 10^{-7} and K_2 2.2×10^{-12}, soluble in water to 0.0156% and in alcohol to 0.66% and chloroform to 20%. It readily forms salts with strong acids and strychnine sulfate $(C_{21}H_{22}N_2O_2)_2 \cdot H_2SO_4 \cdot 5H_2O$ is soluble in water to 2.9%. The structure of strychnine was elucidated by Woodward who also synthesized the compound(*160*). Strychnine is most effective against house mice and other rodents (rat oral LD_{50} 5 mg/kg), although it is generally toxic to warm-blooded animals.

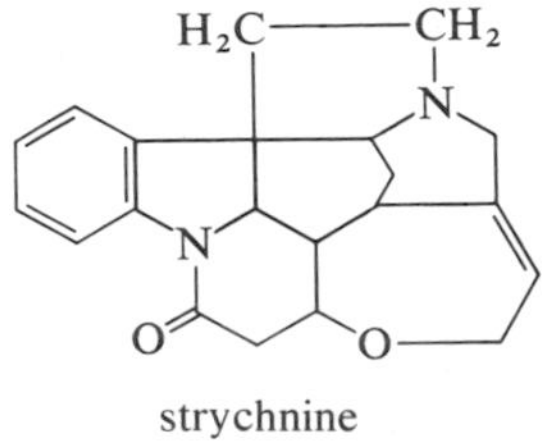

strychnine

Sodium fluoroacetate, FCH_2COONa, has been extensively used as a rodenticide under the designation "1080." The parent fluoroacetic acid, FCH_2COOH, b.p. 167°C, m.p. 31°C, occurs naturally in the South African plant "gifblaar" or *Dichapetalum cymosum* family Dichapetalaceae and in the Australian *Acacia georginae*(*115*). Sodium fluoroacetate is widely used for the control of rats, mice, ground squirrels, and predators but because of its very high general toxicity its use is limited to professional pest control operators. Sodium fluoroacetate is used principally for cereal and water baits to control predators and ground squirrels.

Hydroxycoumarin rodenticides were developed as the result of researches by Campbell and Link(*23*) on the nature of hemorrhagic disease produced in cattle feeding on fermented sweet clover. The toxic principle was found to be *dicoumarol* or 3,3'-methylenebis-(4-hydroxycoumarin), m.p. 287–293°C. This compound when ingested chronically at 2 mg per day by rats produced fatal hemorrhages due to interference with the action of Vitamin K(*31*) but is of low toxicity in acute dosage (rat oral LD_{50}, 540 mg/kg). As a result of studies with analogs (see Sec. B) the related compound *warfarin* or 3-(α-acetonylbenzyl)-4-hydroxycoumarin, m.p. 161°C, was developed as a practical anticoagulant rodenticide

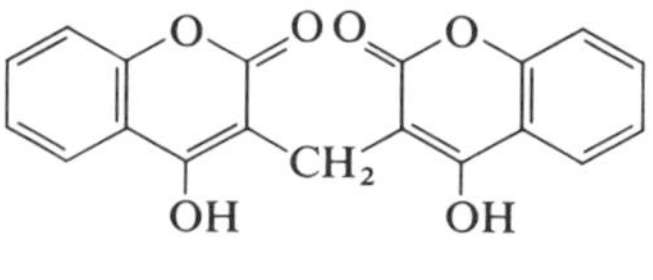

dicoumarol

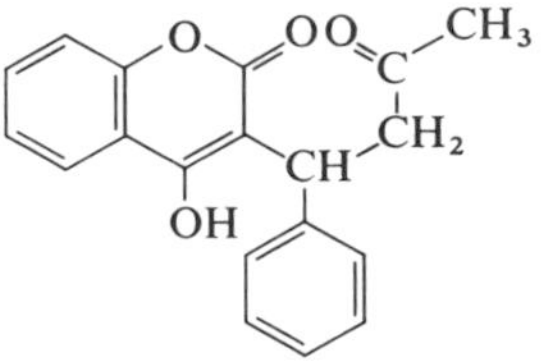

warfarin

which was completely acceptable to rats. Warfarin is almost insoluble in water and is moderately soluble in alcohols. It forms a sodium salt which is water soluble. Warfarin has an oral LD_{50} to the rat of 58♀ and 323♂ mg/kg. Related anticoagulant compounds which have been developed as rodenticides include the 4-chlorobenzyl derivative (coumachlor or TOMORIN, m.p. 169–171°C) and the 3-furyl derivative (*coumafuryl*, or FUMARIN, m.p. 121–123°C, rat oral LD_{50}, 200 mg/kg) which have very similar properties.

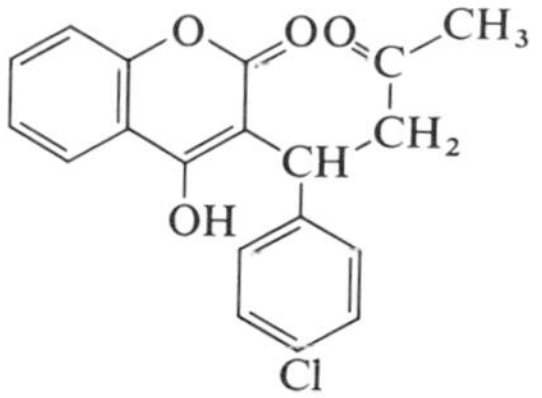

coumachlor

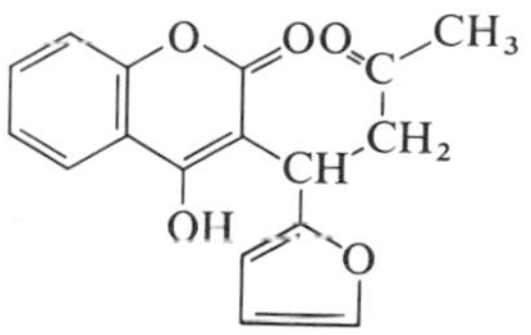

coumafuryl

Certain substituted indandiones also interfere with blood coagulation and *pindone* (PIVAL) or 2-pivalyl-1,3-indandione, m.p. 108.5–110.5°C, a bright yellow crystalline compound insoluble in water, has been developed as a rodenticide. It has a rat oral LD_{50} of 50 mg/kg. The analogous 2-diphenylacetyl-1,3-indandione, or *diphacinone*, m.p. 145–147°C, is also a commercial rodenticide with very similar properties.

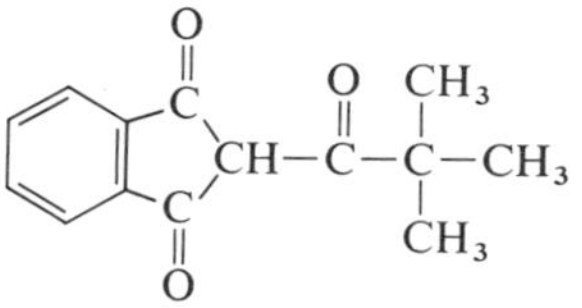

pindone

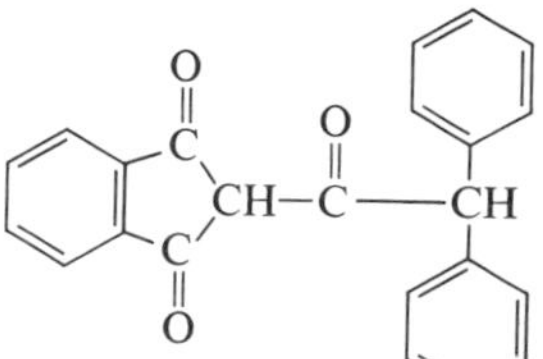

diphacinone

α-*Naphthylthiourea* or antu, m.p. 198°C, is soluble in water to 0.06 g/100 ml and in acetone to 2.4 g and triethylene glycol to 8.6 g/100 ml. It was described by Richter(*118*) as the result of investigations of bitter substances such as phenylthiourea. This compound proved to be toxic to rats with an oral LD_{50} of 3 to 4 mg/kg but its bitter taste resulted in bait rejection. Antu, oral LD_{50}, 7 mg/kg, was found to be especially toxic to the Norway rat and less effective to other species. It causes increased permeability of lung capillaries resulting in pleural effusion and death from drowning pulmonary edema.

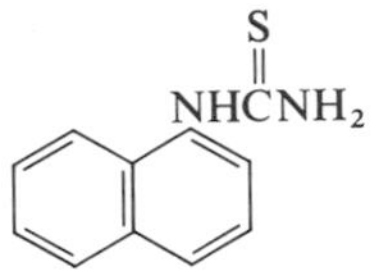

$$\text{(naphthyl)}NHC(=S)NH_2$$

$$Cl\text{-(phenyl)}N=NNHC(=S)NH_2$$

α-naphthylthiourea    MURITAN

A related derivative of phenylthiourea, *p*-chlorophenyldiazothiourea or MURITAN has been used as a rodenticide in Germany(*87*).

Tetramethylene disulfotetramine, m.p. 255°C dec., the condensation product of sulfamide and formaldehyde has been described as a rodenticide by Hecht and Henecka(*57*). This compound has an oral LD_{50} to the mouse of 0.2 to 0.25 mg/kg. Spencer(*134*) has suggested its use in treating coniferous seeds for reforestation projects to reduce loss to rodents. The compound is translocated to seedlings rendering them unpalatable to field mice.

A highly selective anticholinesterase compound, GOPHACIDE, or O,O-di-(*p*-chlorophenyl) N-acetamidino phosphoramidothioate has an oral LD_{50} to the gopher of 1.0 mg/kg but is much less toxic to predators and domestic animals. It is marketed as a 2% bait concentrate for the control of the pocket gophers, family Geomyidae.

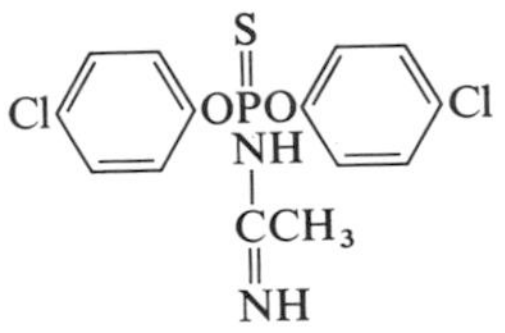

$$Cl\text{-(phenyl)}OPO\text{(phenyl)}Cl$$

GOPHACIDE

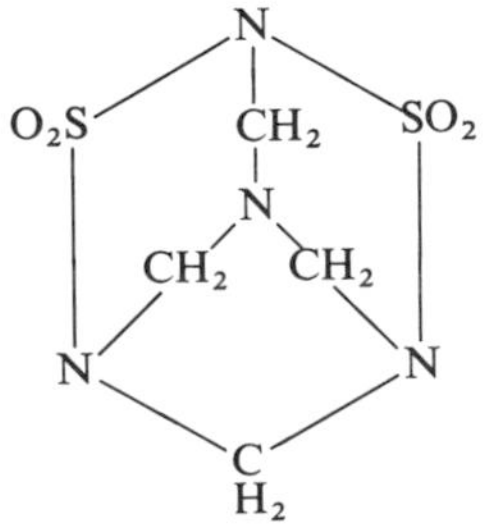

tetramethylene
disulfotetramine

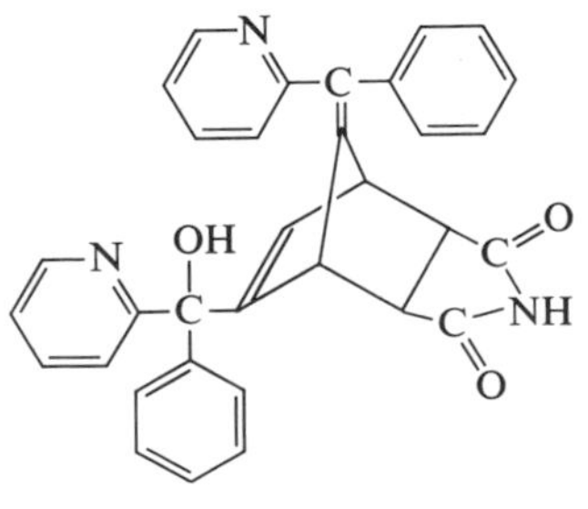

norbormide

Norbormide (RATICATE) or 5-(α-hydroxy-α-phenyl-α-2-pyridyl)-methyl-7-(phenyl-2-pyridylmethylene)-5-norbornene-2,3-dicarboximide is a highly selective rodenticide described by Crabtree et al. (27). It has an oral LD_{50} of 7.4 mg/kg to the rat but is generally nontoxic at 1000 mg/kg to other animals including the mouse, coyote, and goat.

Crimidine or CASTRIX, or 2-chloro-4-dimethylamino-6-methylpyrimidine, m.p. 87°C, has been developed for use on poison grain for mouse control in Germany. It is almost insoluble in water and is soluble in most organic solvents. It has a rat oral LD_{50} of 1.25 mg/kg, and produces violent convulsions. However, it decomposes readily in the body and thus poisoned rats are stated not to be dangerous to predators (88).

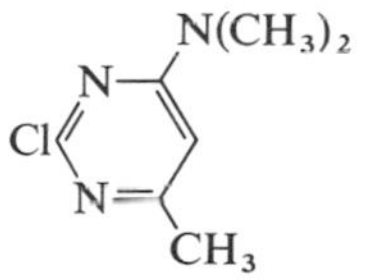

crimidine

The insecticide *endrin* (Sec. 1-4, K) has been used as a spray or dust in orchards to kill mice. It has a rat oral LD_{50} of 7.5 ♀ and 17.8 ♂ mg/kg.

Several inorganic compounds are used as rodenticides. *Barium carbonate*, $BaCO_3$, soluble in water to 0.0024%, rat oral LD_{50} 800 mg/kg, causes hemorrhages of the gastrointestinal tract and kidneys and muscular paralysis. *Thallium sulfate*, Tl_2SO_4, m.p. 632°C, is soluble in water to 4.87% at 20°C, and has a rat oral LD_{50} of 25 mg/kg. It is used as a slow acting rat, rodent, and ant poison, *Zinc phosphide*, m.p. 420°C, water insoluble, rat oral LD_{50} 47 mg/kg, is used principally to kill field mice feeding on orchard trees.

B. Structure–Activity Relationships

1. Fluoroacetic Acid. The simple derivatives of fluoroacetic acid are generally highly toxic to a wide variety of vertebrates and invertebrates. The mode of action (Sec. C) involves in vivo conversion of fluoroacetate to fluorocitric acid so that fluoroacetic acid esters, fluoroacetamide, 2-fluoroethanol, and other simple compounds which are hydrolyzed or oxidized in vivo to fluoroacetate are also highly toxic. The even-numbered ω-fluorinated carboxylic acids are also converted in vivo to fluoroacetate by β-oxidation and are also highly toxic. This biochemical process also produces highly toxic compounds from even number ω-fluoro- alkyl chlorides, mercaptans, amides, and nitriles as shown in Table 1-26 (115).

2. 4-Hydroxycoumarins. Following the identification of dicoumarol as the causative agent in spoiled sweet clover hemorrhagic disease,

TABLE 1-26

TOXICITY OF FLUOROACETATE DERIVATIVES TO MICE
(115)

Compound	Intraperitoneal LD_{50}, mg/kg
FCH_2COOH	8–10
FCH_2COOCH_3	6–7
FCH_2CONH_2	15 (rat oral)
FCH_2CH_2OH	10
FCH_2CH_2Cl	> 100
$F(CH_2)_4Cl$	1.2
$F(CH_2)_5Cl$	32
$F(CH_2)_6Cl$	5.8
$F(CH_2)_5SH$	>100
$F(CH_2)_6SH$	1.2
$F(CH_2)_5NH_2$	50
$F(CH_2)_6NH_2$	0.9
$F(CH_2)_4CN$	1.0
$F(CH_2)_5CN$	50
$F(CH_2)_7COOH$	0.64
$F(CH_2)_8COOH$	>100

Overman et al. (111) evaluated 109 derivatives for their ability to increase
the prothrombin time in rabbits. Some of the most active compounds
are compared with dicoumarol in Table 1-27. In the dicoumarol series,
activity was decreased by substituting the methylene group between the
two dicoumarol moieties with short aliphatic chains or by a phenyl

TABLE 1-27

RELATIVE HYPOPROTHROMBINEMIA-INDUCING CAPACITY OF 4-
HYDROXYCOUMARINS TO THE RABBIT (111)

	Index
3,3′-Methylenebis-(4-hydroxycoumarin)	100
3,3′-Ethylidenebis-(4-hydroxycoumarin)	24
3,3′-Propylidenebis-(4-hydroxycoumarin)	32
3,3′-Butylidenebis-(4-hydroxycoumarin)	8.6
3,3′-Benzylidenebis-(4-hydroxycoumarin)	0.4
3,3′-Thiobis-(4-hydroxycoumarin)	5.3
4-Hydroxycoumarin	0.12
3-(α-Phenyl-β-acetylethyl)-4-hydroxycoumarin (warfarin)	21
3-[α-(p-Methoxyphenyl)-β-acetylethyl]-4-hydroxycoumarin	50
2-Methyl-2-methoxy-4-phenyl-5-oxodihydropyrano-(3,2-c)	
(1)-benzopyran	60
Indanedione	0

group. Replacement of methylene by sulfur retained about 0.05 of the activity. The most interesting synthetic derivatives were the 3-substituted 4-hydroxycoumarins that contained a keto-group in a 1,5-arrangement with respect to the 4-hydroxy group. These compounds have a structural configuration closely resembling that of dicoumarol (compare with warfarin in Sec. 1), and their activity is comparable.

Although indanedione was inactive as an anticoagulant, substitution in the 2-position with an acyl group produces the active compounds pindone and diphacionon. The relationship of the three $C=O$ groups of these compounds suggests that they are competitive inhibitors for vitamin K which is the prosthetic group in the formation of prothrombin (Sec. C).

3. Arylthioureas. Williams and Smith(*157*) have studied the remarkable differences in toxicity between the monoarylthioureas, which are highly toxic, and the corresponding diarylthioureas, which are virtually nontoxic (Table 1-28). These differences seem to be related to differences

TABLE 1-28
TOXICITY OF ARYLTHIOUREAS TO RATS(*157*)

	Oral LD_{50}, mg/kg
Phenylthiourea	3.1
α-Naphthylthiourea	2.5
o-o-Tolylurea	1.5
Diphenylthiourea	>1500
Di-α-naphthylthiourea	>1500
Di-o-tolylthiourea	>2000

in metabolism in the rat, diphenylthiourea being hydroxylated and excreted as the glucuronide while phenylthiourea is converted to phenylcyanamide and then to phenylcarbamic acid with the liberation of H_2S by desulfuration.

C. Mode of Action

1. Strychnine. Strychnine is well known to produce tonic convulsions and lowers synaptic resistance in the nerve cord by a poorly understood inhibition of the inhibitory cells(*31*).

2. Red squill. Red squill contains scillaren and other cardiac glycosides which produce convulsions and respiratory failure in rats.

3. Zinc phosphide. Zinc phosphide reacts with the hydrochloric acid of the rodent stomach to release phosphine gas PH_3 which is extremely reactive and poisonous.

4. Thallium sulfate. Thallium sulfate is a general cellular poison and resembles the arsenicals. Presumably it inactivates —SH groups or other reactive centers in enzymes, producing a variety of neurological, circulatory, and gastrointestinal symptoms.

5. Fluoroacetate. Fluoroacetate is rapidly converted in vivo, by lethal synthesis, to fluoroacetyl-CoA which condenses with oxaloacetate to form fluorocitric acid. Fluorocitric acid enters the tricarboxylic acid cycle and becomes a powerful competitive inhibitor of the enzyme *cis*-aconitase which catalyzes the conversion of citric acid to isocitric acid prior to oxidative decarboxylation(*115*). This action may be depicted as

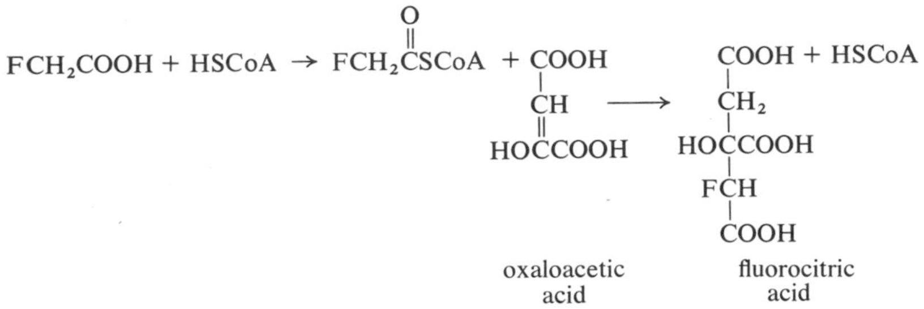

$$
FCH_2COOH + HSCoA \rightarrow FCH_2\overset{\overset{O}{\|}}{C}SCoA +
\begin{matrix} COOH \\ | \\ CH \\ \| \\ HO\overset{}{C}COOH \end{matrix}
\longrightarrow
\begin{matrix} COOH \\ | \\ CH_2 \\ | \\ HO\overset{}{C}COOH \\ | \\ F\overset{}{C}H \\ | \\ COOH \end{matrix}
+ HSCoA
$$

oxaloacetic acid   fluorocitric acid

6. Arylthioureas. Arylthioureas are suggested by Williams and Smith (*157*) as exerting their toxic action through metabolic release of hydrogen sulfide by a desulfuration reaction:

$$C_6H_5NHCSNH_2 \longrightarrow C_6H_5NHCN + H_2S$$

H_2S was found to have a rat LD_{50} of 0.27 to 0.55 mg/kg intravenously, a value compatible with the LD_{50} of phenylthiourea, which would liberate 0.7 mg/kg of H_2S. Further study of this mode of action is in progress.

7. Hydroxycoumarin and Indandione. Derivatives of hydroxycoumarin and indandione have structures resembling vitamin K and apparently

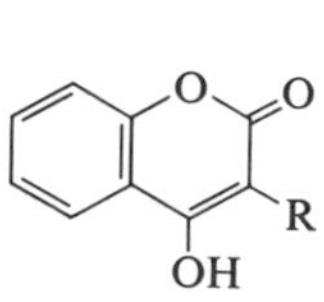

4-hydroxycoumarins 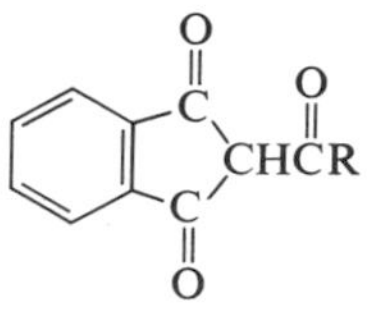   2-acylindandiones 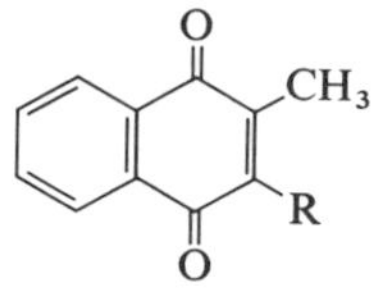   vitamin K act as competitive inhibitors with vitamin K which is the prosthetic group for in vivo synthesis of prothrombin. Thus the presence of the anticoagulants decreases the amount of prothrombin in the blood. Prothrombin is the precursor of thrombin, which catalyzes the conversion of fibrinogen to fibrin, a skeleton of protein fibers that promote coagulation. Many factors are involved in this complicated process of coagulation (67).

D. Biological Activity

The use of rodenticides like all other pesticides varies with the habits and habitat of the pest. In addition, as summarized in Sec. A, the individual rodenticides differ substantially in their spectrum of activity in rodents and in their selectivity with regard to man and his domestic animals and to wildlife.

Baits are the most practical means of using rodenticides and involve mixing the compound with ground food, cut sections of fruits and vegetables, or meat. Impregnation on grains of corn or wheat is especially effective for mice. The concentrations which are customarily used in baits are: antu 1 to 3%, warfarin 0.025%, pindone 0.025%, strychnine 0.5 to 1%, zinc phosphide 0.5 to 1%. Tetramethyl disulfotetramine has been suggested for use at 0.05%.

In situations where large scale control programs are carried out, as in plague control, it is practical to dust the burrows and runways of rats with preparations of the rodenticide in talc or other diluent.

Endrin has been used as a dust for applications to orchards to control mice.

Sodium fluoroacetate is used in water or cereal baits for control of ground squirrels and predators.

E. Toxicity

The rodenticides that are specifically designed to kill mammalian pests present a direct toxic hazard to man and his domestic animals when distributed carelessly or improperly, or left within the reach of children.

The rat acute oral LD_{50} values for these compounds are given in Sec. A. Hayes(55) describes the estimated fatal doses for man, the symptoms of intoxication, and the antidotal treatments as summarized below. The original publication should be consulted for details.

sodium fluoroacetate — 0.5 to 2 mg/kg; epileptiform convulsions and ventricular fibrillation.

thallium sulfate — 4 to 8 mg/kg; gastrointestinal and neurological symptoms; gastric lavage with administration of dithizon chelating agent.

warfarin — 1 to 2 mg/kg per day for several days; inhibition of prothrombin formation and capillary damage leading to hematoma and hemorrhages; vitamin K in massive doses.

endrin — 1 mg/kg; dizziness followed by epileptiform convulsions; phenobarbital in large doses.

strychnine — 0.5 to 1 mg/kg; violent tonic convulsions with death from asphyxia; short acting barbiturate (*93*).

crimidine -- severe convulsions originating in central nervous system; pentobarbital (*93*).

REFERENCES

1. R. Andersen, R. Behrens, and A. Linck, *Weeds*, **10**, 4 (1962).
2. R. Andersen, A. Linck, and R. Behrens, *Weeds*, **10**, 1 (1962).
3. A. Anderson, *Proc. Intern. Congr. Entomol.*, *10th*, **2**, 227 (1958).
4. B. Anderson, V. Bachman, S. McLane, and E. Dean, *Weeds*, **5**, 135 (1957).
5. L. Audus (ed.), *The Physiology and Biochemistry of Herbicides*, Academic, New York, 1964.
6. R. Barlow, *Introduction to Chemical Pharmacology*, Methuen, London, 1955.
7. W. Barthel, *Advan. Pest Control Res.* **4**, 33 (1961).
8. D. Barton, O. Jeger, V. Prelog, and R. Woodward, *Experientia*, **10**, 81 (1954).
9. M. Bell, J. Johnson, B. Wildi, and R. Woodward, *J. Am. Chem. Soc.*, **80**, 1001 (1958).
10. J. Booer, *Ann. Appl. Biol.*, **31**, 340 (1944).
11. W. Boon, *Focus*, Imperial Chemical Industries, 1964.
12. P. Brian, *Ann. Botany* (*London*), **13**, 59 (1949).
13. P. Brian, and H. Hemming, *Ann. Appl. Biol.*, **32**, 214 (1945).
14. R. Brian, "The classification of herbicides and types of toxicity," in *The Physiology and Biochemistry of Herbicides* (L. Audus, ed.), Academic, London, 1964.
15. R. Brian, R. Homer, J. Stubbs, and R. Jones, *Nature*, **181**, 446 (1958).
16. M. Brooks, *Nature*, **170**, 1022 (1952).
17. J. Brown and A. Boyle, *Phytopathology*, **34**, 760 (1944).
18. W. Brown, *J. Hort. Sci.*, **13**, 247 (1935).
19. W. Brown, *Ann. Appl. Biol.*, **34**, 422 (1947).
20. H. Bucha, *Science*, **137**, 537 (1962).
21. H. Bucha, and C. Todd, *Science*, **114**, 493 (1951).
22. K. Büchel, *Angew. Chemie, Intern. Ed.*, **4**, 12 (1965).
23. H. Campbell, and K. Link, *J. Biol. Chem.*, **138**, 21 (1941).
24. W. Carter, *Science*, **97**, 383 (1943).
25. D. Cation, *Phytopathology*, **43**, 468 (1953).
26. W. Chefurka, *Enzymologia*, **18**, 209 (1957).
27. D. Crabtree, W. Robison, and V. Perry, *Pest Control*, **32** (5), 26 (1964).
28. A. Crafts, *Advan. Pest Control Res.*, **1**, 39 (1957).
29. A. Crafts, *The Chemistry and Mode of Action of Herbicides*, Wiley (Interscience), New York, 1961.
30. S. Crowdy, and D. Pramer, *Chem. Ind.* (London), **1955**, 160.

31. W. Cutting, *Handbook of Pharmacology*, 2nd ed., Appleton-Century Crofts, New York, 1962.
32. A. Dimond, J. Heuberger, and J. Horsfall, *Phytopathology*, **33**, 1095 (1943).
33. W. Ennis, *Botany Gaz.*, **109**, 473 (1948).
34. E. Felber and C. Hamner, *Botany Gaz.*, **110**, 324 (1948).
34a. R. Foster, C. Teesdale, and G. Poulton, *Bull. World Health Organ.*, **22**, 543 (1960).
35. L. Frick, and W. de Jiminez, *Bull. World Health Organ.*, **34**, 429 (1964).
36. G. Friessen, *Planta*, **8**, 666 (1929).
37. J. Fukami, *Natl. Inst. Agr. Sci. Japan, Misc. Publ.*, **7**, (1962).
38. T. Fukuto, *Advan. Pest Control Res.*, **1**, 147 (1957).
39. T. Fukuto, and R. Metcalf, *J. Agr. Food Chem.* **4**, 930 (1956).
40. T. Fukuto, R. Metcalf, and M. Winton, *J. Econ. Entomol.*, **52**, 1121 (1959).
41. A. Gast, E. Knüsli, and H. Gysin, *Experientia*, **11**, 107 (1955).
42. J. Getzin, *J. Econ. Entomol.*, **58**, 158 (1965).
43. O. Giannotti, R. Metcalf, and R. March, *Ann. Entomol. Soc. Am.*, **49**, 588 (1956).
44. G. Godfrey, J. Oliviera, and H. Hoshino, *Phytopathology*, **24**, 1332 (1934).
45. M. Goldsworth and S. Gertler, *Plant Disease Reporter, Suppl.*, **182**, 89 (1949).
46. R. Goodman, "The influence of antibiotics on plants and plant disease control," in *Antibiotics, Their Chemistry and Non-medical Uses*, (H. Goldberg, ed.), D. Van Nostrand Co., New York, 1959, Chap. 4.
47. R. Goodman, *Advan. Pest Control Res.*, **5**, 1 (1962).
48. C. Goring, *Advan. Pest Control Res.*, **5**, 47 (1962).
49. J. Grove, J. MacMillan, T. Mulholland, and M. Thorold-Rogers, *J. Chem. Soc.*, **1952**, 3977.
50. R. Gruenhagen, P. Wolf, and E. Dunn, *Contrib. Boyce Thompson Inst.*, **16**, 349 (1951).
51. H. Gysin, *Chimia (Switz.)*, **8**, 205 (1954).
52. H. Gysin and E. Knüsli, *Advan. Pest Control Res.*, **3**, 289 (1960).
53. P. Hamm, and A. Speziale, *J. Agr. Food Chem.*, **4**, 518 (1956).
54. C. Hamner and H. Tukey, *Science*, **100**, 154 (1944).
55. W. Hayes, *Clinical Handbook on Economic Poisons*, U.S. Dept. Health, Education, and Welfare, Public Health Service, 1963.
56. D. Heath, *Organophosphorus Poisons*, Pergamon, New York, 1961.
57. G. Hecht and H. Henecka, *Angew. Chem.*, **61**, 365 (1949).
58. W. Hester, U.S. Pat. 2,317,765 (April 27, 1943).
59. L. Hiltner, *Pratsi Bl. Pfl. Bau.*, **18**, 65 (1915).
60. J. Hilton, J. Ard, L. Janson and W. Gentner, *Weeds*, **7**, 391 (1959).
61. P. Hochstein and C. Cox, *Am. J. Botany*, **43**, 437 (1956).
62. O. Hoffman and A. Smith, *Science*, **109**, 588 (1949).
63. F. Hopkins, E. Morgan, and C. Lutwak-Mann, *Biochem J.*, **32**, 1829 (1938).
64. J. Horsfall and A. Dimond (eds.), *Plant Pathology*, Academic, New York, 1960.
65. O. Johnson, N. Krog and J. Poland, *Pesticides*, Parts 1 and 2, *Chem. Week*, May, June, 1963.
66. J. Kagy, *J. Econ. Entomol.*, **34**, 660 (1941).
67. P. Karlson, *Introduction to Modern Biochemistry*, Academic, New York, 1963.
68. T. Kato, K. Ueda and K. Fujimoto, *Agr. Biol. Chem.* Tokyo **28**, (12), 914 (1964).
69. P. Kearney, C. Harris, D. Kaufman and T. Sheets, *Advan. Pest Control Res.*, **6**, 1 (1965).
70. C. Kearns, L. Ingle and R. Metcalf, *J. Econ. Entomol.*, **38**, 661 (1946).

71. W. Kellerman and W. Swingle, *Bull. Kansas Agr. Exptl Sta.*, **12**, 27 (1890).
72. W. Kenrick, *The New American Orchardist*, 1833.
73. A. Kittleson, *Science*, **115**, 84 (1952).
74. A. Kittleson, *J. Agr. Food Chem.*, **1**, 677 (1953).
75. H. Klöpping and G. Van der Kerk, *Rec. Trav. Chim.*, **70**, 917 (1951).
76. M. Kolbezen, R. Metcalf, and T. Fukuto, *J. Agr. Food Chem.*, **2**, 864 (1954).
77. E. Kornfeld, R. Jones and T. Parke, *J. Am. Chem. Soc.*, **71**, 150 (1949).
78. W. Lange and G. Kreuger, *Ber.*, **65**, 1598 (1932).
79. J. Leaper and J. Bishop, *Botany Gaz.*, **112**, 250 (1951).
80. H. Linser, *Angew. Chem.*, *Intern. Ed. Engl.*, **5**, 776 (1966).
81. H. Lipke and C. Kearns, *Advan. Pest Control Res.*, **3**, 253 (1960).
82. R. Ludwig and G. Thorn, *Advan. Pest Control Res.*, **3**, 219 (1960).
83. R. Lukens and H. Sisler, *Phytopathology*, **48**, 235 (1958).
84. P. Marsh and M. Butler, *Ind. Eng. Chem.*, **38**, 701 (1946).
85. P. Marsh, M. Butler and B. Clark, *Ind. Eng. Chem.*, **41**, 2176 (1949).
86. P. Marth and J. Mitchell, *Botany Gaz.*, **106**, 224 (1944).
87. H. Martin, *The Scientific Principles of Crop Protection*, 4th ed., Arnold, London, 1959.
88. H. Martin, *Guide to the Chemicals used in Crop Protection*, 4th ed., Queen's Printer, Ottawa, Canada, 1961, Suppl. 1965.
89. H. Martin and H. Shaw, British Intell. Obj. Sub-Comm., Final Rept. 1095, H. M. Stationery Office, London, 1946.
90. C. McBeth and G. Bergeson, *Plant Disease Reporter*, **39**, 223 (1955).
91. S. McCallan, Mechanisms of toxicity with special reference to fungicides, *Proc. Intern. Conf. Plant Protection*, 2nd, Butterworth's, London, 1957.
92. G. McWhorter and J. Holstun, *Weeds*, **9**, 592 (1961).
93. *Merck Index*, 7th ed., Merck, Rahway, New Jersey, 1960.
94. C. Metcalf, W. Flint, and R. Metcalf, *Destructive and Useful Insects*, 4th ed., McGraw-Hill, New York, 1962.
95. R. Metcalf, *Organic Insecticides*, Wiley (Interscience), New York, 1955.
96. R. Metcalf, *World Rev. Pest Control*, **3** (1), 28 (1964).
97. R. Metcalf, "Methods of estimating effects," in *Research in Pesticides*, (C. Chichester, ed.), Academic, New York, 1965, Pt. 1, p. 17.
98. R. Metcalf, unpublished data.
99. R. Metcalf and T. Fukuto, *J. Agr. Food Chem.*, **13**, 220 (1965).
100. R. Metcalf and G. Georgiou, *Bull World Health Organ.*, **27**, 251 (1962).
101. P. Millardet, *J. d'Agr. Prat.*, **2**, 513, 707, 801 (1885).
102. L. Miller, S. McCallan, and R. Weed, *Contrib. Boyce Thompson Inst.*, **17**, 173 (1953).
103. J. Mitchell, W. Zaumeyer and W. Anderson, *Science* **115**, 114 (1952).
104. W. Moje, *J. Agr. Food Chem.*, **7**, 702 (1959).
105. W. Moje, *Advan. Pest Control Res.*, **3**, 181 (1960).
106. W. Moje, J. Martin and R. Baines, *J. Agr. Food Chem.*, **5**, 32 (1957).
107. B. Morgan and R. Goodman, *Plant Disease Rept.*, **39**, 487 (1955).
108. P. Müller (ed.), *DDT Insektizid Dichlorodiphenyltrichlorathan und seine Bedeutung*, Birkhauser Verlag, Basel, Switzerland, 1955.
109. R. O'Brien, *Toxic Phosphorus Esters*, Academic, New York, 1960.
110. M. O'Connor, R. Buck, and C. Fellers, *Ind. Engng. Chem.*, **27**, 1377 (1935).
111. R. Overman, M. Stahmann, C. Huebner, W. Sullivan, L. Spero, D. Doherty, M. Ikawa, L. Graf, S. Roseman and K. Link, *J. Biol. Chem.*, **153**, 5 (1944).
112. A. Oxford, H. Raistrick and P. Simonart, *Biochem. J.*, **33**, 240 (1939).

113. J. Pappas and G. Carman, *J. Econ. Entomol.*, **48**, 698 (1955).

114. J. Pappas and G. Carman, *J. Econ. Entomol.*, **54**, 152 (1961).

115. F. Pattison and R. Peters, "Monofluoroaliphatic compounds" in *Pharmacology of Fluoride, Handbook of Experimental Pharmacology*, (F. Smith, ed.), Springer-Verlag, New York, 1966 Chap. 8.

116. H. Reynolds, *Advan. Pest Control Res.*, **2**, 135 (1958).

117. S. Rich, "Fungicidal chemistry," in *Plant Pathology*, Vol. 2, (J. Horsfall and A. Dimond, eds), Academic, New York, 1960 p. 588.

118. C. Richter, *J. Am. Med. Assoc.*, **129**, 927 (1945).

119. L. Riehl and R. Wedding, *J. Econ. Entomol.*, **52**, 88 (1959).

120. W. Ripper, *Advan. Pest Control Res.*, **1**, 305 (1957).

121. J. Robertson, *Trans. London Hort. Soc.*, **5**, 178 (1824).

122. E. Rogers, F. Koniusky, J. Shavely, and K. Folkers, *J. Am. Chem. Soc.*, **70**, 3086 (1948).

123. W. Roth and E. Knüsli, *Experientia*, **17**, 312 (1961).

124. D. Schoene and O. Hoffman, *Science*, **109**, 588 (1949).

125. D. Schoene, H. Tate and T. Brasefield, *Agr. Chem.*, **4**, (11), 24 (1949).

126. G. Schrader, *Die Entwicklung neuer insektizider Phosphorsaure-Ester*, Verlag Chemie, Weinheim, 1963.

127. P. Schuldt, and C. Wolf, *Contrib. Boyce Thompson Inst.*, **18**, 377 (1956).

128. H. Schulthess, *Abhandl. Zurich Naturforch. Gesell.*, **1**, 498 (1761).

129. W. Shaw and C. Swanson, *Weeds*, **2** (1), 43 (1953).

130. T. Sheets and A. Crafts, *Weeds*, **5** (2), 93 (1957).

131. H. Shirk, *Arch. Biochem. Biophys.*, **51**, 258 (1954).

132. R. Slade, W. Templeman and W. Sexton, *Nature*, **155**, 497 (1945).

133. S. Soloway, *Advan. Pest Control Res.*, **6**, 85 (1965).

134. D. Spencer, *J. Forestry*, **52**, 824 (1954).

135. A. Stoll and J. Renz, *Helv. Chim. Acta*, **25**, 43 (1942).

136. A. Stoll and J. Renz, *Helv. Chim. Acta*, **26**, 648 (1943).

137. K. Sund and H. Little, *Science*, **132**, 622 (1960).

138. K. Sund, E. Putula and H. Little, *J. Agr. Food Chem.*, **8**, 210 (1960).

139. C. Tanner, W. Greaves, W. Orrell, N. Smith and R. Wood, British Intell. Obj. Sub-comm., Final Rept. 1480, H. M. Stationary Office, London, 1947.

140. W. Templeman and W. Sexton, *Nature*, **156**, 630 (1945).

141. W. Ter Horst and E. Felix, *Ind. Eng. Chem.*, **35**, 1255 (1943).

142. H. Thompson, C. Swanson and A. Norman, *Botany Gaz.*, **107**, 476 (1946).

143. G. Thorne, *Principles of Nematology*, McGraw-Hill, New York, 1961.

144. W. Tisdale and I. Williams, U.S. Pat. 1,972,961, Sept. 11, 1934.

145. G. Van der Kerk, "Chemical structure and fungicidal activity of dithiocarbamic acid derivatives" in *Plant Pathology, Problems and Progress, 1908–1958*, (C. Holton, G. Fischer, R. Fulton, H. Hart and S. McCallan, eds.), Univ. Wisconsin Press, Madison, 1959, Chap. 26.

146. G. Van der Kerk, *Soc. Chem. Ind., Mono.* **15**, London, 1961.

147. J. Van Overbeek and R. Blondeau, *Weeds*, **3** (1), 55 (1954).

148. J. Van Overbeek, "Mode of action of herbicides," in *The Physiology and Biochemistry of Herbicides*, (L. Audus, ed.), Academic, New York, 1964, Chap. 13.

149. H. Veldstra, *Ann. Rev. Plant Physiol.*, **4**, 151 (1953).

150. R. Wain, *Advan. Pest Control Res.*, **2**, 263 (1957).

151. S. Wakesman, *Antibiot. Chemotherapy*, **3**, 333 (1953).

152. S. Wakesman, *Science*, **118**, 259 (1953).

153. K. Watanabe, T. Tanaka, K. Fukuhara, N. Miyairi, H. Yonehara and H. Umezawa, *J. Antibiotics (Tokyo)*, **10A**, 39 (1957).

154. R. Wellman and S. McCallan, *Contrib. Boyce Thompson Inst.*, **14**, 151 (1946).

155. B. West and F Wolf, *J. Gen. Microbiol.*, **12**, 396 (1955).

156. A. Whiffen, N. Bohonos and R. Emmerson, *J. Bacteriol.*, **52**, 610 (1946).

157. R. Williams and R. Smith, "Biochemistry of rodenticides," in *Drugs and Enzymes*, (B. Brodie and J. Gillette, eds.), Macmillan, New York, 1965, p. 331.

158. F. Winteringham, A. Harrison and P. Bridges, *Biochem. J.*, **61**, 359 (1955).

159. D. Woodcock, "Relation of chemical structure to fungicidal action," in *Plant Pathology, Problems and Progress 1908–1958*, (C. Holton, G. Fischer, R. Fulton, H. Hart and S. McCallan, eds.), Univ. Wisconsin Press, Madison, 1959, Chap. 25.

160. R. Woodward, *J. Am. Chem. Soc.*, **76**, 4749 (1954).

161. I. Yamamoto, *Advan. Pest Control Res.*, **6**, 231 (1965).

162. H. Yonehara and S. Takeuchi, *J. Antibiotics (Tokyo)*, **11**, 122 (1958).

163. G. Zentmyer, *Phytopathology*, **33**, 1121 (1943).

164. G. Zentmyer and S. Rich, *Phytopathology*, **46**, 33 (1956).

165. P. Zimmerman and A. Hitchcock, *Contrib. Boyce Thompson Inst.*, **12**, 321 (1942).

Chapter 2

METABOLISM OF INSECTICIDES AND FUNGICIDES

T. Roy Fukuto and James J. Sims

UNIVERSITY OF CALIFORNIA

RIVERSIDE, CALIFORNIA

2-1   INTRODUCTION

The advent of organic compounds for use in the control of pests has opened up areas of research in which there has been intense activity in the past two decades. Perhaps the most important of these areas has been the study of the metabolism of pesticides in plants and animals. The bulk of the work dealing with pesticide metabolism has been with insecticides, although there has been considerable recent work in herbicide metabolism (see Chap. 4). This chapter therefore is devoted largely to insecticide metabolism and also includes the limited information available on fungicide metabolism. Insofar as possible, the fate and degradation pathways of the more commercially important insecticides and fungicides in crop plants, domestic animals, soils and waters, and the nature of the metabolites in food for human or animal consumption are discussed.

Investigations concerned with pesticide metabolism are necessary for several reasons. From a practical viewpoint, information on the chemical behavior and reactions of insecticides or fungicides in biological systems is essential for the rational assessment of hazards arising from the use of these compounds for pest control. Clearly, the identification and establishment of the toxicological properties of the metabolic products produced in plants and animals exposed to pesticides are mandatory before residual hazards may be assessed. In addition to this more practical demand for metabolism studies, such investigations, particularly in animals, are basic to our understanding of the mode of action of these compounds. They are indeed of paramount importance for the elucidation of the intoxication and detoxication processes that occur in animals, plants, and microorganisms. By understanding these basic processes, it will be possible on a more sophisticated basis to invent new and better compounds with favorable biological activity and selectivity, i.e., compounds effective against pests but innocuous to other living things. Further, this information is needed to understand the manner in which resistant species of pests develop their capacity to avoid the biological action of pesticidal chemicals.

For supplementary information regarding insecticide and fungicide metabolism, the reader is referred to several review articles and books that have been published in recent years(1–8). This chapter discusses the metabolism of organic insecticides in the following order: organophosphorus esters, carbamate esters, chlorinated hydrocarbons, botanicals, and nitrophenols. The last section is devoted to fungicides. Of the classes of compounds listed above, the organophosphorus esters have received the greatest attention and therefore are considered first.

2-2 ORGANOPHOSPHORUS INSECTICIDES

A. Parathion and Related Compounds

The elucidation of the metabolism and fate of the organophosphorus insecticides in plants and animals has been indispensable in establishing the mode of action of these compounds, particularly the phosphorothionate esters. It has been well established that phosphorothionate esters such as parathion, O,O-diethyl O-p-nitrophenyl phosphorothioate, are generally poor inhibitors of the cholinesterase enzymes(5,6,9) because of the deactivating effect of the sulfur atom. On the other hand, paraoxon, diethyl p-nitrophenyl phosphate, the corresponding phosphate ester of parathion, is a very strong anticholinesterase agent, and it is evident that the toxic action of parathion and related compounds is attributable to the action of the corresponding phosphate ester formed by in vivo desulfuration of the thionate ester. Thus, the apparent anomaly between poor enzyme inhibition by phosphorothionate esters and high toxicity is explained by evidence presented below, which shows that these esters are metabolized in vivo to highly potent cholinesterase inhibitors.

Pure parathion has been demonstrated to be almost devoid of anticholinesterase activity, and the high inhibitory activity of the technical material has been proven to be attributable to the presence of impurities, particularly the S-ethyl isomer, iso-parathion(10–12). Other highly insecticidal organophosphorus esters related in structure to parathion, such as

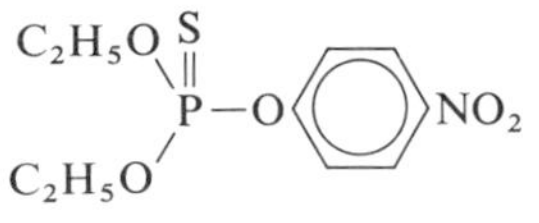

parathion 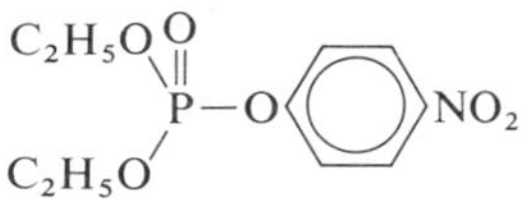   paraoxon 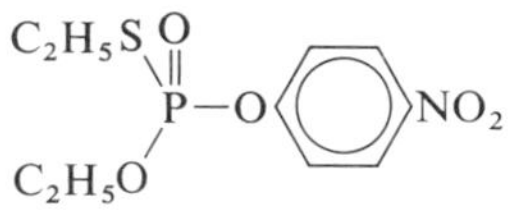   iso-parathion

EPN (O-ethyl O-p-nitrophenyl phenylphosphonothioate) and SUMITHION (O,O-dimethyl O-3-methyl-4-nitrophenyl phosphorothioate), also have been shown to be poor anticholinesterase agents when in the pure state (12, 13). The first evidence that parathion is metabolized in vivo into a strong cholinesterase inhibitor was presented by Chamberlain and Hoskins(14) who found that an active anticholinesterase agent was produced by incubating parathion with the intact ventral nerve cord of Periplaneta americana. Independently and almost simultaneously, Diggle and Gage(10, 11) demonstrated that parathion is converted in vivo in the rat or in vitro by rat blood and rat liver slices into an active cholinesterase inhibitor of then unknown structure. Subsequently, this substance was isolated by column chromatography from acetone extracts of parathion-treated rat livers, and it was demonstrated to be identical to paraoxon by comparison of their behavior on paper chromatographs, ultraviolet absorption spectra, hydrolytic stability, and anticholinesterase activity(15). The conversion of parathion into paraoxon and methyl parathion into methyl paraoxon was demonstrated also at about the same time in various tissues of *Periplaneta americana* by Metcalf and March (12) by use of paper chromatographic techniques. These investigators found that intact tissue and oxygen were needed for paraoxon formation, indicating that the metabolic transformation is oxidative in nature. By comparison of the inhibitory behavior of parathion, paraoxon, and the S-ethyl isomer to rat brain acetyl cholinesterase and aliesterase both in vitro and in vivo, evidence was obtained that indicated that the inhibitory activity of parathion in vivo is attributable largely to the action of paraoxon formed by enzymic conversion of P=S to P=O and not to the S-ethyl isomer. Although the level of paraoxon formed in the animal by metabolic conversion of parathion is low compared to other metabolic products, it is quite certain that this is the primary avenue for intoxication by a phosphorothionate ester. In this regard, Kubistova(16) later demonstrated that at least 4.5% of the LD_{50} dose administered to rats is converted to paraoxon in 1 hr. Of this amount, the liver is responsible for 64%, the gut 21%, lungs 13%, and the kidneys 1.5%. A more recent study on rat liver microsomal oxidation of methyl parathion(17) has shown that about 2.8% of the parent compound is converted into methyl paraoxon. In this case the presence of the phosphate ester was established unequivocally by isolation by column chromatography and comparison of infrared spectra.

Since the pioneering work of Diggle and Gage(10,11) and Metcalf and March(12), others have confirmed and expanded upon these earlier findings regarding phosphorothionate activation to phosphates(18–22). The more recent investigations concerning this activation phenomenon have concentrated more on the biochemical aspects of the reaction. Although there is some disagreement on the specific cofactors needed for in vitro activation with isolated mammalian tissue preparations, it seems that a pyridine nucleotide, oxygen, and perhaps Mg^{2+} are necessary. In mammals, the liver appears to be the best source for active tissue, and activity is found primarily in the microsomal fraction. The mammalian tissue cofactor requirements appear to differ among different investigators. Davison(23) and Murphy and DuBois(22) found NAD superior to NADP for parathion and azinphosmethyl [O,O-diethyl S-4-oxo-1,2,3-benzotriazin-3(4H)-ylmethyl phosphorodithioate] activation. With the latter compound, others(24) have found that the combination of NAD and NADP is superior to either cofactor used alone. In more recent work, Neal and DuBois(25) have demonstrated with EPN that conversion to the phosphonate analog and subsequently to hydrolyzed products by rat liver homogenate is catalyzed twice as fast with NADP than with NAD. No increase in reaction was obtained with the reduced form of either cofactor. With washed liver microsomes, however, the reduced form of NADP (NADPH) was necessary for EPN activation. Further, no requirements for magnesium, calcium, or manganese ions could be demonstrated. The requirement of oxygen was clearly demonstrated since the rate in the absence of oxygen was one-sixth of that in the presence of air. The conversion of phosphorothionates to phosphates by liver microsomes has been shown to be inhibited by general inhibitors, e.g., SKF 525A and piperonyl butoxide, of microsomal systems responsible for the oxidation of a variety of drugs(18,19,26).

In insects, the fat body and gut appear to be the principal activating tissue and activity also appears to reside in the microsomal fraction. Homogenization of tissue destroys activating ability and activity is restored by addition of cofactors. O'Brien(27) demonstrated that the addition of NAD, magnesium ion, and nicotinamide to cockroach tissue homogenates restores the ability of this system to oxidize malathion to malaoxon. In contrast to this, Nakatsugawa and Dahm(19) have shown that activation of azinphosmethyl by cockroach tissue requires NADPH and oxygen. NADH, NAD, and NADP were less effective, and magnesium ion enhanced activation slightly. The same investigators also showed(20) that NADPH and oxygen were needed for parathion activation to paraoxon. Investigations by Fukami and Shishido(28) on parathion activation by the microsomal fraction of whole body homogenates of the rice stem borer larvae have shown NADP more effective than NAD in effecting paraoxon formation. Thus, there appears to be some disagreement on the cofactor requirements necessary for activation in insects also.

The foregoing discussion on phosphorothionate metabolism is concerned with the activation processes that occur in animals, an important aspect of the intoxication mechanism of these compounds. However, it is clear that activation consists of only a very small part of the overall metabolism picture. With the organophosphorus esters, degradation or detoxication by hydrolytic reactions is by far the major route through which these compounds are metabolized.

The enzymatic hydrolysis of organophosphorus esters such as parathion has been attributed to the phosphatases, defined as enzymes that catalyze the hydrolysis of esters of phosphoric acid. In view of the wide variety of organophosphorus esters and the different functional groups that have been demonstrated to be degraded in biological systems, it is reasonable to assume that different phosphatases must be involved in hydrolysis of organophosphorus insecticides. Schmidt and Laskowski (29) in a review chapter have classified phosphatases into three groups, phosphomonoesterases, phosphodiesterases, and phosphoanhydrases. The first two catalyze the hydrolysis of mono and diesters of phosphoric acid but do not hydrolyze anhydride bonds. Phosphatases that hydrolyze P—O—p-nitrophenyl bond in paraoxon are classified in the third category because of the high lability of this bond. Enzymes that hydrolyze the P—O—alkyl bond of a tertiary ester, however, do not fall into any category. It appears that phosphatases act by cleavage of the P—O (or P—S) bond(30) and probably involve nucleophilic attack on the phosphorus atom by a group in the enzyme. However, it is possible that C—O bonds may also be cleaved, e.g., in the enzymatic demethylation of methyl parathion, in which the net result is the hydrolysis of a phosphorus ester bond. C—O bond cleavage, however, may be attributable to the action of an enzyme other than a phosphatase, e.g., by an O-dealkylation reaction in which the methyl group is first oxidized and then removed(31).

In vitro degradation of parathion type organophosphorus compounds by enzymes from mammalian tissue has been studied by several different investigators. Aldridge(32) was the first to report on an enzyme present in mammalian serum that rapidly hydrolyzed p-nitrophenyl acetate and paraoxon to give p-nitrophenol. The enzyme was labeled A-esterase, distinguishing it from B-esterase, which hydrolyzed the acetate but was inhibited by small amounts of paraoxon. It also was shown that the liver was the most active detoxifying organ in mammals. Based on kinetic analysis, a mechanism for the hydrolysis of paraoxon by A-esterase was suggested that is closely analogous to the inhibition of esterases except for a facile decomposition of the phosphorylated enzyme. More recently, Main(33,34) has demonstrated that the hydrolysis of p-nitrophenyl acetate is not attributable entirely to the enzyme that hydrolyzes paraoxon. Paraoxonase obtained from sheep serum was purified 385-fold and shown to be much lower in its activity in hydrolyzing p-nitrophenyl acetate compared to the crude enzyme. The rapid hydrolysis of the acetate by crude enzyme is attributed to another enzyme labeled D-esterase. The purified paraoxonase was found to hydrolyze diisopropyl phosphorofluoridate (DFP), but curiously enough it did not hydrolyze tetraethyl pyrophosphate (TEPP). In addition to A-esterase, human serum albumin also has been demonstrated to be capable of catalyzing the hydrolysis of paraoxon(35). For further information regarding the nature and distribution of the different A-esterases, the reader is referred to other more extensive reviews(5,6,36,37).

Enzymes that degrade parathion-like molecules also are present in insects. Homogenates or enzyme preparations from insects have been shown to be capable of hydrolyzing both phosphate and phosphorothionate esters but usually at rates somewhat slower than that shown by mammalian tissue preparation. An "aromatic esterase" present in bee abdomen brei has been found to hydrolyze a variety of organophosphorus esters including parathion and paraoxon(38). Parathion was hydrolyzed twice as fast as paraoxon. The homogenate of house flies has been shown to hydrolyze paraoxon slowly at 10^{-2} M by spectrophotometric analysis, although no hydrolysis occurred at 10^{-3} M(39). However, by using less sensitive manometric methods, the hydrolysis of neither parathion nor paraoxon could be demonstrated by the homogenates of five different insects or rat liver homogenate, an indication of the slowness of the reaction. Others also have reported inability to demonstrate in vitro hydrolysis of phosphorothionate insecticides by insect homogenates, but this is probably attributable also to the insensitive techniques used(40).

Recently, Matsumura and Hogendijk(41) have reported the isolation of a detoxifying esterase present in house flies capable of hydrolyzing parathion at the P—O—nitrophenyl bond. This enzyme, present in parathion- and diazinon-resistant and susceptible flies, was partially purified by precipitation and chromatographic procedures and shown to catalyze the hydrolysis of parathion and diazinon to diethyl phosphorothioic acid and the corresponding hydroxy compound. Labeled as a "thionase," the enzyme is relatively poor in hydrolyzing paraoxon and appears to be specific for phosphorothionates, in direct contrast to the A-esterase or paraoxonase isolated from mammalian tissues(32,39). Parathion-degrading ability of the resistant flies both in vivo and in vitro is greater than that found with susceptible flies and suggests that resistance to parathion and diazinon [O,O-diethyl O-(2-isopropyl-6-methyl-4-pyrimidyl) phosphorothioate] is attributable to direct degradation of these compounds by this enzyme. Oppenoorth and van Asperen(42–44), however, have demonstrated with the same strain of flies that organophosphorus resistance is attributable to increased degradation of paraoxon and diazoxon [O,O-diethyl O-(2-isopropyl-6-methyl-4-pyrimidyl) phosphate], the corresponding phosphate esters, presumably by a modified aliesterase present in greater amounts in resistant flies. Homogenates of organophosphorus-resistant strains of flies were found to contain less aliesterase than is present in susceptible flies and the change of aliesterase in susceptible flies into a phosphatase isoenzyme (or modified aliesterase) was postulated. This enzyme change appears to be attributable to a single gene mutation and results in a phosphatase that shows a high affinity to paraoxon and diazoxon hydrolysis.

Mechanistically, the phosphatases appear to behave in a manner similar to other esterases that are inactivated by organophosphorus esters by reaction to form a dialkyl phosphorylated enzyme. The difference in enzymatic properties, however, is attributable to rapid breakdown of the phosphorylated phosphatase to produce dialkyl phosphoric acid and the free enzyme.

The foregoing discussion has been concerned with enzyme systems that catalyzed the cleavage of the P—O—aryl bond in organophosphorus ester as shown below.

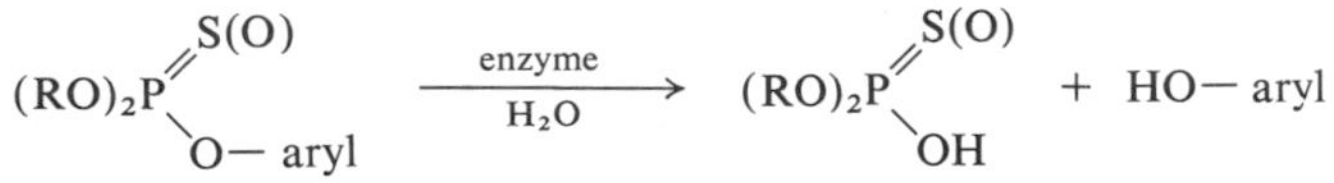

$$(RO)_2P\overset{\displaystyle S(O)}{\underset{\displaystyle O-aryl}{}} \xrightarrow[H_2O]{enzyme} (RO)_2P\overset{\displaystyle S(O)}{\underset{\displaystyle OH}{}} + HO-aryl$$

In addition to this type of degradation, the enzymatic hydrolysis of the P—O—alkyl linkage also has been demonstrated in mammals and insects with both phosphates and phosphorothionates according to the equation below.

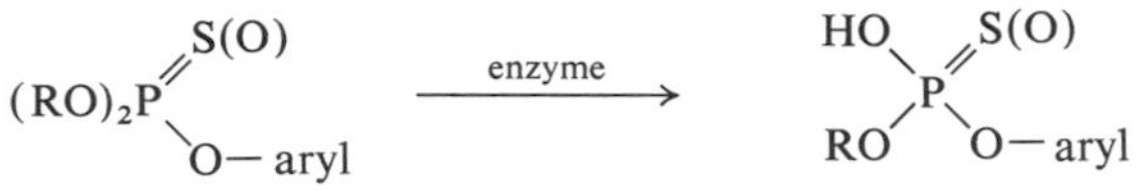

$$(RO)_2P\overset{\displaystyle S(O)}{\underset{\displaystyle O-aryl}{}} \xrightarrow{enzyme} \overset{\displaystyle HO}{\underset{\displaystyle RO}{}}P\overset{\displaystyle S(O)}{\underset{\displaystyle O-aryl}{}}$$

Plapp and Casida(45) first demonstrated the importance of P—O—alkyl cleavage as a metabolic pathway in animals, particularly in mammals. These investigators examined the metabolism of a variety of different organophosphorus insecticides including ronnel (O,O-dimethyl O-2,4,5-trichlorophenyl phosphorothioate), CHLORTHION [O-(3-chloro-4-nitro-phenyl) O,O-dimethyl phosphorothioate], methyl parathion, parathion, paraoxon, and diazinon in the rat and cockroach. By means of ion exchange chromatography, substantial amounts of the dealkylated organophosphorus compound were found in the urine of rats after oral administration. In general, the dimethyl esters were more susceptible to dealkylation than the diethyl esters, particularly at the higher dosages. For example, 42% of ronnel at the 100 mg/kg dosage was metabolized to desmethyl ronnel compared to 6% at the 2 mg/kg dosage. Less dealkylation occurred in the American cockroach, with the exception of methyl parathion, where 23% desmethyl product was observed compared to 9% in the rat.

O-dealkylation of methyl parathion, parathion, SUMITHION, and methyl paraoxon by various subcellular fractions of rat tissues, American cockroach, and larvae of the rice stem borer also has been demonstrated by Fukami and Shishido(46,47). Rat liver supernatant was particularly active in effecting O-demethylation, converting 80–90% of the parent dimethyl phosphorothionate to the desmethyl derivative. The mitochondria fraction was less active, converting about 20–40% of the parent ester to the demethylated product. With the rice stem borer the supernatant, mitochondrial, and microsomal fractions were roughly equal in effecting O-demethylation. The rat liver supernatant showed a pH optimum of about 8.5 and was inhibited by cupric ion and p-mercuribenzoate, a phenomenon typical with many phosphatases.

The enzymatic demethylation of DDVP (2,2-dichlorovinyl dimethyl phosphate) by rat tissue homogenates and rat and rabbit plasma also has been demonstrated(48). With rat liver homogenate, ammonium sulfate fractionation showed that the enzymes hydrolyzing the P—O—methyl linkage precipitated predominantly between 60–80% saturation, while the enzyme hydrolyzing the P—O—vinyl group precipitated between 40–60% saturation. Thus, it appears that the phosphatases responsible for hydrolyzing the dichlorovinyl and methyl moieties are separate enzymes.

Another metabolic reaction peculiar to parathion and other nitrophenyl containing phosphorus esters is the reduction of nitro to an amino group. This reaction has been shown to take place in the rumen and is probably attributable to organisms present in this organ. The production of both amino parathion (O,O-diethyl O-4-aminophenyl phosphorothioate) and amino paraoxon (corresponding phosphate) has been demonstrated by incubating parathion with bovine rumen fluid(49–51). The formation of amino parathion also has been demonstrated in soil and is believed to be caused by yeast(52).

Surprisingly few investigations have been carried out on the metabolism of parathion and related compounds in plants in spite of its wide usage on a variety of crops. Although parathion is not generally regarded as a systemic insecticide, early work by several different investigators (53–55) showed that the leaves of plants grown in parathion-treated soil were toxic to different insects. However, it was not established whether the toxicity was attributable to translocated parathion, fumigant effects, or the impurities present in technical parathion. In some rather definitive experiments, David and Aldridge(56) have demonstrated that parathion is taken up slowly by cabbage and wheat plants and is converted to paraoxon. No parathion was found in the fluid of plants, and the systemic action was attributed to the movement of paraoxon in the plant. The possibility of fumigant effects emanating from the soil was ruled out, as was also the conversion of parathion to paraoxon in soil. Thus, it appears that an activation process occurs in the roots of plants where parathion is converted to paraoxon and it is the latter that is translocated. In vitro experiments by Knaak et al.(57) have shown that peroxidase isolated from green bean hypocotyls effectively catalyzed the conversion of parathion to paraoxon in the presence of dihydroxymaleic acid, which serves as a hydrogen donor. In addition, this enzyme catalyzed the hydrolysis of both parathion and paraoxon. Although this work does not rule out other enzymes for the activation process, it does indicate that peroxidase in the bean plant effects parathion to paraoxon conversion.

The complete elucidation of all metabolic pathways for relatively few organophosphorus compounds has been established in animals or in plants, owing to the difficult nature of the work. Two compounds that have recently received wide attention because of their unusual relative toxicities to mammals are methyl parathion and SUMITHION. SUMITHION, or O,O-dimethyl O-3-methyl-4-nitrophenyl phosphorothioate, is different from methyl parathion only in that it contains a methyl group in the 3-position of the benzene ring. Both compounds are about equally toxic to insects, but SUMITHION is approximately 54 times less toxic to the white mouse than is methyl parathion. A recently completed and rather thorough investigation of the comparative metabolism of these two compounds has shown that the main metabolic pathways for methyl parathion and SUMITHION in the white mouse and housefly are those given below for methyl parathion(58,59). In general, the same reactions also apply to parathion, the ethyl analog(1).

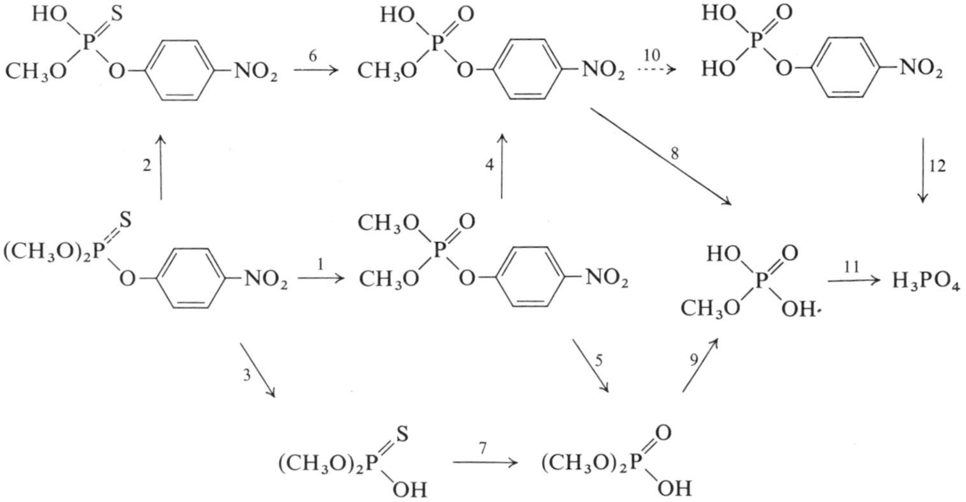

Both SUMITHION and methyl parathion, after oral administration to mice, are rapidly metabolized and excreted, practically all of it in the urine. The compounds shown in the scheme above, with the possible exception of *p*-nitrophenyl dihydrogen phosphate (product from step 10), were positively identified as metabolic products in mouse urine by column, thin layer, and paper chromatographic methods. The same metabolites were also produced by the housefly. Although the metabolic pathways in the mouse are similar for both methyl parathion and SUMITHION, there were distinct differences in the relative amounts of the various metabolites formed which may account, in part, for the difference in toxicity of these two compounds. With methyl parathion, the major metabolic product, constituting approximately 53% of the radioactive metabolites in the urine at the 3 mg/kg dosage, was dimethylphosphoric acid. At the 17 mg/kg dosage, a value which approaches the LD_{50} level of 23 mg/kg, 32%, was found. In comparison, SUMITHION at 3 mg/kg gave 32% dimethylphosphoric acid and at 850 mg/kg (LD_{50} 1250 mg/kg) 3% was found. These values indicate that the P=S to P=O activation occurs much more readily with methyl parathion than with SUMITHION. Another striking difference between these two compounds was their relative rates of demethylation to the desmethyl derivatives as indicated by pathway 2. With methyl parathion at 17 mg/kg dosage, the amount of desmethyl methyl parathion was 19% compared to 66% for desmethyl SUMITHION at the 850 mg/kg dosage. This indicates that the demethylation step plays a much more important role in the detoxication of SUMITHION than with methyl parathion. The differences in these two factors, greater P=S to P=O conversion with methyl parathion and faster demethylation with SUMITHION, must in large part be responsible for the selective action of SUMITHION. This conclusion is supported to some extent by the work of others(60) who found after incubation of these compounds with mouse liver slices that SUMITHION was more rapidly degraded and that methyl parathion was converted to an active cholinesterase inhibitor while SUMITHION was not.

B. Demeton and Related Compounds

The discovery of systemically active organophosphorus insecticides stimulated great interest in the metabolism of these compounds for several reasons. Systemically active insecticides may be separated into two distinct classes, those that exert their toxic action by being absorbed by the plant (or animal) and distributed unchanged, as appears to be the case with certain derivatives of 2-fluoroethyl alcohol(61), or those that are acted upon by the plant and converted to other metabolic products. In general, the organophosphorus systemics fall into the latter category, and there are numerous cases in which the metabolic products are intrinsically more toxic than the parent substance. Obviously the identification and knowledge of the chemical and toxicological properties of these metabolic poisons are essential for assessing the hazards of residues resulting from the use of these compounds on edible crops. Further, information accrued from metabolic investigations is invaluable in establishing the mode of action of these insecticides and for a clearer understanding of the biochemical systems involved in metabolism.

1. Demeton. The active ingredients in commercial demeton consist of a pair of structural isomers, O,O-diethyl O-2-(ethylthio)ethyl phosphorothioate (herein referred to as PS-demeton) and O,O-diethyl S-2-(ethylthio)ethyl phosphorothioate (PO-demeton). The chemical and physical properties of these isomers have been described in detail, and they appear to differ in spite of their similar chemical structure(*62–64*). The systemic action of commercial demeton is not attributable to the action of the unchanged isomers, but to metabolic products that are toxic to insects and mammals. Preliminary investigations by Heath et al. (*64*) showed that when the roots or foliage of brassicaceous plants are treated with either PS- or PO-demeton, essentially none of the starting isomers is found in the plant after a few days. By countercurrent partition methods, at least three metabolites were isolated from PO-demeton, and the properties of these metabolites were examined, but the identity of these compounds was not established.

A more detailed study, described in a series of papers on demeton metabolism(*65–69*), has shown that the isomers are converted to oxidation products that are more potent inhibitors of acetyl cholinesterase than the parent compounds. By comparing the biological and chemical properties, as well as the relative behavior on paper chromatograms of the hydrogen peroxide and perbenzoic acid oxidation products of the isomers with the metabolites obtained from the cotton plant, it was shown that each isomer was first converted to the sulfoxide and this in turn was converted to the sulfone. Unequivocal identification of four metabolites was subsequently accomplished(*68,69*) by isolating each metabolite in the pure state through the combined use of paper and column chromatography, and solvent partition, and comparing their infrared spectra with that of the synthetic sulfoxides and sulfones. The metabolic pathway of demeton metabolism known to occur in the cotton plant is depicted below.

Evidence also was obtained indicating the presence of at least one and possibly two additional metabolites from PS-demeton. Because of the somewhat greater polar characteristics of these metabolites, as estimated by their chromatographic behavior, it was concluded that these were the sulfoxide and sulfone of the phosphate, obtained by

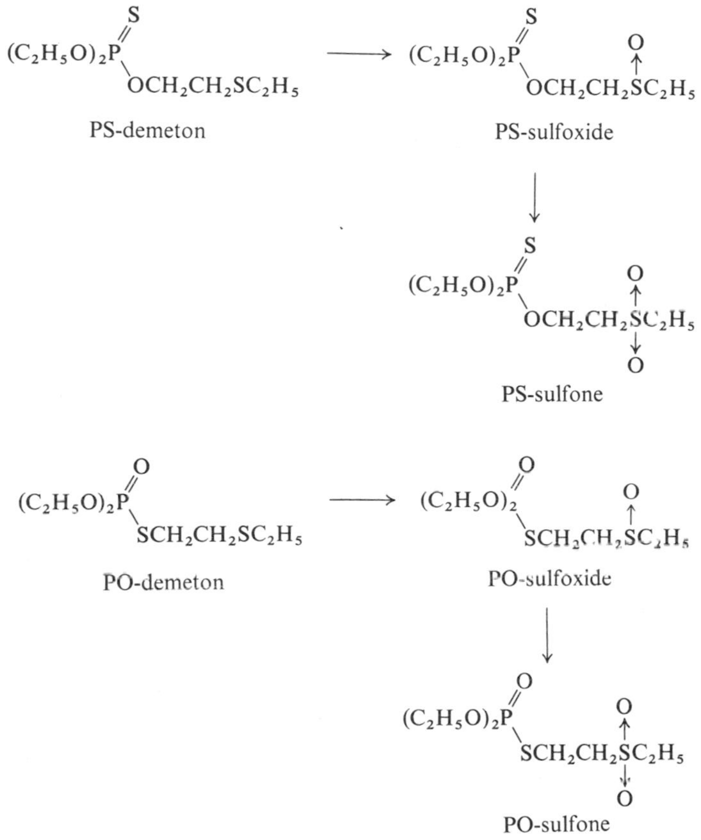

$$\underset{\text{PS-demeton}}{(C_2H_5O)_2\overset{\displaystyle S}{\overset{\|}{P}}\diagdown_{OCH_2CH_2SC_2H_5}} \longrightarrow \underset{\text{PS-sulfoxide}}{(C_2H_5O)_2\overset{\displaystyle S}{\overset{\|}{P}}\diagdown_{OCH_2CH_2\overset{\displaystyle O}{\overset{\uparrow}{S}}C_2H_5}}$$

$$\underset{\text{PS-sulfone}}{(C_2H_5O)_2\overset{\displaystyle S}{\overset{\|}{P}}\diagdown_{OCH_2CH_2\underset{\downarrow O}{\overset{\displaystyle O}{\overset{\uparrow}{S}}}C_2H_5}}$$

$$\underset{\text{PO-demeton}}{(C_2H_5O)_2\overset{\displaystyle O}{\overset{\|}{P}}\diagdown_{SCH_2CH_2SC_2H_5}} \longrightarrow \underset{\text{PO-sulfoxide}}{(C_2H_5O)_2\overset{\displaystyle O}{\overset{\|}{P}}\diagdown_{SCH_2CH_2\overset{\displaystyle O}{\overset{\uparrow}{S}}C_2H_5}}$$

$$\underset{\text{PO-sulfone}}{(C_2H_5O)_2\overset{\displaystyle O}{\overset{\|}{P}}\diagdown_{SCH_2CH_2\underset{\downarrow O}{\overset{\displaystyle O}{\overset{\uparrow}{S}}}C_2H_5}}$$

desulfuration of the PS-demeton and metabolic oxidation to the sulfoxide and sulfone or by direct desulfuration of PS-sulfoxide or PS-sulfone.

The metabolism of the demeton isomers also was investigated in the white mouse and American cockroach(70) and the routes of metabolism and the toxic metabolites formed were shown to be similar to that occurring in plants. Both PS- and PO-demeton are rapidly metabolized, degraded, and eliminated in these animals. The principal route of elimination in the mouse was through the urine, 50 to 70% of an orally administered dose being eliminated in 24 hr, 90% of which was in the form of hydrolyzed nontoxic products. The remaining 10% of the urine-excreted compounds was chloroform-partitioning substances which by paper chromatography and cholinesterase inhibition measurements were believed to consist of the P-O-demeton, its sulfoxide and sulfone,

P-S-demeton and its sulfoxide, and the desulfurated P-S-demeton O,O-diethyl O-2-(ethylthio)ethyl phosphate. Evidence also was obtained that indicated that other metabolites, related to those mentioned above and that behaved similarly on the paper chromatographic systems used, were present. Although the rates of metabolism, degradation, and elimination were somewhat slower in the cockroach than in the mouse, in gross aspects the metabolic pathways were the same.

Similar studies with the methyl analog of PO-demeton [O,O-dimethyl S-2-(ethylthio)ethyl phosphorothioate or methylisosystox] were carried out by Mühlmann and Tietz(71). These investigators showed that methylisosystox, like PO-demeton, is converted in different plants to the corresponding sulfoxide and sulfone. Identification of these metabolites was accomplished by isolating the pure metabolites by counter-current partition methods. The sulfone was a crystalline solid, m.p. 50°C, identical to the synthetic sulfone. The sulfoxide isolated from plants was shown to have identical solvent partition and paper chromatographic properties as the sulfoxide synthesized in the laboratory. Further, hydrogen peroxide oxidation of the plant isolated sulfoxide resulted in excellent yield of the crystalline sulfone.

Of considerable importance is the behavior of demeton type compounds in water. P=S demeton isomerization to P=O demeton is strongly catalyzed by water and other polar solvents via a cyclic sulfonium ion intermediate(72–74). Methyl demeton appears to be considerably more reactive than demeton in water and, in addition to isomerizing, reacts with itself to form sulfonium ions of greatly increased toxicity and anticholinesterase activity as shown below(75).

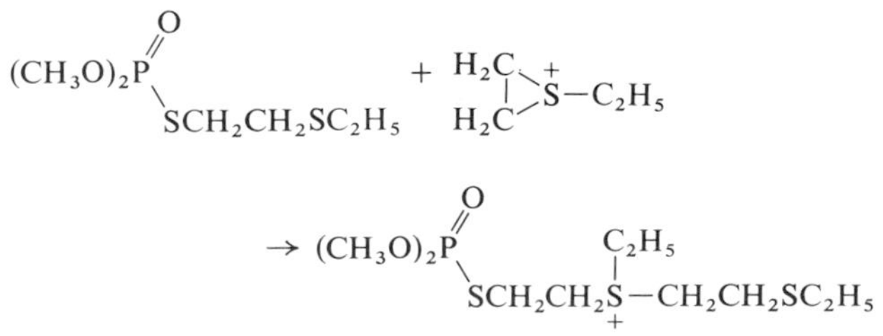

$$
(CH_3O)_2P\overset{O}{\underset{SCH_2CH_2SC_2H_5}{\big\|}} \quad + \quad \underset{H_2C}{\overset{H_2C}{\diagdown}}\overset{+}{S}-C_2H_5
$$

$$
\rightarrow \quad (CH_3O)_2P\overset{O}{\underset{SCH_2CH_2\overset{+}{S}-CH_2CH_2SC_2H_5}{\big\|}}\overset{C_2H_5}{} \tag{1}
$$

$$
(CH_3O)_2P\overset{O}{\underset{SCH_2CH_2SC_2H_5}{\big\|}} \quad \longrightarrow \quad (CH_3O)_2P\overset{O}{\underset{SCH_2CH_2\overset{+}{S}-C_2H_5}{\big\|}}\overset{CH_3}{}
$$

$$
+ \quad \overset{CH_3O}{\underset{^-O}{\diagup}}P\overset{O}{\underset{SCH_2CH_2SC_2H_5}{\diagdown}} \tag{2}
$$

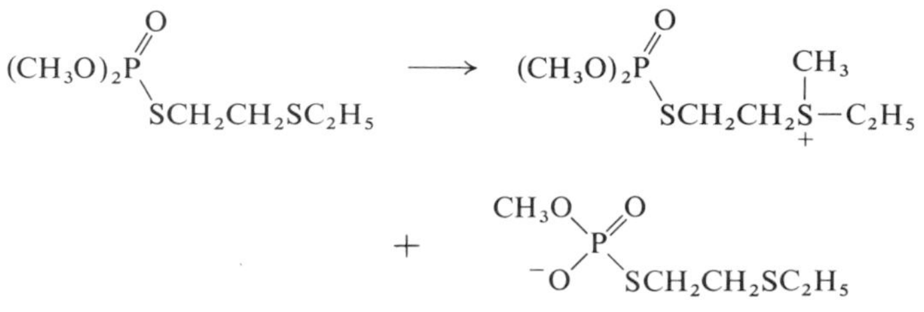

The intravenous LD_{50} to rats of the two sulfonium ions are 0.007 and 0.062 mg/kg, respectively (72,75).

2. Disulfoton and Phorate. The cotton plant metabolism of two other commercially important systemic insecticides closely related to demeton, disulfoton [DI-SYSTON or O,O-diethyl S-2-(ethylthio)ethyl phosphorodithioate], and phorate [THIMET or O,O-diethyl S-(ethylthio)methyl phosphorodithioate] also has been investigated (76). Both compounds were metabolized in a similar manner, i.e., oxidation of the thio ether moiety to the corresponding sulfoxide and sulfone and desulfuration of the P=S moiety to P=O, producing a phosphorothiolate ester. The metabolic scheme is shown below.

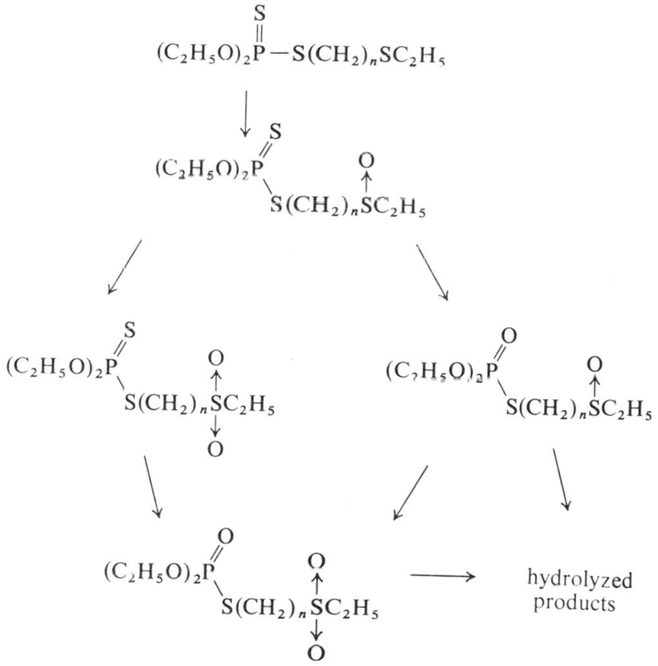

$$(C_2H_5O)_2\overset{\overset{S}{\|}}{P}-S(CH_2)_nSC_2H_5$$

$$(C_2H_5O)_2\overset{\overset{S}{/\!/}}{P}\diagdown \underset{S(CH_2)_n\overset{\overset{O}{\uparrow}}{S}C_2H_5}{}$$

$$(C_2H_5O)_2\overset{\overset{S}{/\!/}}{P}\diagdown \underset{S(CH_2)_n\overset{\overset{O}{\uparrow}}{\underset{\downarrow}{S}}C_2H_5}{\overset{}{O}} \qquad (C_2H_5O)_2\overset{\overset{O}{/\!/}}{P}\diagdown \underset{S(CH_2)_n\overset{\overset{O}{\uparrow}}{S}C_2H_5}{}$$

$$(C_2H_5O)_2\overset{\overset{O}{/\!/}}{P}\diagdown \underset{S(CH_2)_n\overset{\overset{O}{\uparrow}}{\underset{\downarrow}{S}}C_2H_5}{\overset{}{O}} \longrightarrow \text{hydrolyzed products}$$

In this scheme when n is 1, the compound is phorate; when n is 2, the compound is disulfoton. It was shown that conversion of disulfoton to its sulfoxide occurred immediately upon absorption into the plant and at no time could disulfoton be detected. On the other hand, small amounts (less than 8%) of unchanged phorate were found to be present immediately after absorption, this eventually disappearing after about 50 hr. Figure 2-1 is a graph showing relative amounts of disulfoton metabolites as fractions of total chloroform-partitioning ^{32}P in extracts of isolated

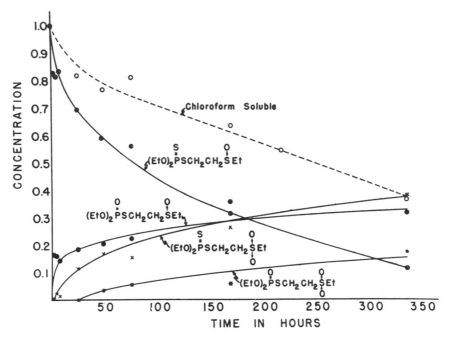

Fig. 2-1. Graph showing relative amounts of disulfoton metabolites as fractions of total chloroform-partitioning ^{32}P in extracts of isolated cotton leaves at various intervals after treatment.

cotton leaves at various intervals after treatment. The values for the chloroform-soluble material represent intact ester, the rest of the ^{32}P remaining in the aqueous phase as ionic products attributable to hydrolysis. The graph clearly shows the rapid accumulation of disulfoton sulfoxide and the gradual appearance of the secondary oxidation products. Phorate gave somewhat similar plots except for the presence of unchanged phorate in the early stages. Although the earlier work with demeton gave strong but inconclusive evidence for the conversion of P=S to P=O esters, there is little doubt that desulfuration of disulfoton and phorate sulfoxide to the respective thiolate esters takes place in the cotton plant. Thus, it is evident that in plants there are at least three distinct metabolic reactions with compounds of this type, i.e., oxidation of thioethers to sulfoxide and sulfones, desulfuration of P=S to P=O, and hydrolysis of ester bonds.

Figure 2-1 points out the importance of the identification and quantification of metabolites with regard to the assessment of residues. It is quite evident, in the early stages after absorption by the plant, that disulfoton sulfoxide is the predominant metabolite. However, after 14 days, the PO sulfoxide and disulfoton sulfone were the principal metabolites. Since these metabolites possess different toxicological properties, information concerning their relative proportions in plants is obviously essential before residual hazards may be assessed.

Further work with disulfoton(77) showed that the rate of metabolism of disulfoton to its oxidation products is increased 1.9-fold for each 10°C rise in temperature between a temperature range of 37° and 100°F. It was further shown that metabolism of disulfoton occurs at a rate two to three times faster in tomato leaves at 70°F. than with cotton leaves, and at an intermediate rate in a variety of other plants including alfalfa, bean, brussels sprout, cabbage, corn, and lemon.

A recent report(78) on disulfoton metabolism in the boll weevil (*Anthonomus grandis*), bollworm (*Heliothis zea*), isolated cotton leaves, and the rat corroborates the results reported above. In addition to the various intact oxidation products, a number of hydrolysis products also were isolated and identified. The hydrolytic degradation of the oxidative metabolites was considerably faster in the boll weevil and bollworm than in the cotton leaf. In each case, however, the principal hydrolytic products were diethyl phosphoric and diethyl phosphorothioic acids. No phosphorodithioic acid was found, indicating that hydrolysis occurs through cleavage of the P—S bond as exemplified below with the sulfoxide derivative.

$$(C_2H_5O)_2P \overset{\overset{S \;(O)}{\parallel}}{\underset{\overset{SCH_2CH_2SC_2H_5}{\uparrow}}{\diagdown}} \qquad \overset{O}{\uparrow} \qquad \longrightarrow \qquad (C_2H_5O)_2P \overset{\overset{S \;(O)}{\parallel}}{\underset{OH}{\diagdown}}$$

$$\underset{HOH}{\overset{\diagup}{\uparrow}}$$

In addition to these two acids, five other degradation products of minor importance also were detected.

Bowman and Casida(79) have examined phorate metabolism in the pea and cotton plant as soil and as seed treatments. Phorate in the cotton plant two weeks after seed treatment was found to be completely metabolized to oxidized or hydrolyzed products. By column chromatography an alpha- and a beta-fraction, consisting of 83% and 17%, respectively, of the chloroform-soluble radioactive metabolites, was obtained. Infrared spectra, solvent partitioning behavior, and anticholinesterase data showed that the alpha-fraction was 61% phorate sulfoxide and 39% phorate sulfone. The beta-fraction consisted of 70% PO-phorate sulfoxide and

30% PO-phorate sulfone. In absolute quantities, the two-week cotton plant contained no unchanged phorate, 11.6 ppm phorate sulfoxide, 7.4 ppm phorate sulfone, 1.7 ppm PO-phorate sulfoxide, and 1.1 ppm PO-phorate sulfone. In addition, the metabolism of O,O-diethyl S-(isopropylthio)methyl phosphorodithioate, a closely related compound of strong systemic activity, was shown to follow a metabolic route similar to phorate.

The formation of phorate sulfoxide and sulfone, PO-phorate sulfoxide and sulfone by incubation of phorate with rat liver slices(80) and with the yeast-like plant *Torulopsis*(81) also has been demonstrated. In contrast, in the few insects that have been examined, including the German cockroach, milkweed bug, and Rhodnius bug, the evidence indicates that THIMET is converted to the sulfoxide and sulfone, but not to the corresponding PO analogs(82).

3. Carbophenothion. Carbophenothion [O,O-diethyl S-(*p*-chlorophenylthio)methyl phosphorodithioate] is another thioether containing compound closely related in structure to phorate. When applied to lettuce, carbophenothion is converted into five oxidative metabolites within a period of 4 hr, and they persisted for up to 21 days(83). By paper chromatographic comparison with known oxidation products of carbophenothion, the various metabolites were identified as carbophenothion sulfoxide and sulfone, the oxygen analog of carbophenothion, its sulfoxide and sulfone. Thus, the metabolic pathways for carbophenothion are identical to those occurring with other thioether containing insecticides.

C. Fenthion and Related Compounds

1. Fenthion. The metabolism of fenthion [O,O-dimethyl O-(4-methyl-thio)-*m*-tolyl phosphorothioate] and related compounds is qualitatively similar to that described for demeton, disulfoton, and other thioether containing compounds. As has been demonstrated with thioether containing insecticides, fenthion is metabolized in plants initially to the sulfoxide(84–88). The metabolic conversion of the thioether group to the sulfoxide is rapid and subsequent metabolism to the sulfone and desulfuration to the phosphate sulfoxide and sulfone is slower. In addition to the aforementioned compounds, the presence of the S-methyl isomer, its sulfoxide and sulfone also have been demonstrated in plants (84,85). The S-methyl isomer, however, is a known contaminant in technical fenthion, and it is possible that isomerization occurred prior to plant treatment.

The transformation of fenthion to its sulfoxide and sulfone appears to be a facile process and even occurs upon ordinary exposure of a thin film to the atmosphere and sunlight and is apparently a function of light intensity (86). There is a probability that fenthion is converted to oxidation products on leaf surfaces before absorption and metabolism by the plant. Fenthion readily isomerizes to the S-methyl compound with heat, substantial isomerization occurring after 1 hr and 100°C (89). The isomerization probably occurs through a sulfonium ion intermediate.

Fenthion sulfoxide and sulfone and fenthion phosphate sulfoxide or sulfone have been identified in houseflies after topical application of fenthion (86). At least five intact ester derivatives of fenthion have been isolated from the boll weevil, German cockroaches, and rats (90).

The various reactions giving rise to intact fenthion metabolites are given below.

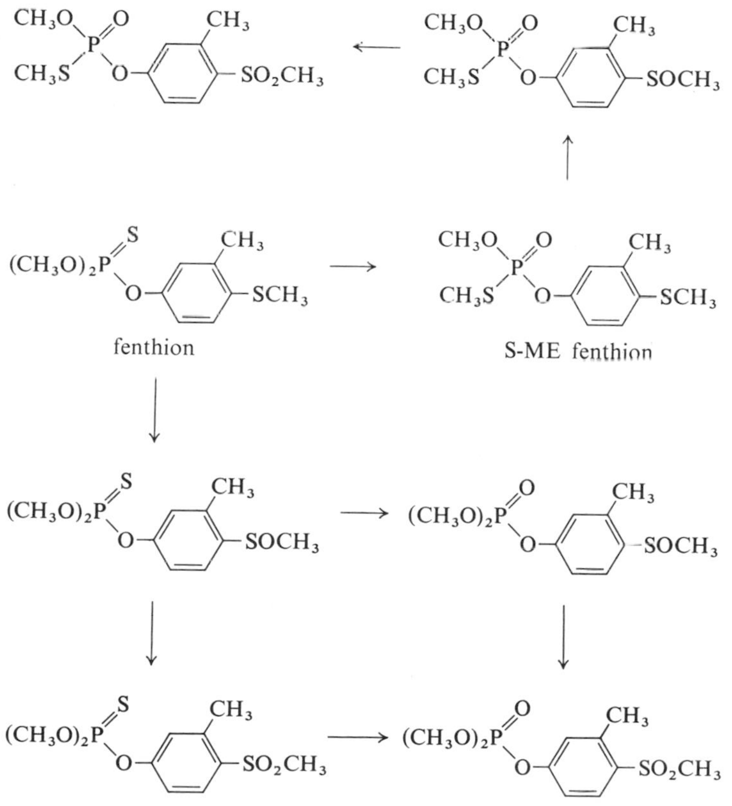

The metabolism of O,O-dimethyl and O,O-diethyl O-(p-methylsulfinyl)-phenyl phosphorothioate (Bayer 25198 and Bayer 25141, respectively), two compounds closely related to fenthion but containing the methylsulfinyl instead of the methylthio moiety, has been investigated in cotton plants and the American cockroach(91,92). In general, the metabolic pathways are similar to those described above for fenthion.

2. Malathion. Malathion, owing to its generally good insecticidal properties and its low mammalian toxicity, is probably the most widely used organophosphorus insecticide today. Because of its biological selectivity, information regarding the comparative metabolism of this important compound in insects and mammals has been of great interest. Malathion or O,O-dimethyl S-[1,2-bis(ethoxycarboxy)ethyl] phosphorodithioate is a molecule containing several labile functional groups and consequently its metabolism in animals is expected to be complex. Early investigations by March et al.(93) showed that malathion metabolism in the hen and white mouse is indeed complex. In the hen and mouse, where metabolism appeared to be the same, at least seven metabolites were detected in excreta by paper chromatography. In comparison, the paper chromatogram of malathion metabolites isolated from an extract of American cockroach gut contained only two spots. High insecticidal activity of malathion was attributed in part to slower rates of detoxication in insects compared to mammals. Although the chemical structures of the metabolites were not established, a metabolic route consistent with the biological and chemical properties of the metabolites was suggested.

In vitro studies have shown that malathion is rapidly activated to malaoxon (phosphorothioate analog) by PS to PO conversion by the action of mouse liver slices in the presence of necessary cofactors(94). Further, malathion was degraded by the same system in the absence of cofactors at a rate greater than the activation process. Isolated cockroach gut also activated malathion to malaoxon, but in this case degradation was slow, leading to a steady accumulation of malaoxon. Since malaoxon is a strong cholinesterase inhibitor and malathion is weak, the biological specificity of the latter was attributed to the buildup of malaoxon in insects. Subsequent in vivo investigations have shown that malathion degradation is much more extensive in the mouse than in insects with correspondingly lower malaoxon formation, confirming earlier in vitro results(40). By using high specific activity malathion and ion exchange chromatography, a total of 11 metabolites were isolated from the German cockroach, American cockroach, and the common housefly, and seven metabolites from the white mouse. The principal metabolites isolated from the mouse were the monoethyl ester of malathion (68%), dimethyl phosphorothioic acid (13%), dimethyl phosphorodithioic acid (5%), and 10% of an unknown metabolite. The relative amounts of the various metabolites obtained were similar between the roaches, which in turn were slightly different from flies. In roaches, the main metabolites were dimethyl phosphoric acid, dimethyl phosphorothioic acid, and dimethyl phosphorodithioic acid obtained by hydrolysis of the P—S—C linkage, monoethyl ester of malathion, and completely deethylated malathion obtained by hydrolysis of the carboethoxy group. The combined percentages of the phosphorus acids and carboxylic acids were approximately equal, and their sum accounted for about 70–80% of the total metabolites. In houseflies dimethyl phosphorothioic acid was by far the predominant metabolite (30–40%), with lesser amounts of dimethyl phosphoric acid (20–30%). Further, the level of malaoxon was substantially greater in the cockroach than in the mouse at any time after injection, e.g., at one hour after injection there was ten times more malaoxon in the roach than in the mouse per gram of animal.

The relative proportion of metabolites indicates greater PS to PO activation in insects compared to mammals, followed by hydrolysis of the P—S—C linkage. This is especially true in the housefly. In mammals degradation via hydrolysis of the carboethoxy moiety predominates, leading to ionic substances that are rapidly excreted. Thus, the biological specificity of malathion between insects and mammals is readily explained by the difference in rates and routes by which malathion is metabolized. Hydrolysis of the phosphate linkages is presumed to be attributable to phosphatase activity, and carboethoxy hydrolysis has been attributed to "carboxyesterase" activity.

Further studies of in vitro malathion metabolism by different kinds of rat tissue have shown that the various tissues produce similar metabolites in comparable proportions, with the carboethoxy ester hydrolysis products appearing in substantially greater amounts than products from phosphate ester hydrolysis(95). The tissues of rats treated with EPN, a compound known to enhance the toxicity of malathion(96), were generally less effective in activating and degrading malathion. The evidence indicated that EPN, however, was more effective in inhibiting the carboxyesterase activity. Later in vivo studies with the rat and dog (97) showed that EPN affects metabolism by inducing a change from carboxy ester hydrolysis to phosphate hydrolysis. Also, it was in this later study that the identity of desmethyl malathion was first demonstrated. In malathion treated rats this compound accounts for about 11% of the total metabolites found in rat urine.

The metabolic pathway for malathion in animals therefore may be depicted as follows:

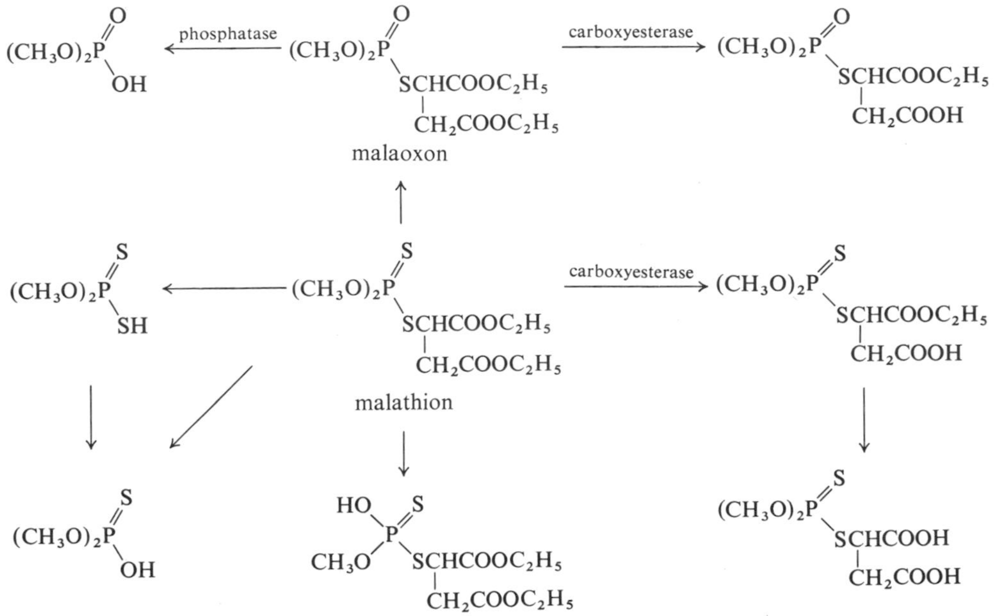

$$(CH_3O)_2\overset{O}{P}{}OH \xleftarrow{\text{phosphatase}} (CH_3O)_2\overset{O}{P}\begin{matrix}SCHCOOC_2H_5\\ |\\ CH_2COOC_2H_5\end{matrix} \xrightarrow{\text{carboxyesterase}} (CH_3O)_2\overset{O}{P}\begin{matrix}SCHCOOC_2H_5\\ |\\ CH_2COOH\end{matrix}$$

malaoxon

$$(CH_3O)_2\overset{S}{P}{}SH \xleftarrow{} (CH_3O)_2\overset{S}{P}\begin{matrix}SCHCOOC_2H_5\\ |\\ CH_2COOC_2H_5\end{matrix} \xrightarrow{\text{carboxyesterase}} (CH_3O)_2\overset{S}{P}\begin{matrix}SCHCOOC_2H_5\\ |\\ CH_2COOH\end{matrix}$$

malathion

$$(CH_3O)_2\overset{S}{P}{}OH \qquad \underset{CH_3O}{\overset{HO}{}}\overset{S}{P}\begin{matrix}SCHCOOC_2H_5\\ |\\ CH_2COOC_2H_5\end{matrix} \qquad (CH_3O)_2\overset{S}{P}\begin{matrix}SCHCOOH\\ |\\ CH_2COOH\end{matrix}$$

The metabolic routes given here are consistent with the metabolites found in insects or mammals. Since at least two or more metabolites have not been identified, the metabolic picture is incomplete. This scheme, however, accounts for the majority of the metabolites found. Although no metabolite for malaoxon with a hydrolyzed carboethoxy group has been identified, carboxyesterase hydrolysis of malaoxon undoubtedly must occur in view of the great difficulty encountered in detecting malaoxon in animal tissue.

The metabolism of malathion in the cow also has been studied by analysis of the urine and is substantially the same as that found in other mammals in that hydrolysis of the carboxy ester group predominates over other pathways(98). The fecal matter from a cow given malathion orally, however, contained mainly dimethyl phosphoric acid and dimethyl phosphorothioic acid, suggesting strong phosphatase action in the rumen. Also noteworthy is that 7% of the total metabolites in feces was chloroform soluble, of which 85% was malathion and 12% malaoxon. The milk contained a small amount of malathion metabolites (9.2% of total dose after seven days); of this, only 29% was extractable out of milk and partitioned in favor of water over benzene, indicating the absence of either malathion or malaoxon.

3. Dimethoate. Because of its high insecticidal activity and low mammalian toxicity, dimethoate [O,O-dimethyl S-(N-methylcarbamoylmethyl) phosphorodithioate] has been widely used in insect control. Dimethoate is an extremely versatile compound and may be used as a systemic insecticide not only in plants but for control of animal parasites. The remarkable properties of this compound have stimulated extensive investigations on its metabolism in insects, mammals, and plants.

In gross aspects, the metabolism of dimethoate is similar in insects, mammals, and plants. The presence of the oxygen analog, O,O-dimethyl S-(N-methylcarbamoylmethyl) phosphorothioate, has been demonstrated unequivocally in all three systems and evidently is the metabolite responsible for the toxic action of dimethoate(99–102). This metabolite is found in highest levels in insects, particularly in those highly susceptible to dimethoate. For example, in the highly susceptible housefly as much as 31% of the applied dose was recovered as oxygen analog after injection(103) and 16% after topical application(102). Large amounts of oxygen analog have been found in locusts poisoned with radiolabeled dimethoate(104). Although the oxygen analog of dimethoate appears to be the principal intermediate responsible for toxicity, there is a distinct possibility that other metabolites are formed that also possess anticholinesterase activity. Since two of the several unidentified metabolites that have been isolated from different insects have been shown to possess chromatographic properties similar to dimethoate and its oxygen analog, it is tempting to suggest that their structures are N-desmethyl dimethoate and N-desmethyl oxygen analog, formed by oxidative desmethylation of the parent compounds(101,102). The formation of the desmethyl derivatives is analogous to the metabolic transformation that takes place with AZODRIN (see section on AZODRIN and BIDRIN). The possibility that desmethyl oxygen analog is responsible in part for toxicity is supported by the fact that N-desmethyl dimethoate is quite toxic to houseflies(102).

The hydrolysis of the amide moiety to the carboxyl derivative also appears to take place in insects. The presence of O,O-dimethyl S-carboxymethyl phosphorodithioate and possibly the corresponding oxygen analog has been demonstrated in a variety of insects(100–102). Because of their ionic nature, these compounds probably do not contribute to toxicity. The equations below depict the possible metabolic transformations involving intact esters that may occur in insects. In addition to these esters, the presence of a number of hydrolysis products has been established in dimethoate treated insects. These include desmethyl dimethoate (**V**), dimethyl phosphoric acid, dimethyl phosphorothioic acid, dimethyl phosphorodithioic acid, and phosphoric acid(*100,101*).

Qualitatively, the metabolism of dimethoate in mammals is essentially identical to that occurring in insects. There are, however, quantitative differences. In general, dimethoate is much more rapidly degraded in mammals and eliminated via the urine(*105–107*). For example, 87–90% of the oral dose given to cattle was found in the urine after 24 hr, mainly as hydrolysis products(*105*). The principal metabolite in the urine immediately after oral administration of dimethoate to the cow(*106*) and sheep(*107*) has been shown to be the deamidated carboxymethyl dithioate (**IV**). This metabolite steadily declined after several hours with concomitant increase in dimethyl phosphoric acid. Dimethoate oxygen analog was found also, but only in minor amounts. Other metabolites identified were dimethyl phosphorodithioic acid, dimethyl phosphorothioic acid, monomethyl phosphoric acid, and desmethyl dimethoate (**V**). Several unidentified metabolites also were isolated; two of these were suspected to be the carboxymethyl thioate (**III**) and desmethyl carboxymethyl dithioate, whose structure is given below.

Substantial amounts of unchanged dimethoate were found in cattle blood shortly after administration by intramuscular injection or orally. Considerably less but significant amounts of a strong cholinesterase inhibitor, presumed to be dimethoate oxygen analog, were also found(105).

Dimethoate is 20-fold more toxic to the pheasant and in general is more toxic to avians(104). In the pheasant, the oxygen analog is formed and tends to accumulate, because of much slower degradation to nontoxic products, in line with the higher toxicity of dimethoate to avians.

In vitro studies have shown that dimethoate degradation in mammals occurs principally in the liver(100). With whole liver homogenates or liver microsomes, dimethoate is degraded to dimethyl phosphorodithioic acid or carboxymethyl dithioate (IV). Guinea pig liver gave only dimethyl phosphorodithioic acid; on the other hand, liver from castrated sheep produced only carboxymethyl dithioate. The two products presumably are attributable to phosphatase and amidase action, respectively, in the two different livers.

The metabolism of dimethoate has been studied in cotton plants (101, 107a, 108), corn, pea, and potato seedlings(108) and olive trees (109). The metabolite accumulating in the cotton plant in largest amount is the carboxymethyl dithioate (IV) after dimethoate absorption through the roots or into excised leaves(101,107a). However, by foliar application, desmethyl dimethoate (V) is found as the principal metabolite, and apparently no carboxymethyl dithioate is present within the plant(108). In general, the various metabolites found in plants were the same as those identified in insects and mammals. Dimethoate oxygen analog (II) was found in significant amounts in all plants studied. Thus, in the living systems examined, the metabolic reactions that take place include activation of P=S to P=O by desulfuration, deamidation by amidase action, hydrolysis of P—O and P—S bonds by phosphatase action, and possibly N-dealkylation by oxidases. The various points of attack in the molecule are shown below.

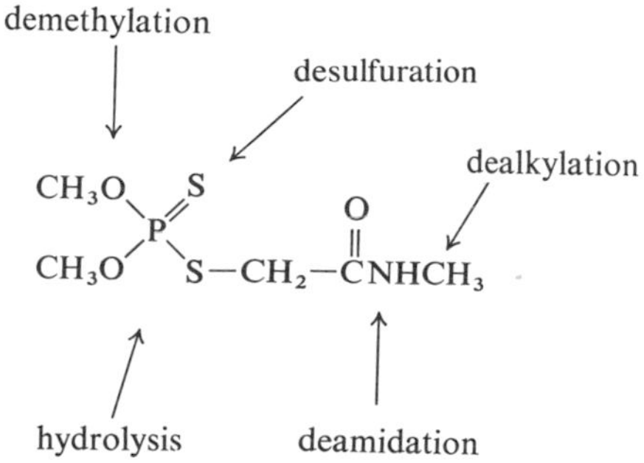

4. Diazinon. Surprisingly little information is available on diazinon metabolism [O,O-diethyl O-(2-isopropyl-4-methyl-6-pyrimidyl) phosphorothioate] despite its wide usage as an insecticide. Pure diazinon is a poor anticholinesterase, but is readily converted to a strong inhibitor upon storage or heating, particularly in the presence of a trace of moisture (*110*). The thermal decomposition of pure diazinon is quite complex; among the many products obtained are pyrophosphates that probably are responsible for increase in anticholinesterase activity (*111*).

In an effort to obtain an explanation for its selective action to insects, the metabolism of diazinon has been investigated in houseflies, the American cockroach, and the mouse (*99*). Metabolic conversion of diazinon to the oxygen analog, diethyl 2-isopropyl-4-methyl-6-pyrimidyl phosphate, has been demonstrated in these animals and selective action appears to be attributable to the relatively higher level of the oxygen analog in insects compared to the mouse. No attempts were made to identify other metabolic products in the study.

Diazinon metabolism has been investigated also in a goat (*112*) and a cow (*113*). Radioactivity was found in the milk, urine, and feces of the goat after oral administration of ^{32}P diazinon. From measurements of the anticholinesterase activity and organic ^{32}P content in these fractions, it was concluded that no diazinon was present and that the radioactivity was attributable to hydrolysis products. Obviously this information is subject to severe criticism because of the poor anticholinesterase activity of diazinon. More definitive experiments with the cow (*113*) using chromatographic techniques have shown the presence of significant amounts of unchanged diazinon in milk, urine, and blood. The hydrolysis products were demonstrated to be diethyl phosphoric acid and diethyl phosphorothioic acid.

The presence of diazinon oxygen analog has been demonstrated in spinach and snap beans sprayed with diazinon (*114*). Also, 2-isopropyl-4-methylpyrimidin-6-ol has been isolated from tomato plants treated with ^{14}C pyrimidine ring-labeled diazinon, indicating direct hydrolysis of the phosphorus ester moiety as an important metabolic pathway.

5. Coumaphos. The metabolism of coumaphos or O,O-diethyl O-(3-chloro-4-methylumbelliferone) phosphorothioate has been studied extensively in mammals because of its wide usage in controlling animal parasites. When administered orally to mammals, coumaphos is rapidly detoxified and eliminated in the urine and feces, mainly as detoxication products (*114a, 115, 116*). Diethyl phosphorothioic acid was the principal metabolite found in the urine of orally treated rats, consisting of greater than 50% of the excreted metabolites. Found in lesser amounts were phosphoric acid, diethyl phosphoric acid, and either desethyl coumaphos or the oxygen analog of desethyl coumaphos(115). Monoethyl phosphoric acid also has been reported in rat urine(116). In contrast, the feces contained significant amounts of intact ester, either as coumaphos or coumaphos oxygen analog, in addition to hydrolytic products. In gross aspects the metabolic pathway in the cow, goat, and mouse is similar to that in the rat(115, 117–120).

Metabolism and elimination of coumaphos are considerably slower following dermal application and appear to be a function of absorption and translocation rates in the animal. For example, rats treated orally eliminated 70% of the applied dose in seven days compared to about 5% after dermal treatment(116). Further, dermal treatment resulted in approximately equal amounts of phosphoric, diethyl phosphoric, and diethyl phosphorothioic acids in the urine, but none of the desethyl products could be detected(115).

Tissues from animals treated with coumaphos have been shown to contain largely degradation products. However, significant quantities of organosoluble substances, indicating the presence of intact ester, have been found in liver and kidney of dermally treated cattle(118). Also, organosoluble substances were observed in the milk of dermally treated cows, the highest level of 0.25 ppm found at the 5-hr sample after application of a 0.75% spray(119, 120). The organosoluble portion declined rapidly during the first 48 hr and continued to decline during the next 14 days. The substances were not identified, but were presumed to be either coumaphos or its oxygen analog.

The excreta from hens dusted with coumaphos have been shown to contain unchanged coumaphos, its oxygen analog, O,O-diethyl phosphorothioic acid, and diethyl phosphoric acid(121). The same substances were found after oral treatment(122).

Considerably less information is available on coumaphos metabolism in insects(116, 119). In vitro studies with the house fly and cattle grub (Hypoderma bovis) homogenates have shown that coumaphos is activated to a cholinesterase inhibitor, presumably to the oxygen analog, in these systems, but no degradation occurs. In comparison, tissue homogenates from the ox or rat, particularly the liver, contain systems that degrade coumaphos. These findings are consistent with the selective toxicity of coumaphos.

From the information presented above, the metabolism of coumaphos may be summarized by the following equations.

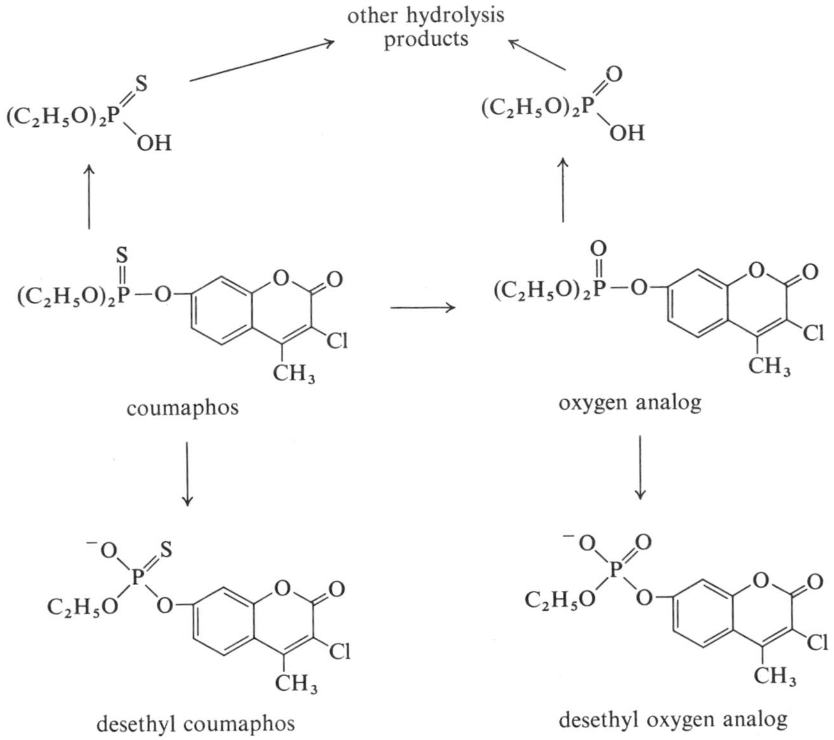

coumaphos

oxygen analog

desethyl coumaphos

desethyl oxygen analog

6. Azinphosmethyl. Despite the wide usage of azinphosmethyl [O,O-dimethyl S-4-oxo-1,2,3-benzotriazin-3(4H)-ylmethyl phosphorodithioate] on a variety of crops, there is relatively little information concerning its metabolism. In vitro studies with liver homogenates from rats, mice, and guinea pigs(123, 124) have demonstrated that azinphosmethyl, like other phosphorothionates, is converted into a strong cholinesterase inhibitor, presumably to the oxygen analog of azinphosmethyl. The rate of this desulfuration reaction was shown to be highest when the liver homogenate is fortified with NAD. The anticholinesterase metabolite, in the presence of liver homogenate, is rapidly degraded by hydrolysis. The activation of azinphosmethyl to an anticholinesterase and simultaneous degradation of the metabolic product also have been demonstrated in tissue preparations from the American cockroach(125). With cockroach tissue homogenates, NADPH was the most effective cofactor, and NADH, NAD, and NADP were moderately effective. The addition of sodium fluoride to the tissue preparation inhibited hydrolytic degradation of the metabolite. The reactions that occur with mammalian liver and insect tissue are shown.

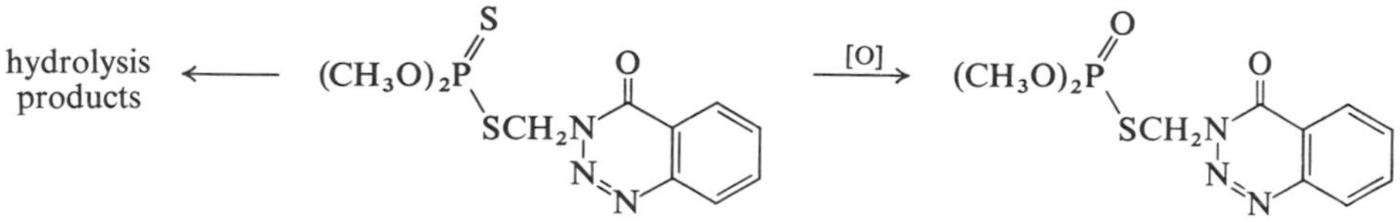

hydrolysis products ← $(CH_3O)_2P$ (=S) SCH_2N ... →[O]→ $(CH_3O)_2P$ (=O) SCH_2N ...

Azinphosmethyl oxygen analog is an extremely potent anticholinesterase with an I_{50} value of approximately $3 \times 10^{-9} M$ against fly-head cholinesterase (125).

[32]P-Azinphosmethyl given orally to a cow is metabolized into at least six products which are found in the urine, probably all in the hydrolyzed state (126). By use of azinphosmethyl labeled with ^{14}C in the methylene carbon, evidence was obtained that indicated the presence of four phosphorus-free metabolites in milk. Of these, one metabolite of unknown structure is responsible for approximately 85–90% of the total amount.

7. Ronnel. The metabolism of ronnel, O,O-dimethyl O-2,4,5-trichlorophenyl phosphorothioate, has been investigated in the rat, cow, and housefly (127). Both ronnel and its phosphate analog are rapidly degraded in the rat, over 60% of the administered dose being excreted as hydrolyzed products in rat urine after two days. Excretion from the cow was somewhat slower, taking seven days for about 50% elimination. Two primary pathways of approximately equal importance for ronnel metabolism were demonstrated, the hydrolysis of the P—O—phenyl bond and hydrolysis of the P—O—methyl bond to secondary esters as shown below.

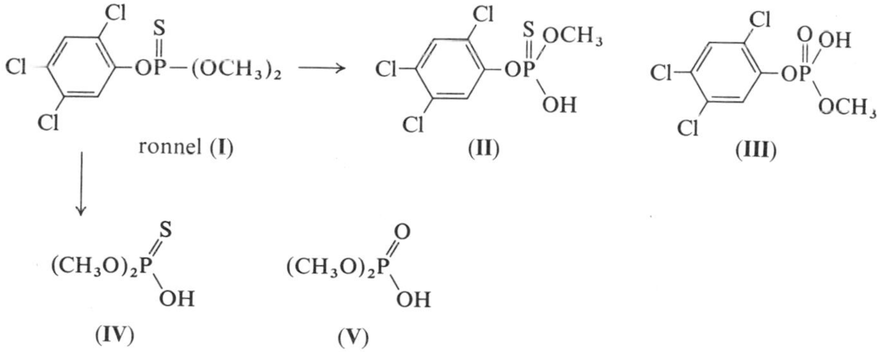

ronnel (I) → (II) (III)

(IV) (V)

The compounds shown above were all identified with reasonable certainty, and compound **V** and the demethylated products (**II** and **III**) were found in the greatest quantities. The metabolism sequence depicted above has considerable significance in that it was the first reported example demonstrating dealkylation as an initial metabolic step. In addition to hydrolytic degradation, the activation of ronnel to its phosphate derivative was indicated by large depressions in the cholinesterase activity of the blood of the cow.

8. IMIDAN. The metabolism of IMIDAN O,O-dimethyl S-phthalimido-methyl phosphorodithioate) has been studied in the cotton plant(*128*), rat(*129*), cattle(*130*), and in soil(*131*). In plants and mammals the inital step in metabolism appears to proceed via cleavage of the sulfur—methylene or nitrogen—methylene bond. Cleavage at the sulfur—methylene bond is supported by the tentative identification of N-hydroxymethyl-phthalimide as one of the metabolites from cotton plants.

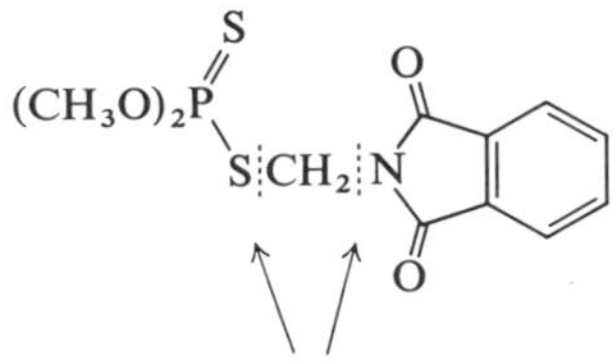

cleavage

This compound, however, could not be detected among the metabolites isolated from cattle. There was no evidence for the presence of phthali-mide in cotton, but chromatographic results indicated its possible pre-sence in cattle. Both in the plant and in mammals the conversion of IMIDAN to its oxygen analog appears to contribute only slightly to the overall metabolic processes compared to hydrolytic reactions. Since all of the work was carried out with ^{14}C carbonyl-labeled IMIDAN, the fate of the phosphorodithioate moiety was not determined. The principal metabolites were demonstrated to be either phthalamic or phthalic acid, presumably arising from phthalimide or N-hydroxymethylphthalimide.

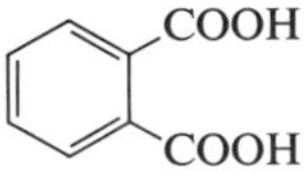

phthalic acid

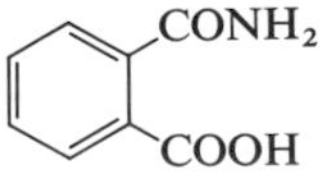

phthalamic acid

Among other minor constituents indicated from chromatographic evi-dence were benzoic acid and *p*-hydroxybenzoic acid.

In the rat, IMIDAN is rapidly degraded after oral treatment by the aforementioned metabolic reactions and eliminated principally in the urine(*129*). Only very small amounts (2–3%) of the applied radioactivity was found in tissues, and essentially no radioactive carbon dioxide was expired by the rat.

IMIDAN is rapidly broken down in moist soil, probably by hydrolytic reactions(131). The implication that degradation is hydrolytic is supported by a half-life of IMIDAN of 19 days in dry soil compared to three days in moist soil.

9. Trichlorfon (DYLOX). Trichlorfon (dimethyl 2,2,2-trichloro-1-hydroxyethylphosphonate) has been examined for metabolism in insects and mammals because of its unusual structure, its high selectivity to insects, and low mammalian toxicity.

Under alkaline conditions trichlorfon is rapidly converted into dichlorvos or DDVP (dimethyl 2,2-dichlorovinyl phosphate) as shown below (132–134).

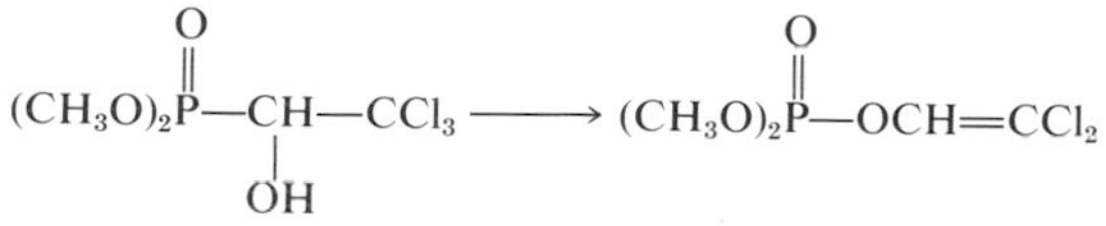

$$(CH_3O)_2\overset{\displaystyle O}{\overset{\|}{P}}-\underset{\underset{\displaystyle OH}{|}}{CH}-CCl_3 \longrightarrow (CH_3O)_2\overset{\displaystyle O}{\overset{\|}{P}}-OCH=CCl_2$$

trichlorfon   dichlorvos

This is an unusual reaction in that it represents a case in which a phosphonate ester is rearranged to a phosphate ester. Dichlorvos has been found to be an impurity present in technical trichlorfon. The rearrangement is rapid in strong alkali and even occurs at a reasonable rate at pH as low as 7.0, the half-life of the reaction at this pH being in the order of about six hours(135). Hence, one may surmise that the alkalinity of the water used in trichlorfon crop treatment will have a large bearing on its performance in insect control.

The metabolism of trichlorfon has been investigated in the cow(136) and a dog(137). In the cow, ^{32}P-labeled trichlorfon administered orally is rapidly metabolized and eliminated in the urine, about 65% of the administered dose being excreted in 12 hr. Chromatographic analysis of the urine showed the presence of at least three compounds, 75% of which consisted of a compound of unknown structure. Of the total radioactiviy, less than 1% appeared to be unchanged trichlorfon and about 17% was tentatively identified as dimethyl phosphoric acid. Less than 0.2% of the administered dose was found in the milk. Of this, only 10% partitioned in favor of chloroform, indicating that the radioactivity was largely in the form of hydrolyzed ionic products. The blood contained substantial amounts of radioactivity, a maximum of 15.1 μg equivalents 2 hr after administration, decreasing gradually to about 1 μg equivalents after 24 hr. At 2 hr, approximately 7.5% of the radioactivity in the blood was in the form of trichlorfon.

Investigation with a dog treated intravenously with trichlorfon has shown that the urine contained 63% of the original compound in the form of the glucuronide of trichloroethanol, indicating cleavage of the carbon—phosphorus bond. Approximately 77% of the ^{32}P in urine collected during the first six hours was in the form of hydrolyzed products of unknown structure and 23% passed through as unchanged trichlorfon.

The rate of conversion of trichlorfon to dichlorvos under different pH conditions has been determined(135, 138). By chloride ion titration and radiotracer analysis it has been shown that even at pH 6.0 trichlorfon rearranges slowly to dichlorvos and at pH 7.0 and 8.0 the rearrangement is rapid. These results have been supported by other investigations in which the reaction was followed by a manometric method. It has been shown that trichlorfon does not inhibit chymotrypsin at pH 5.0, while dichlorvos inhibits the enzyme under the same conditions. Similar results were found with fly-head cholinesterase. These results indicate that the toxic action of trichlorfon is exerted through in vivo formation of dichlorvos. That dichlorvos is the active intermediate in trichlorfon intoxication is substantiated further by experiments in which in vivo formation of DDVP in house flies feeding on ^{32}P trichlorfon was demonstrated. This conclusion is in disagreement with the findings of others (137) who found no evidence of in vivo dichlorvos formation in flies treated topically with trichlorfon.

Because of its ineffectiveness against the cotton leaf worm late-instar, *Prodenia litura*, the metabolism of trichlorfon has been investigated in this insect to provide a basis for its low insecticidal activity(139, 140). Trichlorfon is absorbed and rapidly metabolized by the late-instar of the cotton leaf worm. By means of ^{32}P- and ^{14}C-labeled material, the excretion of 40–45% of the topically applied dose within 20 hr has been demonstrated. The evidence, although not conclusive, indicates the identity of the principal metabolite as the glucuronide of 1-hydroxy-3,3,3-trichloroethylphosphonic acid, the completely demethylated product. This substance represented 68% of the recovered metabolites; the other metabolites consisted of dimethyl phosphoric acid (27%) and monomethyl phosphoric acid (5%). Approximately 10–14% of the recovered ^{14}C-metabolites was eliminated as carbon dioxide. The low toxicity of trichlorfon to the cotton leaf worm late-instar therefore is attributed in part to its rapid metabolism.

From cotton plants treated with ^{32}P-trichlorfon by root absorption from an aqueous solution, dimethyl and monomethyl phosphoric acid and inorganic phosphate have been identified as degradation products(141). However, in view of earlier discussion on the rapid rearrangement of trichlorfon to dichlorvos in water it is possible that the actual material being absorbed is dichlorvos. Trichlorfon itself is not readily absorbed by cotton leaves.

Recent studies in the rat indicate that O-demethylation of trichlorfon is an important metabolic route in mammals(*142, 143*). As much as 27% of the injected dose of methoxy labeled ^{14}C-trichlorfon has been recovered as ^{14}CO$_2$ within 10 hr, presumably by O-demethylation to form methyl alcohol and subsequent oxidation to carbon dioxide. The urine contained dimethyl phosphoric acid as the principal metabolite (60–70%), monomethyl phosphoric acid, and about 10% of an unknown metabolite. The latter was proven not to be monodemethylated trichlorfon. The complete absence of desmethyl trichlorfon is difficult to explain in view of the large amounts of carbon dioxide eliminated. The possibility exists that ^{14}CO$_2$ originates from dimethyl phosphoric acid, but this is unlikely since it has been shown that dimethyl phosphoric acid is quantitatively excreted in the urine when injected into the rat(*144*). From in vitro investigations with rat brain homogenates, however, desmethyl trichlorfon has been isolated as the major metabolite along with lesser amounts of monomethyl phosphoric acid and a compound believed to be the trichlorohydroxyethylphosphonic acid(*145*).

10. Dichlorvos or Dimethyl 2,2-dichlorovinyl phosphate (DDVP). Dichlorvos is an important commercial insecticide with low residual activity because of its hydrolytic instability in an aqueous environment. It can be prepared directly from chloral and trimethyl phosphite or can be obtained by treating trichlorfon with a base. The rate at which dichlorvos hydrolyzes in water under different pH conditions has been determined(*135*), the first-order hydrolysis constants at pH 7.0 and 8.0 reported as 1.5×10^{-3} and 2.3×10^{-3} min^{-1}, respectively. This corresponds to a half-life of 462 min at pH 7.0 and 301 min at pH 8.0.

The in vivo metabolism in mammals and in vitro enzymatic metabolism of dichlorvos have been studied by Casida and co-workers(*146, 147*). In vitro studies with rabbit and rat tissue preparations have shown that dichlorvos is split initially either through the dichlorovinyl ester linkage or the methyl ester linkage. Liver preparations from the rabbit and rat were most effective in catalyzing hydrolysis, although homogenates of the kidney, spleen, and plasma also were active. These same preparations also were effective in hydrolyzing desmethyl dichlorvos. The general in vitro metabolic pathways of dichlorvos by the various tissue preparations are depicted below. The major proportion of the dichloroacetaldehyde obtained from vinyl linkage cleavage appears to be reduced to dichloroethanol with a very small amount going to dichloroacetic acid.

In vivo investigations with rats, cows, and a goat have proven that the same metabolites obtained by in vitro methods also occur in the intact

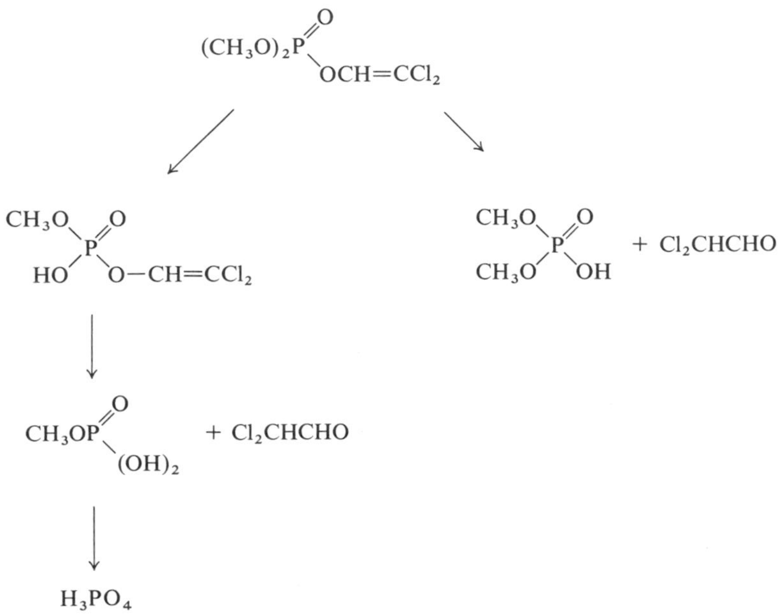

$$(CH_3O)_2P \begin{matrix} O \\ \diagdown \\ OCH=CCl_2 \end{matrix}$$

$$\begin{matrix} CH_3O & O \\ & P \\ HO & O-CH=CCl_2 \end{matrix} \qquad \begin{matrix} CH_3O & O \\ & P \\ CH_3O & OH \end{matrix} + Cl_2CHCHO$$

$$CH_3OP \begin{matrix} O \\ \diagdown \\ (OH)_2 \end{matrix} + Cl_2CHCHO$$

$$H_3PO_4$$

animal. Dichlorvos is rapidly hydrolyzed in mammals and eliminated as ionic products in the excreta. For example, a cow given 20 mg/kg of dichlorvos orally, eliminated 91% of the administered dose after 128 hr, 40% in the urine and 51% in the feces. Over 70% is eliminated within 32 hr after treatment. Low levels of organosoluble residues of DDVP were found in the milk of both cows and goats, the amount varying with the method of administration and the dosage. For the first and second 12-hr periods after oral administration, the levels of organo-soluble insecticide in parts per billion were 0.46 and 0.39 at the 1.0 mg/kg dosage and 7.3 at the 20 mg/kg dosage.

11. Naled (DIBROM). Naled (1,2-dibromo-2,2-dichloroethyl dimethyl phosphate) is prepared by bromination of dichlorvos. In mammals, naled is degraded almost completely to hydrolysis products(147). The metabolic products found in cow urine after oral administration have been identified as mono and dimethyl phosphoric acid, desmethyl dichlorvos, and inorganic phosphates. These results indicate that demethylation is one of the initial hydrolytic reactions occurring in naled metabolism. Other metabolites which have been found from naled include dichlorvos, dichlorobromoacetaldehyde, dichloroacetaldehyde, and certain amino acid conjugates of the degraded products. Naled reacts rapidly with sulfhydryls leading to dichlorvos and other degradation products(148).

12. Mevinphos (PHOSDRIN). Mevinphos, a short-residual insecticide, consists of the cis and trans isomers of 3-(dimethoxyphosphinyloxy)-crotonate. Of the two isomers the *cis*-crotonate or α-isomer is the more biologically active molecule(*149, 150*).

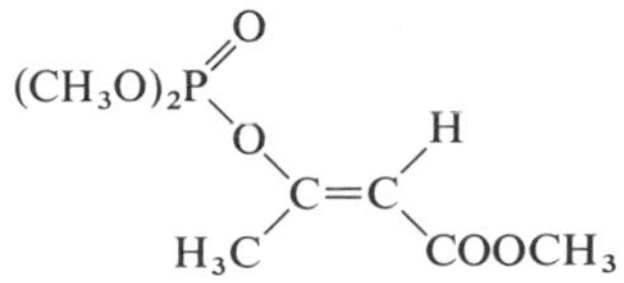

cis-crotonate (**I**) 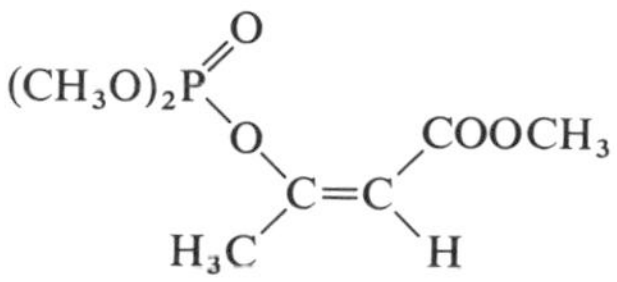   *trans*-crotonate (**II**)

In the pea and bean plant, the isomeric cis-crotonate (**I**) is degraded more rapidly than the trans-crotonate (**II**)(*151*). (The reader is reminded that the cis and trans assignments refer to the position of the methyl and carbomethoxy moieties about the double bond and is opposite to the assignment given in the reference.) The half-life of the cis-isomer is approximately 20 hr, the trans-isomer about 48 hr. The metabolic degradation of mevinphos (either cis or trans) appears to take place via two routes as shown below for the cis-crotonate(*152*).

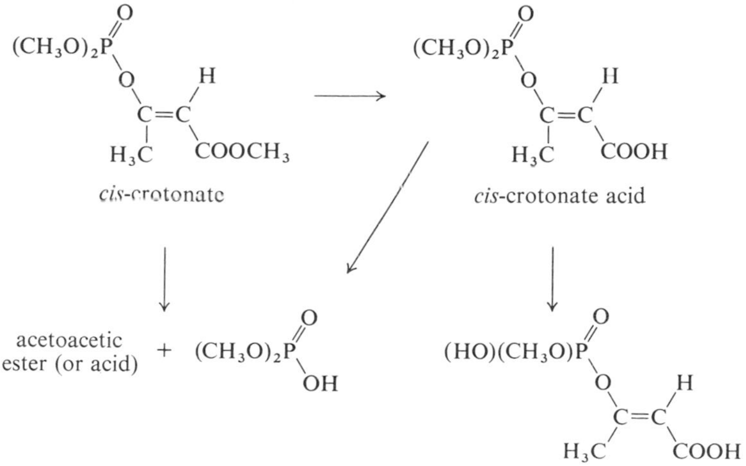

cis-crotonate   cis-crotonate acid

acetoacetic ester (or acid)

Undoubtedly other metabolites also are formed in the plant, but those shown above have been isolated and identified with reasonable certainty after treatment of the plant with either the *cis*-crotonate or *cis*-crotonate acid.

13. Phosphamidon. The metabolism of phosphamidon (2-chloro-2-diethyl-carbamoyl-1-methylvinyl dimethyl phosphate) has been studied only in plants(153, 154). Desethyl phosphamidon (2-chloro-2-ethyl-carbamoyl-1-methylvinyl dimethyl phosphate), α-chloroacetoacetic acid diethylamide, α-chloroacetoacetic acid ethylamide, dimethyl phosphoric acid, and unchanged phosphamidon were isolated from bean plants treated with ^{14}C-labeled phosphamidon. No other metabolites could be detected. Desethyl phosphamidon was isolated in milligram quantities and unequivocally identified. The metabolic pathway for phosphamidon in the bean plant may be suggested as follows.

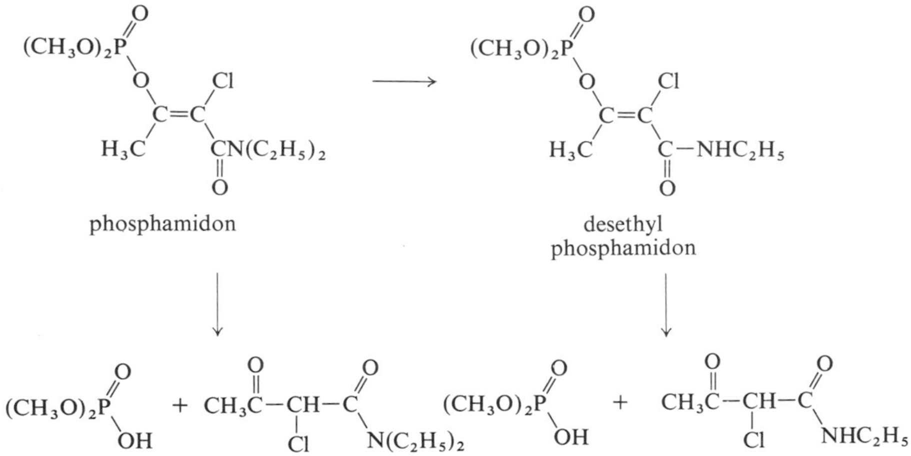

The metabolism of phosphamidon is unusual in that facile deethylation of the diethylamido moiety occurs with no demethylation at one of the methoxy groups.

Environmental factors such as temperature and, to a lesser extent, light influence plant metabolism of phosphamidon to desethyl phosphamidon with increased rates at higher temperatures(155). The degradation of desethyl phosphamidon is faster than that of phosphamidon since it never appears alone in plants, but always in the presence of phosphamidon. Desethyl phosphamidon is a stronger anticholinesterase and generally is more toxic to insects than phosphamidon.

14. BIDRIN. The metabolism of BIDRIN (3-hydroxy-N,N-dimethyl-*cis*-crotonamide dimethyl phosphate) has been studied in insects, mammals, plants, and in soil(156–158). In gross aspects, BIDRIN metabolism is the same in different living systems, depicted in part by the scheme below in which only the intact biologically active esters are shown.

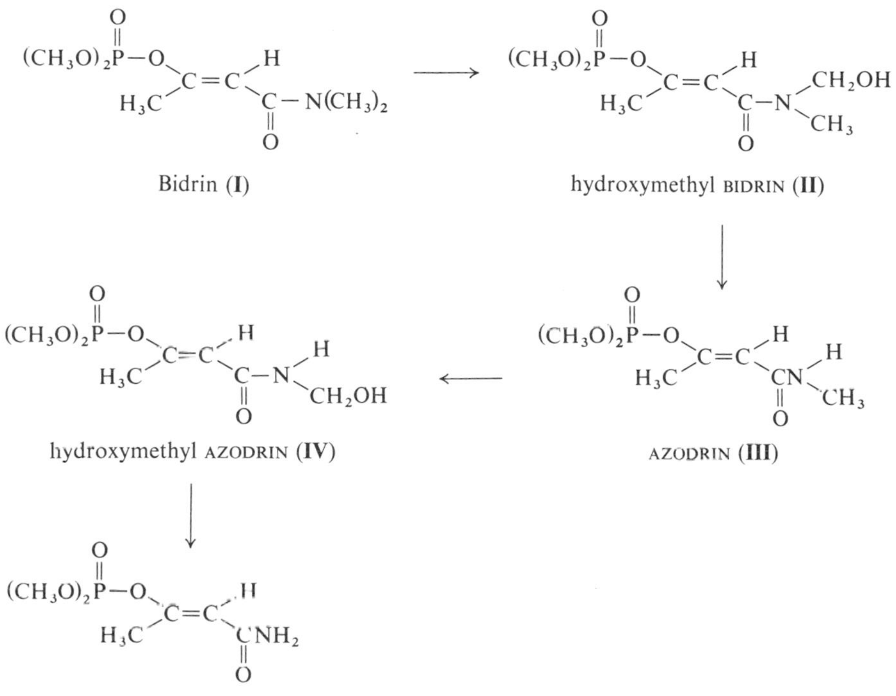

Bidrin (**I**)

hydroxymethyl BIDRIN (**II**)

hydroxymethyl AZODRIN (**IV**)

AZODRIN (**III**)

desmethyl AZODRIN (**V**)

Initial metabolism studies in cotton plants showed that BIDRIN is converted into at least three compounds in which the phosphate ester linkage is intact(*156*). Two of the three metabolites proved to be hydroxymethyl BIDRIN (**II**) and N-desmethyl BIDRIN or AZODRIN (**III**). The third metabolite labeled unknown A was found to be more polar in its behavior on paper chromatograms than either BIDRIN or hydroxymethyl BIDRIN but was less polar than the hydrolysis products. Unknown A proved to be a weak inhibitor of boll weevil cholinesterase ($I_{50} = 1 \times 10^{-4} M$), but treatment with hydrochloric acid gave a product with strong anticholinesterase activity whose chromatographic behavior was similar to hydroxymethyl BIDRIN. Hydroxymethyl BIDRIN, AZODRIN, and BIDRIN are equally strong in inhibiting boll weevil cholinesterase ($I_{50} = 8 \times 10^{-6} M$). Hydrolysis products found in the cotton plant were O-desmethyl BIDRIN (**VI**), O-desmethyl BIDRIN acid (**VII**), dimethyl phosphoric, monomethyl phosphoric, and phosphoric acid. Curiously, no O-desmethyl AZODRIN was detected.

Boll weevils treated with ^{32}P-labeled BIDRIN gave O-desmethyl BIDRIN

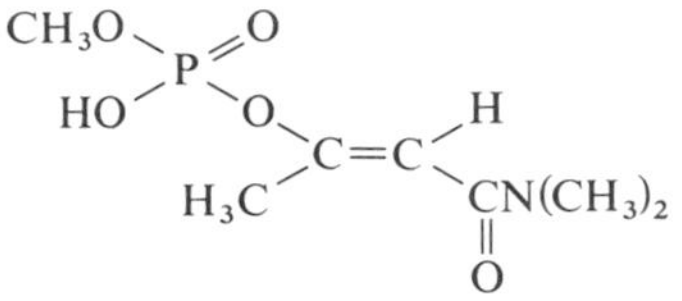

CH₃O and P=O structures with chemical diagrams:

O-desmethyl BIDRIN (VI) 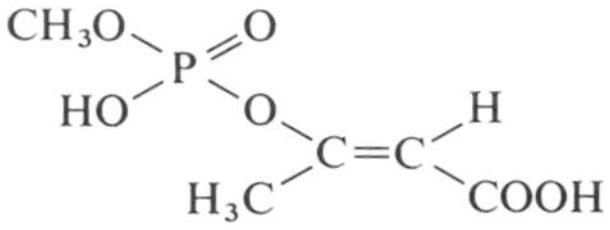   O-desmethyl BIDRIN acid (VII)

(VI), dimethyl phosphoric, monomethyl phosphoric, and phosphoric acid as metabolites along with unchanged BIDRIN. Rats also produced these same metabolites, but, in addition, hydroxymethyl BIDRIN (II) and AZODRIN (III) were found.

Subsequent investigations by others(157) have confirmed most of the earlier work. Studies using the snap bean plant, several different mammals, and insects have demonstrated the presence of the aforementioned metabolites. However, in addition to these, hydroxymethyl AZODRIN (IV) and desmethyl AZODRIN (V) also were found among the metabolites possessing the intact phosphate ester. In mammals, including rats, mice, dogs, rabbits, and a goat, the metabolites consisted mainly of hydrolysis products, but substantial amounts of AZODRIN and smaller amounts of II, IV, and V also were found. Further, trace amounts of N-methylacetoacetamide and 3-hydroxy-N,N-dimethylbutyramide were indicated in the urine of BIDRIN treated rats.

The evidence indicates a total of four intact phosphate esters as metabolites (II, III, IV, V). All four compounds are strikingly similar in their anticholinesterase activity (I_{50} range with fly-head cholinesterase 1×10^{-7} to $6 \times 10^{-8} M$). The LD_{50} values (mg/kg) against the mouse for I, II, III, IV, and V are 14, 18, 8, 12, and 3, respectively; with the houseflies, 38, 14, 6.4, 30, and 1. The LD_{50} values show that the metabolites AZODRIN (III) and desmethyl AZODRIN (V) are more toxic to the mouse and to the house fly and undoubtedly contribute heavily to the overall toxicity of BIDRIN. This is another example in which the living organism produces by metabolic conversion compounds intrinsically more toxic than the parent compound. The metabolism of demeton to its oxidation products is an example cited earlier.

The metabolism of BIDRIN to II and III also has been demonstrated in house flies by others(158). Co-treatment of the flies with the methylenedioxyphenyl synergist sesamex, however, markedly lowered the extent to which metabolism occurred and decreased the amount of AZODRIN formed. These results, plus the fact that hydroxymethyl intermediates have been isolated, suggest that N-demethylation takes place through the action of mixed-function oxidase first to form the methylol with subsequent hydrolysis to the N-demethylated product(*31, 159*).

BIDRIN is rapidly degraded in soil and the rate of degradation is directly related to the microbial content(*158*). Sterilization of the soil by heat or by the action of selective biocides drastically lowered the degree of BIDRIN degradation, providing evidence that microorganisms are responsible for the decomposition of BIDRIN in soil.

The discovery of the high insecticidal properties of the BIDRIN metabolite, AZODRIN, has led to its development as a potential insecticide. As expected, AZODRIN is metabolized in insects and rats into hydroxymethyl AZODRIN (**IV**) as the principal unhydrolyzed metabolite and, to a minor extent, to desmethyl AZODRIN (**V**)(*157, 160*). Dimethyl phosphoric acid and O-desmethyl AZODRIN were the major hydrolytic metabolites. An unknown metabolite closely related in properties to the unknown metabolite A from BIDRIN was detected in minor amounts in insects. Acidification of this substance gave a compound that appears to be hydroxymethyl AZODRIN.

15. CIODRIN. CIODRIN (α-methylbenzyl 3-hydroxycrotonate dimethyl phosphate), a compound closely related to mevinphos, is a promising insecticide for the control of insects affecting domestic animals. Technical CIODRIN consists of two isomers, α and β, with α predominating over β. The configurations for the isomers have not been assigned. In the lactating ewe after oral administration, ^{32}P-CIODRIN is rapidly metabolized and eliminated in the urine(*161*), 79% of the eliminated dose being eliminated in two days. At least nine substances were detected in the urine by paper chromatography, but some of these were found to originate from impurities in the labeled preparation. The principal metabolite was dimethyl phosphoric acid consisting of 60–90% of the total radioactivity in the urine. Another important metabolite was 3-(dimethoxyphosphinyloxy)crotonic acid which appeared as 11% of the metabolites after 3 hr, decreasing to 3% after 6 hr. The remaining metabolites were found in lesser amounts and their structures remain unknown. Similar studies with the pure α-isomer of CIODRIN in the lactating goat produced six distinct metabolites in urine, again the principal metabolite consisting of dimethyl phosphoric acid (80–91% of total metabolites) with substantially smaller amounts of the crotonic acid derivative and other metabolites of unknown structure(*161a*). The two metabolic reactions that have been proven to occur with certainty are shown below. Undoubtedly other reactions occur, e.g., demethylation, leading to desmethyl CIODRIN and its subsequent hydrolysis to monomethyl phosphoric acid, analogous to the metabolism of mevinphos.

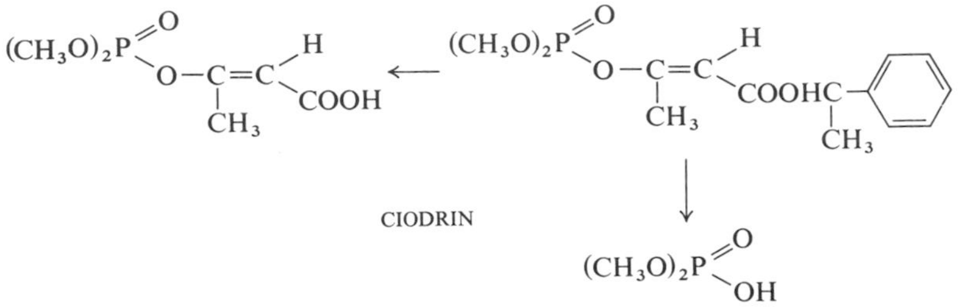

$$(CH_3O)_2P \overset{O}{\diagup} \underset{\underset{CH_3}{|}}{O-C=C} \overset{H}{\diagdown} COOH \quad \longleftarrow \quad (CH_3O)_2P \overset{O}{\diagup} \underset{\underset{CH_3}{|}}{O-C=C} \overset{H}{\diagdown} COOHC \overset{}{\diagdown} \underset{CH_3}{}$$

CIODRIN

$$(CH_3O)_2P \overset{O}{\diagup} OH$$

16. Famphur. Famphur [O-p-(dimethylsulfamoyl)phenyl O,O-dimethyl phosphorothioate] metabolism has been studied in insects and mammals. The principal metabolites found in the urine of famphur treated sheep and a calf were N- and O-desmethyl famphur and hydrolysis products from P—O— phenyl cleavage(162). The organosoluble material present in the blood was shown to be either unchanged famphur or its oxygen analog. Although famphur is quite toxic to the white mouse (LD$_{50}$ 11.6 mg/kg), it is rapidly metabolized after intraperitoneal injection since less than 10% of the total dose is found as organosoluble material in total mouse homogenate after one hour(163). Famphur, its oxygen analog, and a minor amount of N-desmethyl famphur were present. Famphur metabolism in the American cockroach and the milk-weed bug (*Oncopeltus fasciatus*) is markedly slower than in the mouse, but the same organo-soluble metabolites were found in whole insect homogenates. The known metabolic reactions for famphur are shown below.

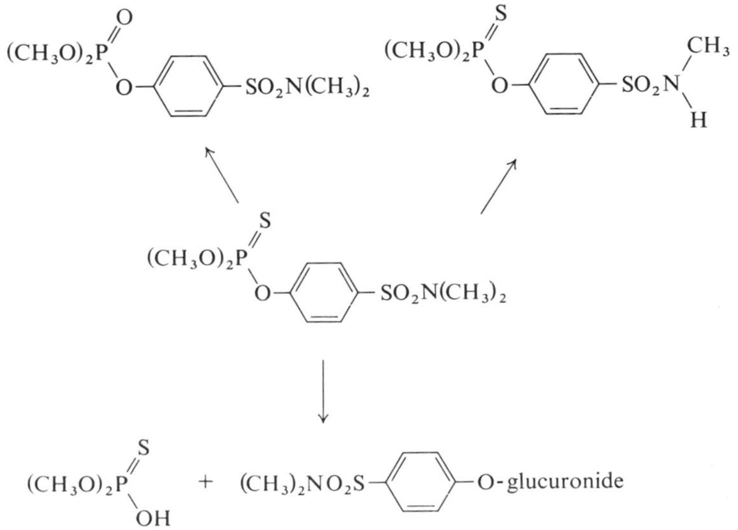

$$(CH_3O)_2P \overset{O}{\diagup} O- \overset{}{\bigcirc} -SO_2N(CH_3)_2 \qquad (CH_3O)_2P \overset{S}{\diagup} O- \overset{}{\bigcirc} -SO_2N \overset{CH_3}{\diagdown} H$$

$$(CH_3O)_2P \overset{S}{\diagup} O- \overset{}{\bigcirc} -SO_2N(CH_3)_2$$

$$(CH_3O)_2P \overset{S}{\diagup} OH \quad + \quad (CH_3)_2NO_2S- \overset{}{\bigcirc} -O\text{-glucuronide}$$

17. Dioxathion (DELNAV). Technical dioxathion (S,S'-*p*-dioxane-2,3-diyl O,O-diethyl phosphorodithioate cis and trans isomers) has been separated by column chromatography into eight different fractions (*164*). The structures of the major constituents of dioxathion are given below. Because of the heterogeneity of this material the complete elucidation of dioxathion metabolism is virtually an impossible task.

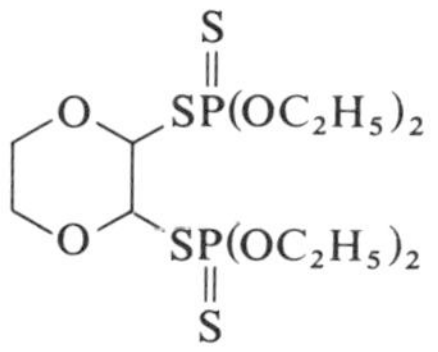

cis and trans isomers

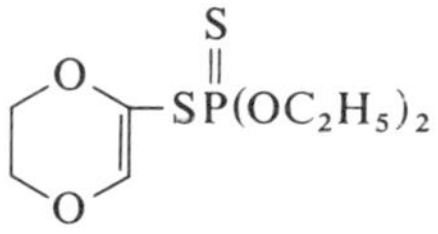

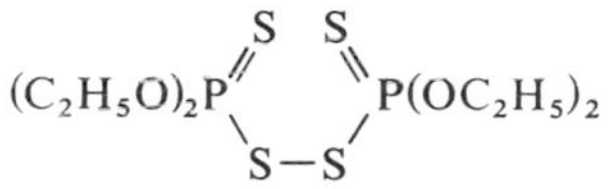

In the American cockroach and by rat liver slices the above constituents of dioxathion apparently are degraded to diethyl phosphoric acid, O,O-diethyl phosphorothioic and dithioic acids (*164*). These three substances also have been found as the major urinary metabolites from cattle treated orally or dermally with dioxathion (*165*). Several other substances also were detected, but it was not possible to ascertain whether these were actual metabolites or originated as impurities in the radiolabeled dioxathion.

18. COLEP. The metabolism of ^{14}C COLEP [O-phenyl O-(*p*-nitrophenyl) methylphonothionate] labeled in the phenyl ring has been investigated in plants and in the rat (*166*). Although the identity of none of the metabolites was established on a firm basis, the evidence indicated that COLEP is metabolized in apple and in cotton plants to O-phenyl methylphosphonic acid, the oxygen analog O-phenyl O-(*p*-nitrophenyl) methylphosphonate, and various conjugates of phenol. Analysis of urine from orally treated rats showed only the presence of phenol conjugates. Obviously, supplementary work with ^{32}P-labeled or methyl-^{14}C COLEP is needed for complete elucidation of the metabolic pathways.

D. Phosphoramides

1. Schradan. The most prominent of the phosphoramidate insecticides is schradan (octamethylpyrophosphoramide), one of the first systemic insecticides discovered. This compound has received considerable attention by different investigators because of its strong systemic activity and insecticidal specificity(167–169), i.e., having the property of being toxic to certain insects, but almost nontoxic to others(167, 170, 171). The intoxication process in insects and mammals has been shown to involve the oxidative metabolism of schradan, a poor anticholinesterase, to one or more metabolites of strong anticholinesterase activity. This activation has been demonstrated in a wide variety of biological tissues in several susceptible insects, including *Anasa tristis* and *Oncopeltus fasciatus*, and nonsusceptible insects *Periplaneta americana* and *Tenebrio molitor* L.(167, 172). Metabolism in mammals appears to occur principally in the liver(173–175). Metabolism of schradan to a more active inhibitor also has been shown to occur in plants(169, 172) and the metabolic products appear to be identical to those occurring in animals. Plant metabolism, however, appears to take place slowly and the hydrolytically unstable metabolites are rapidly destroyed as they are formed.

The activation of schradan to a strong anticholinesterase also has been shown to occur by the action of oxidizing agents such as permanganate, dichromate, hypochlorite, bromine water, and peracetic acid(170, 176–178), and by hydrogen peroxide(173). Comparison of the oxidation products obtained by chemical oxidation and by the action of oxygenated rat liver, insect tissue, and several different plants indicates that schradan is being acted upon by these different systems in a similar manner. By this oxidation process, schradan with a half-life of over 10 years at pH 8 and 25°C is converted into a compound with a half-life of 40 min and 10^5 times greater anticholinesterase activity.

In spite of the considerable amount of work published on schradan metabolism, there appears to be some disagreement concerning the chemical nature of the metabolites. Because of the oxidative nature of the metabolic change, it was suggested at an early stage(179) that schradan is converted to the N-oxide (**II**) which then rearranges to the N-methylol derivative (**IV**). Subsequent investigations by Casida and his group(169, 176, 180) provided evidence that concurred in part with the early hypothesis. The initial metabolite obtained from schradan incubation with rat liver and by a variety of chemical agents was shown to be the N-oxide by infrared spectra and comparison of chemical and biological properties. The infrared spectrum of this oxidation product showed an absorption at 1690 cm^{-1}, very close to the reported absorption of 1680 cm^{-1} for trimethylamine oxide dihydrate. The N-oxide was found to be unstable and prolonged storage or treatment with heat, dilute acid, or alkali converted it to a less polar isomeric product of greater stability and poorer anticholinesterase activity. Plants treated with schradan also have been shown to produce at least two compounds, an unstable and a more stable derivative(*181*). From the available evidence, Casida et al. proposed that schradan was first oxidized to the N-oxide (**II**) and this then rearranged to the N-methoxy derivative (**III**) as shown below.

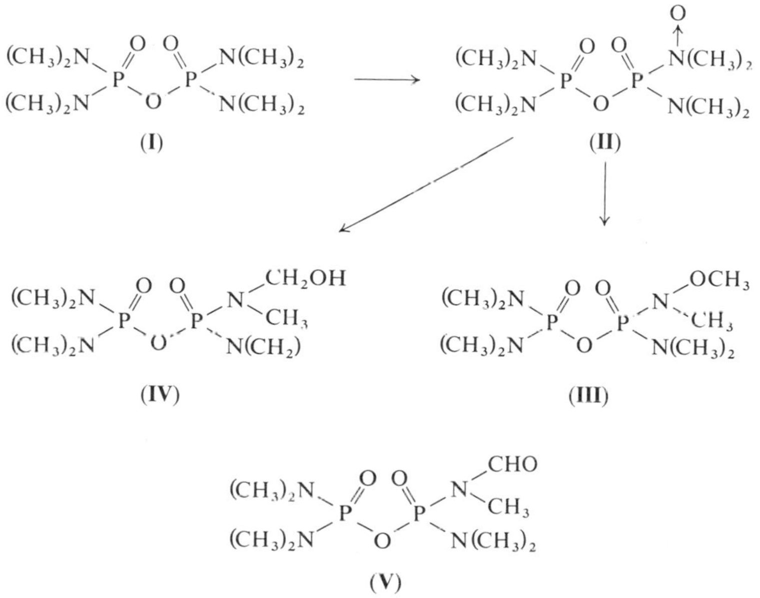

The N-methylol derivative (**IV**) was also considered as a candidate for the isomeride but was ruled out because of the absence of —OH absorption in the infrared. The isomeride gave formaldehyde upon treatment with strong acid, which apparently can occur from either **III** or **IV**, according to the behavior of hydroxylamine ethers(*176*).

Spencer and O'Brien(*178, 182*), however, believe that the initial product obtained in schradan metabolism or oxidation is not the N-oxide (**II**), but is actually the N-methylol derivative (**IV**). They observed that the product that showed infrared absorption at 1690 cm^{-1} was not associated with the active anticholinesterase but with the less active derivative. It was further shown that anhydrous trimethylamine oxide did not absorb in the 1680 cm^{-1} region, nor did N,N-dimethyltryptamine N-oxide. Mild hydrolysis of the active anticholinesterase gave formaldehyde and heptamethylpyrophosphoramide. These investigators concluded that schradan is oxidized to the N-methylol derivative (**IV**) possibly by way of a transitory N-oxide (**II**). The second, more stable component from schradan metabolism or oxidation is believed to be the N-formyl derivative (**V**) produced by further oxidation of **IV**. The N-formyl structure would be expected to absorb in the 1690 cm⁻¹ region. As evidence against compound **III** as the second product, N-methoxy-N,N′,N′-heptamethyl-pyrophosphoramide was synthesized and demonstrated to be different from the actual oxidation product.

The chemical and metabolic conversion of tetramethylphosphoro-diamidofluoride and p-nitrophenyl tetramethylphosphorodiamidate, both weak anticholinesterases, to highly active inhibitors by rat liver slices and by permanganate oxidation has also been demonstrated (*183*). The activation probably occurs in a manner similar to schradan.

2. RUELENE. Among other phosphoramidates of insecticidal impor-tance the metabolism of RUELENE (4-*tert*-butyl-2-chlorophenyl methyl N-methyl-phosphoramidate), a compound which possesses a rather broad spectrum of anthelmintic activity in cattle, sheep, and goats, has been investigated(*184*). Orally treated sheep eliminated 85% of the ³²P-labeled RUELENE in the urine in 24 hr. Paper chromatographic and in-frared analysis of the radioactive metabolites in urine indicated the presence of unchanged RUELENE and the deamidated product, 4-*tert*-butyl-2-chlorophenyl methyl phosphoric acid, as the major products. The chloroform extract of whole blood contained these two compounds as well as traces of three more polar metabolites of unknown structure.

RUELENE is rapidly degraded in poultry, presumably eventually to phosphoric acid, and incorporated in bone tissue(*185*). Of the oral dose given in the feed only 29% was excreted after 21 days. At least six dif-ferent metabolites were isolated in the excreta and the evidence indicated that three others were present. Among those tentatively identified were 4-*tert*-butyl-2-chlorophenyl methyl phosphoric acid and methyl phos-phoric acid. The identity of the two principal metabolites, however, remains unknown. These metabolites were demonstrated to be water-soluble and therefore probably represent nontoxic degradation products.

The metabolism of RUELENE has been examined in a wide variety of insects and appears to be less complex than in mammals(*186*). The prin-cipal metabolite isolated from insects is believed to be 4-*tert*-butyl-2-chlorophenyl dihydrogen phosphate.

2-3 CARBAMATE INSECTICIDES

The carbamate esters are relative newcomers to the family of organic insecticides now in commercial use, and for this reason there is less published information concerning the metabolism of these compounds in animals and plants. However, reports on the partial elucidation of the metabolic pathways of some of the more important carbamates have appeared in the literature during the past several years. The metabolism of a number of new carbamate insecticides is currently under intense investigation in several laboratories.

Current information provides evidence for the operation of two basic but mechanistically different processes for carbamate metabolism in biological systems. These processes can operate separately or together.

In the limited amount of information available it has been found without exception that the carbamate ester linkage is broken with consequent liberation of the phenolic or enolic moiety, either directly by esterase action or indirectly through metabolic conversion of the carbamyl moiety into an unstable intermediate. The hydrolytic process therefore can be depicted generally as follows .

$$
\underset{\substack{O \\ \parallel \\ RO-CNHCH_3}}{}\xrightarrow[\text{esterase}]{} ROH + CH_3NHC\overset{O}{\underset{OH}{\diagup}} \quad (CH_3NH_2 + CO_2) \quad (1)
$$

$$
\xrightarrow{[O]} \left[\underset{\substack{O \\ \parallel \\ ROCNHCH_2OH}}{} \right] \longrightarrow ROH + CO_2 + NH_3 + HCHO
$$
(2)

Direct split of the carbamate linkage may also occur by nonenzymatic base–acid catalyzed hydrolysis, depending upon the intrinsic stability of the carbamate. Direct esterase catalyzed hydrolysis [mechanism(1)] has been shown to occur with carbaryl (1-naphthyl N-methylcarbamate) and p-nitrophenyl N-methylcarbamate by a plasma albumin fraction that can be separated from the aliphatic, aromatic, and cholinesterases from several mammal and avian sources(187). It was further shown that N-methyl-, N-ethyl-, N-propyl-, and N-l-propylcarbamates of p-nitrophenol were hydrolyzed by the same plasma albumin fraction, but not by arylesterase, cholinesterase, chymotrypsin, lipase, papain, pepsin, trypsin, or egg albumin(188). N,N-Dimethylcarbamyl fluoride has been shown to be hydrolyzed by an esterase present in rabbit plasma that also hydrolyzes diisopropyl phosphorofluoridate(189). This compound also was hydrolyzed by an enzyme found in rabbit liver. Subsequent information(190) has shown that neither N,N-dimethylcarbamyl fluoride nor dimethylamine give formaldehyde by the action of rat liver microsomes, suggesting that hydrolysis occurs directly and not via an oxidized intermediate as is discussed below.

In addition to direct enzymatic hydrolysis of the carbamate linkage there is ample evidence supporting the second mechanism, i.e., hydrolysis after prior oxidation of the carbamate moiety. Owing to the prevalence of N-dealkylation reactions of foreign materials in biological systems (*31, 159*), this aspect of carbamate degradation was initially investigated by Hodgson and Casida(*190, 191*). *p*-Nitrophenyl N,N-dimethylcarbamate and a large series of other N,N-dimethylcarbamates of various structures were shown to be oxidized by a rat liver microsomal preparation to intermediates that liberated formaldehyde upon treatment with strong mineral acids. This enzymatic reaction was shown to require NADPH and molecular oxygen. Because of the liberation of formaldehyde, detected by chromotropic acid, the N-methylol derivation shown below was suggested as the structure of the oxidized product.

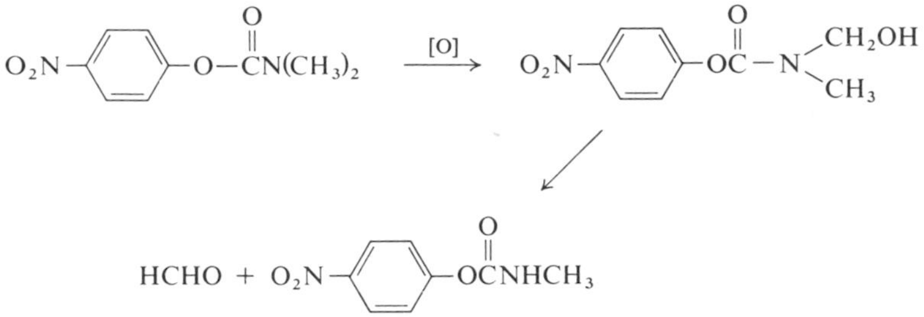

The oxidized product was shown to liberate formaldehyde and *p*-nitrophenyl methylcarbamate in equimolar quantities and was unstable to both acid and base. The formation of this intermediate also was shown to occur in rats given oral dosages of *p*-nitrophenyl N,N-dimethylcarbamate and similar results were indicated in houseflies and the American cockroach. N-Methylcarbamates treated with the microsomal system also produced formaldehyde but in considerably smaller quantities than the dimethylcarbamates examined.

Other evidence supporting N-dealkylation as a degradative pathway is found in $^{14}CO_2$ expiration studies(*192, 193*). Rats treated with various N-methyl labeled carbamates have been found to expire from 25–77% of the applied dose during a 48-hr period, providing direct evidence for N-demethylation. Similar results were obtained with houseflies.

The other metabolic reactions known to occur with carbamate insecticides, e.g., ring hydroxylation, O-dealkylation, etc., are discussed in the following sections dealing with prominent carbamates for which metabolism data are available.

The effect of methylenedioxy synergists such as piperonyl butoxide on carbamate metabolism and mode of action has been adequately discussed(*193a*). The reader is referred also to Chap. 1 of this book.

A. Carbaryl

The metabolism of carbaryl (1-naphthyl N-methylcarbamate) has been examined in insects and mammals by several different investigators (*194–197*). Relatively little work has been reported with plants(*195, 196a*). The earliest report on carbaryl metabolism by Eldefrawi and Hoskins(*194*), using ring-labeled ^{14}C material, in houseflies (*Musca domestica* L.), German cockroach [*Blatella germanica* (L.)], and the large milkweed bug (*Oncopeltus*) showed that carbaryl is rapidly metabolized by the housefly and the roach. In this preliminary study it was shown by paper chromatography that carbaryl was metabolized into at least three major products by houseflies and at least six major products by the German roach. These products were found to be substantially more polar than the original carbaryl. The pyrethrin synergist, sesamex, was found to decrease the metabolic rate, thus providing strong evidence that synergistic enhancement of carbaryl toxicity was attributable to inhibition of the metabolic detoxication reactions. Since ^{14}C-labeled 1-naphthol was found to form the same series of polar products, these investigators concluded that the first step in carbaryl metabolism is the hydrolysis of the carbamate to the naphthol. None of the metabolites were identified, however.

Subsequent work from Casida's laboratory(*195, 196*) has shown conclusively that carbaryl metabolism can occur without prior cleavage of the carbamyl moiety. Preliminary examination of carbaryl metabolism by rat liver microsomes fortified with NADPH, the American cockroach (*Periplaneta Americana*) and the housefly (*Musca domestica* L.) gave at least eight different metabolites, three of which were tentatively identified as 1-naphthyl N-hydroxymethylcarbamate, 4-hydroxy-1-naphthyl N-methylcarbamate, and 5-hydroxy-1-naphthyl N-methylcarbamate (*195*). Two other unidentified metabolites were shown to have the carbamate moiety intact, thus making a total of five metabolites in which the ester linkage is not split. The structure of the metabolites was established by comparing the metabolites with synthetic materials by column and thin-layer chromatography and the use of chromogenic agents. Further work by the same group(*196*) showed that the following metabolites were obtained from the urine of carbaryl treated rabbits and from liver enzyme preparations from rats, mice, and rabbits: (1) 1-naphthyl N-hydroxymethylcarbamate, (2) 4-hydroxy-1-naphthyl methylcarbamate, (3) 5-hydroxy-1-naphthyl methylcarbamate, (4) 5,6-dihydro-5,6-dihydroxy-1-naphthyl methylcarbamate, (5) 1-hydroxy-5,6-dihydro-5,6-dihydroxy-naphthalene, and (6) 1-naphthol. Several other unidentified metabolites were also detected.

Obviously the metabolism of carbaryl, a relatively simple carbamate ester, is quite complex. From current information the metabolic pathway for this compound in vivo in rabbit, houseflies, and cockroaches, and in vitro by rat, mice, and rabbit liver microsomes may be depicted as follows:

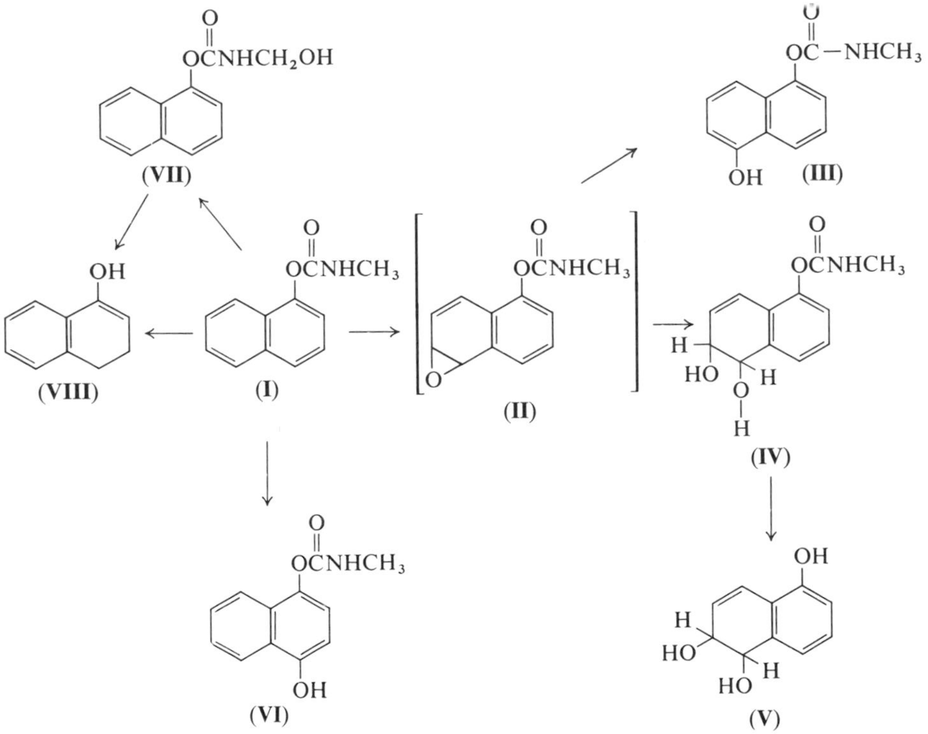

The epoxide **II** postulated as a transitory intermediate, although not isolated or detected, is a reasonable intermediate through which 5-hydroxy-1-naphthyl methylcarbamate (**III**) and 5,6-dihydro-5,6-dihydroxy-naphthyl methylcarbamate (**IV**) can be formed by a hydride ion migration and epoxide opening or by direct epoxide opening by water, respectively.

The metabolites shown in the metabolic scheme above have been identified with reasonable degree of certainty. In addition to these, the urine from carbaryl treated rabbits contained at least four or five other metabolites of secondary importance that were detected by thin-layer chromatography. It is likely that these substances are conjugates of carbaryl metabolites and their structure remains unknown. Preliminary examination indicates that these are conjugated as glucuronides or as ethereal sulfates. Also, the structure of another metabolite with the carbamate moiety missing and three ether-soluble metabolites remains unknown. In some cases, urine from carbaryl treated rabbits contained at least 15 metabolites. In vitro studies with rabbit, mice, and rat liver enzyme preparations have shown the presence of at least 20 metabolites. Seven of the major metabolites, particularly those still in the ester form, were examined for toxicity to houseflies and for anticholinesterase activity. It appears that they are of little toxicological importance compared to the parent compound.

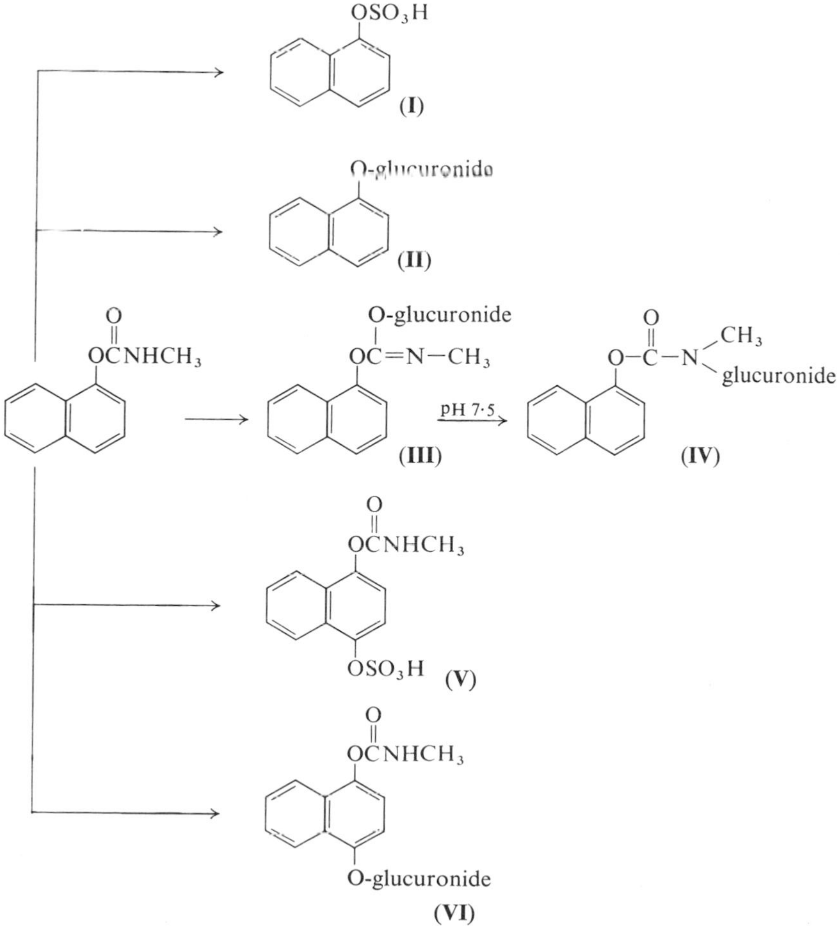

Simultaneous but independent investigations by Knaak et al.(*197*) with rats, guinea pigs, and humans have shown that carbaryl is rapidly metabolized and excreted from animals. After seven days, no ring-labeled or carbonyl-labeled ^{14}C-carbaryl residues could be detected in the rat and the radioactivity was excreted completely either in the urine or as respiratory carbon dioxide. These investigators found approximately eight water-soluble metabolites in the urine of the rat and guinea pig treated orally with either ring, carbonyl, or N-methyl labeled carbaryl. Approximately 50% of the excreted metabolites contained the intact ester linkage, supporting Casida's results for the existence of a nonhydrolytic pathway for carbaryl detoxication. The tentative metabolic scheme as shown above is proposed for the in vivo metabolism of carbaryl in rats and guinea pigs. The structures of the metabolites were established by chromatographic and spectrofluorometric techniques.

B. ZECTRAN

The metabolism of ZECTRAN, 4-dimethylamino-3,5-xylyl N-methylcarbamate, has been examined in broccoli and in the dog(*198,199*). Carbon-14–labeled ZECTRAN applied to the stem of broccoli flowers gave, after 10 days, a number of metabolic products as shown below in the proposed metabolic scheme. The metabolic pathway was established by the combination of paper chromatography and a rather complex and tedious chemical fractionation procedure in which all of the radioactivity was accounted for. Of the total radioactivity, 17.9% was associated with the extracted pulp, presumably by incorporation of breakdown products into lignin. The remaining 82.1% was accounted for as shown in Table 2-1.

TABLE 2-1

DISTRIBUTION OF RADIOACTIVITY IN BROCCOLI TREATED WITH ZECTRAN

Metabolites	% Total radioactivity
I Unchanged ZECTRAN	5.1
II 4-Dimethylamino-3,5-xylenol	1.4
III 4-Dimethylamino-3,5-dimethyl-o-benzoquinone	11.2
IV 2,6-Dimethyl-*p*-benzoquinone	13.6
V Conjugated 4-dimethylamino-3,5-xylenol	23.4
VI Conjugated 2,6-dimethylhydroquinone	27.4
VII Lignin	17.9

Although the initial step in the metabolism of ZECTRAN is indicated as a hydrolytic step to the phenol, this point has not been proven with certainty. The rather large amounts of 4-dimethylamino-3,5-xylenol found as a conjugate in the plant would suggest, however, that hydrolysis of the carbamate, either directly or through an intermediate derived by a change in the carbamyl moiety, is an important step in the degradation process. 4-Dimethylaminoxylenol and 2,6-dimethylhydroquinone in the form of their conjugates (**V** and **VI**) appeared to be the major metabolic products, representing over 50% of the total radioactive residues. The percentages indicated for the quinones **III** and **IV** are probably high since it was shown that both hydroquinone intermediates were converted to the quinones during the isolation procedures.

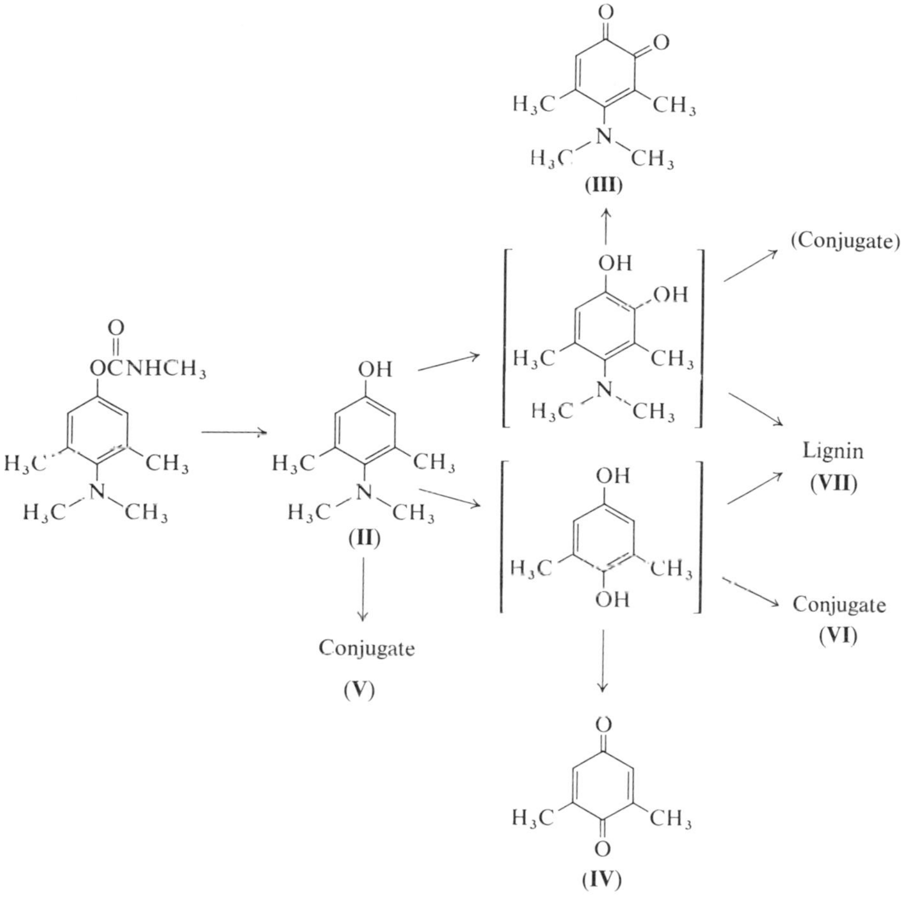

ZECTRAN injected into bean plants produced a number of metabolites, the following were tentatively identified by chromatography: 4-formamido, 4-methylformamido-, 4-methylamino, and 4-amino-3,5-xylyl N-methylcarbamate (*196a*). A closely related compound, MATACIL (4-dimethylamino-3-cresyl N-methylcarbamate), also gave the analogous 3-cresyl derivatives.

The principal metabolite consisting of 89.1% of the total radioactivity in the urine of a dog treated orally with radioactive ZECTRAN was found to be conjugated 4-dimethylaminoxylenol either as the sulfate or glucuronide. Three-fourths of the total radioactivity excreted was found in the urine, the rest in the feces. The other metabolites in the urine were conjugated 2,6-dimethylhydroquinone 5.5%, unidentified, unconjugated nonacidic material 5.1%, 4-dimethylamino-3,5-xylenol 1.9%, and 1.4% unidentified, unconjugated acidic material. The proposed metabolic pathway for ZECTRAN in the dog is shown below.

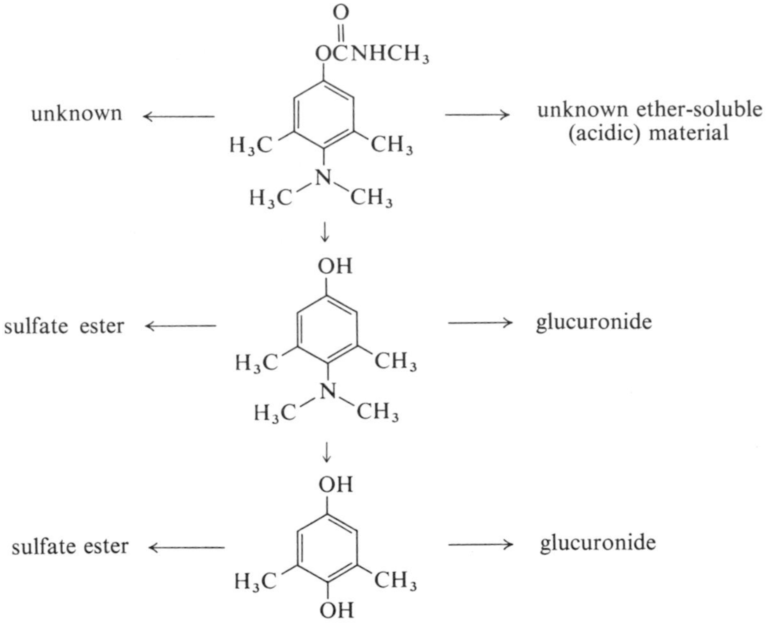

The proposed scheme shows an intial breakdown of ZECTRAN into the phenol, which is then conjugated either as the glucuronide or sulfate, or converted to the hydroquinone by replacement of the dimethylamino group. It was suggested that the latter conversion occurs by oxidative demethylation of the methyl group on nitrogen with subsequent oxidation of the amino phenol to the hydroquinone.

C. Arprocarb (BAYGON)

According to preliminary investigations, arprocarb (2-isopropoxyphenyl N-methylcarbamate) is metabolized into at least four compounds when incubated with rat liver microsomes in the presence of NADPH (195). By spectroscopic and chromatographic methods the principal metabolite was shown to be 2-isopropoxyphenyl N-(hydroxymethyl)-carbamate. The other metabolites were not identified, but one is believed to be a ring hydroxylated derivative.

There is strong evidence that one of the major degradative pathways of arprocarb in insects and mammals is through ether cleavage of the isopropoxy moiety. Houseflies treated with ^{14}C-BAYGON labeled on the 2-carbon of the isopropyl group expired up to 35% of the total dose as ^{14}C-acetone, indicating splitting of the ether bond as shown below (193). Co-treatment of the flies with piperonyl butoxide markedly decreased the amount of ^{14}C expired.

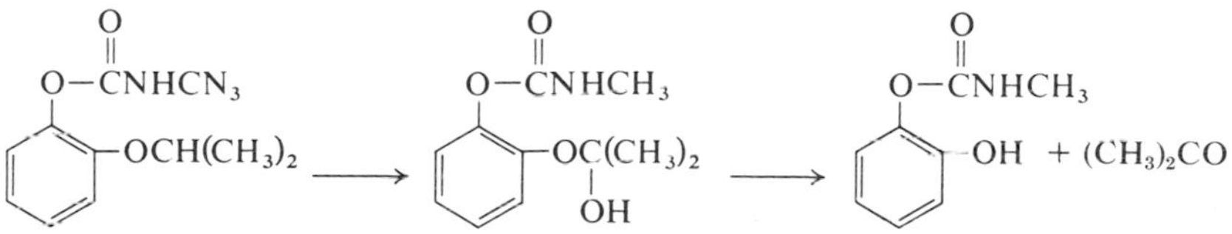

$$\underset{\text{OCH(CH}_3)_2}{\overset{\overset{\text{O}}{\|}}{\text{O}-\text{CNHCN}_3}} \longrightarrow \underset{\overset{\text{OC(CH}_3)_2}{\underset{\text{OH}}{|}}}{\overset{\overset{\text{O}}{\|}}{\text{O}-\text{CNHCH}_3}} \longrightarrow \underset{\text{OH}}{\overset{\overset{\text{O}}{\|}}{\text{O}-\text{CNHCH}_3}} + (\text{CH}_3)_2\text{CO}$$

Microsomal dealkylation of oxygen ethers has been discussed adequately in review articles(31,159). Similar ^{14}C expiration studies have been carried out in the rat using arprocarb labeled on the 1,3-carbon atoms in the isopropyl group(192). Like houseflies, the rat also degrades arprocarb by O-dealkylation with 30% of the applied dose expired as [14]CO$_2$ within a 48-hr period. Arprocarb is rapidly metabolized in rats as 76% of the dose was found in the urine after 24 hr.

In limited experiments, the analysis of urine from humans given small dosages of arprocarb showed that the major portion of the given amount was eliminated as a conjugate of 2-isopropoxyphenol, presumably as the glucuronide(200).

D. Dimetilan

Dimetilan [1-(dimethylcarbamoyl)-5-methyl-3-pyrazolyl dimethylcarbamate], although of high insecticidal activity, is rapidly metabolized in houseflies and cockroaches with a half-life of less than 30 min(201). Hydrolysis of the carbamate moiety in the 3-position is responsible in part for the metabolic products formed, but the majority of the various metabolites isolated is attributable to reactions not involving the 3-ester linkage. Nine metabolites were detected by thin-layer chromatography from extracts of dimetilan-treated German roaches, six from the American cockroach, and four from houseflies. Of the five metabolites isolated from the American cockroach in sufficient amounts for structure analysis, none showed any changes in the 5-methylpyrazole ring since this compound was isolated in each case when the metabolites were subjected to hydrolysis. Also, hydrolysis of the metabolites liberated the following combination of degradation products: (1) dimethylamine and formaldehyde, (2) dimethylamine, formaldehyde, and methylamine, (3) dimethylamine alone, and (4) formaldehyde alone. Attempts to isolate and identify methylamine were made only in a single case, but it is probable that this compound was present also in cases where formaldehyde was liberated. These results indicate that dimetilan metabolism occurs principally through oxidative conversion of the methyl groups on either of the carbamyl moieties to form the N-methylol derivatives. Although none of the metabolites was identified with certainty, the evidence indicates that metabolism occurs according to the following reactions.

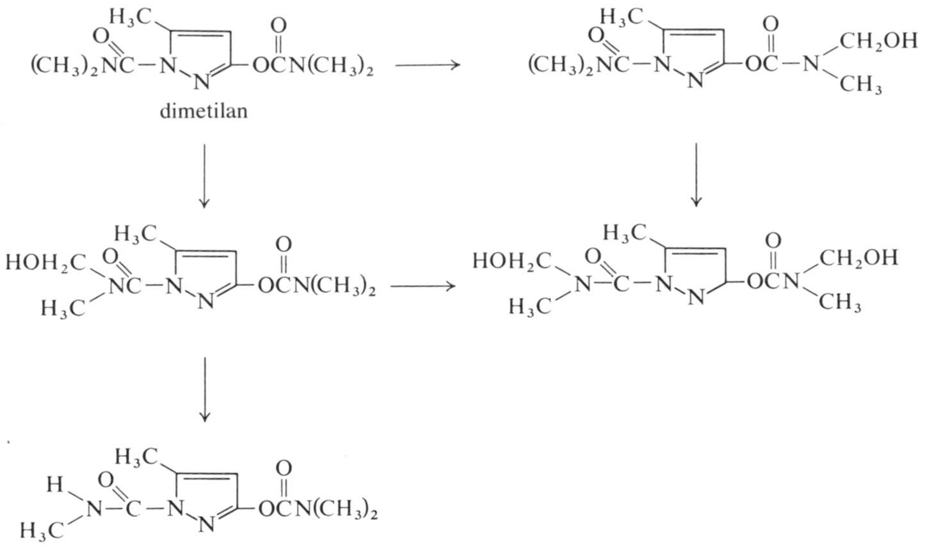

Several of the metabolites were highly toxic to both houseflies and the American cockroach and must be responsible in part for the insecticidal activity of dimetilan.

E. TEMIK

TEMIK [2-methyl-2-(methylthio) propionaldehyde O-methylcarbamoyl oxime] is a novel carbamate insecticide in that it is a carbamate ester of an oxime(202). Oxime carbamates recently have received wide attention as potential insecticides.

The metabolism of TEMIK in cotton plants and housefly is similar to the oxidative and hydrolytic pathways that are found in the thioether containing organophosphorus compounds(203). The thioether moiety in TEMIK is rapidly oxidized to the sulfoxide and the sulfoxide much more slowly to the sulfone. TEMIK sulfoxide is relatively stable in the plant and is the major metabolite present in mature cotton plants two months after systemic treatment. The principal degradative metabolite is TEMIK oxime sulfoxide formed by hydrolysis of TEMIK sulfoxide and consists of 12% of the total metabolites 36 days after treatment. Minor amounts of TEMIK oxime and sulfone oxime also were detected. In the housefly, the metabolism of TEMIK is very similar to that in the cotton plant with rapid formation of TEMIK sulfoxide after topical application. After 24 hr, from 40 to 60% of sublethal dosage was excreted in fly feces as TEMIK sulfoxide and sulfone in a 1 to 3 ratio. No evidence of hydrolytic products was obtained in fecal material. Thus, the metabolic reactions of TEMIK in plants and insects are essentially identical to those found for demeton and disulfoton and are depicted by the equations below.

$$
\begin{array}{ccc}
\underset{\overset{\displaystyle CH_3}{|}}{\underset{\overset{\displaystyle |}{CH_3}}{CH_3SCCH-NOCNHCH_3}} & \longrightarrow & \underset{\overset{\displaystyle CH_3}{|}}{\underset{\overset{\displaystyle |}{CH_3}}{CH_3SO_2CCH=NOCNHCH_3}} \\
\text{TEMIK} & & \text{TEMIK sulfone}
\end{array}
$$

$$
\underset{\overset{\displaystyle |}{CH_3}}{\underset{\overset{\displaystyle CH_3}{|}}{CH_3SOCCH=NOCNHCH_3}}
$$

TEMIK sulfoxide

$$
\begin{array}{ccc}
\underset{\overset{\displaystyle |}{CH_3}}{\underset{\overset{\displaystyle CH_3}{|}}{CH_3SCCH=NOH}} & & \underset{\overset{\displaystyle |}{CH_3}}{\underset{\overset{\displaystyle CH_3}{|}}{CH_3SO_2CCH=NOH}} \\
\text{Oxime} & & \text{oxime sulfone}
\end{array}
$$

$$
\underset{\overset{\displaystyle |}{CH_3}}{\underset{\overset{\displaystyle CH_3}{|}}{CH_3SOCCH=NOH}}
$$

oxime sulfoxide

Except for the formation of the sulfone, TEMIK metabolism in mammals also occurs according to the above equations(204). In the rat, TEMIK is metabolized and excreted in the urine principally as TEMIK sulfoxide and oxime sulfoxide. Minor amounts of unknown polar compounds believed to be acids also were detected. No evidence was obtained for the presence of TEMIK sulfone by any of the analytical procedures. Evidence for the presence of trace quantities of TEMIK oxime and oxime sulfone was found but was inconclusive.

F. MESUROL

MESUROL or 4-(methylthio)3,5-xylyl N-methylcarbamate is a compound closely related to ZECTRAN but contains a methylthio moiety in place of dimethylamino. In the bean plant, MESUROL is rapidly converted to the sulfoxide and to a lesser extent to the sulfone as shown below (196a).

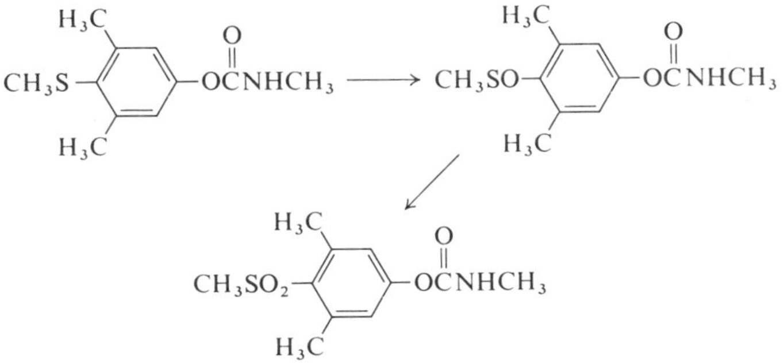

Neither of the oxidation products reached high concentration levels in plants and appears to be degraded rapidly.

2-4 ORGANOCHLORINE INSECTICIDES

A. DDT and Related Compounds

Probably no single compound has been more widely investigated for metabolism than DDT (1,1,1-trichloro-2,2-bis [p-chlorophenyl] ethane), due primarily to concern over the widespread development of resistance to this compound by different animals. The subject of DDT metabolism and that of other organochlorine compounds, and its relationship to insecticide resistance has been reviewed by others(205–208). The reader is referred to these articles for more detailed information regarding earlier work on DDT metabolism although a brief summary is given here.

Since the development of DDT resistance was first discovered in the common housefly, *Musca domestica*, much of the early work concerning DDT metabolism was carried out with this insect. Sternburg et al.(209) were the first to show that DDT-resistant strains of houseflies rapidly converted DDT to DDE [1,1-dichloro-2,2-bis(p-chlorophenyl)ethylene] by simple dehydrochlorination of DDT. In contrast, DDT-susceptible flies could not effect this conversion or only to a small extent. This discovery was confirmed almost simultaneously by Perry and Hoskins (210–212) who demonstrated similar occurrences with DDT-resistant

$$\left(\text{Cl} - \!\!\!\left\langle \bigcirc \right\rangle \!\!\!- \right)_2 \text{CHCCl}_3 \xrightarrow{\ -\text{HCl}\ } \left(\text{Cl} - \!\!\!\left\langle \bigcirc \right\rangle \!\!\!- \right)_2 \text{C}\!\!=\!\!\text{CCl}_2$$

and -susceptible flies of different origin. These findings were supported subsequently by others (213–215) although slight differences in rates of DDT dehydrochlorination were found among the different investigators with their resistant and susceptible strains of flies. Nevertheless, there is little doubt that the conversion of DDT to nontoxic DDE is the major route of metabolism of DDT in the house fly and is the primary reason for DDT resistance in this insect.

In addition to DDE, at least one or more minor metabolites of unknown structure has been discovered in the house fly. The first indication of an additional metabolite was reported by Tahori and Hoskins (216) who showed that a part of the administered dose of DDT did not respond to Schecter–Haller analysis. The difference in applied dose and that found as DDT or DDE was labeled as compound X. Variable amounts of a water-soluble metabolite of unknown structure have been found in the excreta of several different strains of houseflies (217–219), and it is possible that metabolite X and the water-soluble metabolite are the same. The amount of water-soluble metabolite appears to differ with investigators, from about 1% of the applied dose by Perry et al. (218) and up to 80% by Terriere and Schonbrod (219). The latter have demonstrated that the metabolite is a water-soluble conjugate, which upon acidic hydrolysis, gives a product that is extractable from water by carbon tetrachloride. Chromatographic and chemical evidence showed that this substance is neither bis(p-chlorophenyl)acetic acid nor 4,4'-dichlorobenzophenone but may be a phenolic derivative.

The dehydrochlorination of DDT to DDE in the housefly has been proven unequivocally to be enzymatic in nature by the isolation and characterization of the enzyme DDT-dehydrochlorinase (220–222). This enzyme is found in abundance in DDT-resistant house flies and has been isolated, also, from susceptible flies but with much smaller specific activity (223). The enzyme requires glutathione as an activator and is highly specific to DDT analogs with similar steric properties, e.g., analogs substituted in the para positions in the ring such as methoxychlor are dehydrochlorinated, whereas o,p'DDT is not. The unsubstituted 1,1,1-trichloro-2,2-bis(phenyl)ethane was dehydrochlorinated very slowly. DDT-dehydrochlorinase has a pH optimum of 7.4, and at pH 8.5 and 6.5 there is markedly less activity, indicating a rather narrow pH activity range.

The in vivo conversion of DDT to DDE has been demonstrated also in numerous insects, including various species of mosquito larvae

(*224–228*), the Mexican bean beetle(*229*), the European corn borer (*230*), and others(*231*). A summary of DDT metabolism by different insects has been published by Perry(*231*). DDT-dehydrochlorinase has been isolated from the mosquito larvae of *Aedes aegypti*(*225*), *Culex fatigans*(*226*), several resistant and susceptible anophelines(*228*), and the Mexican bean beetle(*229*). The enzymes obtained from the housefly and *Aedes aegyti* are, in general, similar in properties although some slight differences in substrate specificity are observed.

Although houseflies and different species of mosquitoes seem to metabolize DDT to principally a single substance DDE, conversion to other DDT metabolites has been demonstrated in a wide variety of other animals. The hydroxylation of DDT to dicofol or KELTHANE [4,4'-dichloroalpha(trichloromethyl)benzhydrol], in which the hydrogen on the tertiary carbon atom is replaced by a hydroxyl group, has been shown to occur in *Drosophila melanogaster*(*232–234*). Curiously enough, no DDE was found although at least three other unknown metabolites formed in minor proportions were detected by paper chromatography. Two of these were suspected to be *p,p'*-dichlorobenzophenone and 2,2-dichloro-1,1-bis(*p*-chlorophenyl)ethanol(*234*). The American cockroach, *Periplaneta americana*, rapidly absorbs and metabolizes DDT, and at least five metabolites are present in the excreta(*235*). As much as 75% of the DDT applied is excreted within a 24-hr period. Of the five metabolites, DDE and three others were found in minor amounts, from 2–5% each, while the major metabolite found in about 80% appeared to be more polar than DDT as judged from its paper chromatographic behavior. Work by others(*217*) with the American cockroach has demonstrated the presence of at least three metabolites. Paper chromatographic data indicate that one of the metabolites is *p,p'*-dichlorobenzophenone. Similar studies with the Madeira cockroach, *Leucophaea maderae*, have demonstrated the formation of DDE and three other metabolites of unknown structure. Two of the metabolites isolated from roach feces appeared to be substantially more polar than DDT or DDE and constituted the majority of the radioactivity recovered (*230*). The DDT-resistant human body louse, *Pediculus humanus*, has been shown to degrade DDT rapidly into a nontoxic water-soluble material that is unextractable in common organic solvents. This material responded to the Schecter–Haller test and the nitrated product was shown to be extractable from ether into dilute alkali. The properties of the nitrated product were unlike those of nitrated *p,p'*-dichlorodiphenylacetic acid under Schecter–Haller conditions but were similar to those of nitrated *p*-chlorobenzoic acid.

A microsomal enzyme system capable of converting DDT to a substance that behaves chromatographically similar to dicofol has been isolated from the German cockroach, *Blattella germanica* (236). This enzyme is also present in several strains of the housefly, particularly a parathion-resistant strain, and in the American cockroach. The enzyme requires oxygen, NADPH, magnesium ion, and nicotinamide as cofactors.

DDT metabolism in mammals also has been widely investigated. An extensive review of the literature up to 1957 on the fate of DDT in animals has been published by Hayes (237). Investigations with the rabbit have shown that DDT administered orally accumulates in different tissue with the largest amounts found in the bile, brain and cord, blood and liver (238, 239). DDT or its equivalent is slowly eliminated in the urine but substantially larger amounts are found in the feces (240). Analysis of urine from orally treated rabbits has shown that the principal urinary metabolite is bis-(*p*-chlorophenyl)acetic acid (DDA) (241). Since 80–85% of the total organic chlorine was extractable into aqueous alkali, it was concluded that other metabolites and DDT itself were minor

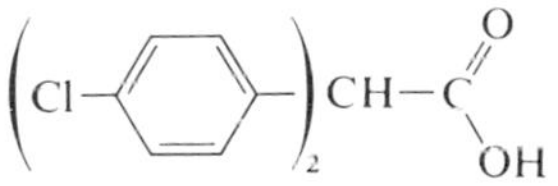

$$\left(Cl-\!\!\left\langle\bigcirc\right\rangle\!\!-\right)_{\!2}\!CH\!-\!C\!\!\begin{array}{c}O\\\diagup\\\diagdown\\OH\end{array}$$

DDA

constituents in urine. Studies with a single human male also showed DDA as the principal metabolite in urine (242). No unchanged DDT could be detected. DDA, also, is believed to be the principal metabolite of DDT administered to rats (243, 244) where it is found primarily in the bile. Most of the DDA is eliminated in the feces with very little found in the urine. Small amounts of DDE are found in both bile and feces and small amounts of DDT are found in the bile. In addition to DDA, an acidic metabolite that produces DDA upon treatment with strong acid has been shown to be present in rat feces. This unknown metabolite is believed to be a conjugate or complex of DDA although it is quite stable to alkaline hydrolysis.

The metabolic pathway for DDT in the rat has been proposed recently from results obtained after heavy DDT administration and analysis of liver and kidney tissue (245). The presence of all but two of the compounds given in the scheme was demonstrated by chromatographic methods and infrared spectroscopy.

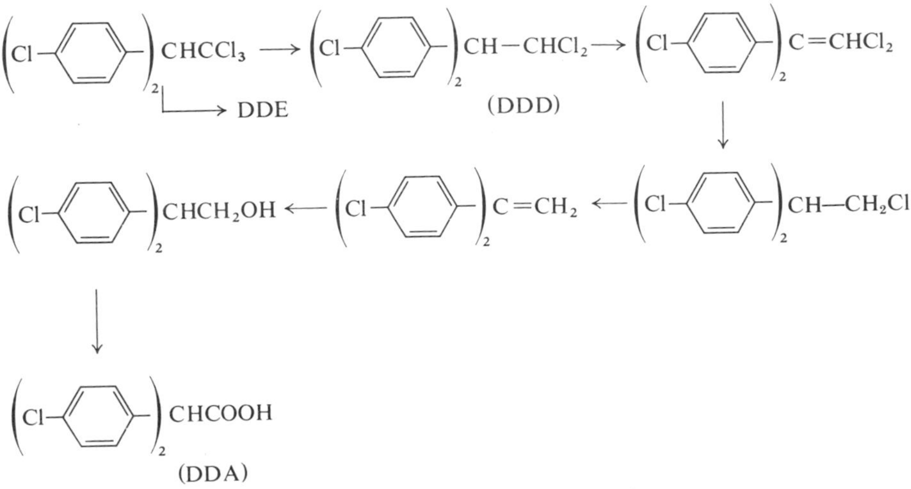

$$\left(Cl-\!\!\left\langle\bigcirc\right\rangle-\right)_2\!\!CHCCl_3 \longrightarrow \left(Cl-\!\!\left\langle\bigcirc\right\rangle-\right)_2\!\!CH-CHCl_2 \longrightarrow \left(Cl-\!\!\left\langle\bigcirc\right\rangle-\right)_2\!\!C=CHCl_2$$

$$\downarrow DDE \qquad\qquad (DDD)$$

$$\left(Cl-\!\!\left\langle\bigcirc\right\rangle-\right)_2\!\!CHCH_2OH \longleftarrow \left(Cl-\!\!\left\langle\bigcirc\right\rangle-\right)_2\!\!C=CH_2 \longleftarrow \left(Cl-\!\!\left\langle\bigcirc\right\rangle-\right)_2\!\!CH-CH_2Cl$$

$$\downarrow$$

$$\left(Cl-\!\!\left\langle\bigcirc\right\rangle-\right)_2\!\!CHCOOH$$

$$(DDA)$$

Recently, there has been increased evidence for the conversion of DDT to DDD, 1,1-dichloro-2,2-bis(p-chlorophenyl)ethane, sometimes referred to as TDE or tetrachlorodiphenylethane, in different living systems. The conversion represents an example of reductive dechlorination.

$$\left(Cl-\!\!\left\langle\bigcirc\right\rangle-\right)_2\!\!CH-CCl_3 \longrightarrow \left(Cl-\!\!\left\langle\bigcirc\right\rangle-\right)_2\!\!CH-CHCl_2$$

Certain microorganisms apparently are very active in metabolizing DDT to DDD. *Serratia* marcescens and another unidentified bacterial strain isolated from the excreta of the stable fly (*Stomoxys calcitrans*), and, under anaerobic conditions, *E. coli* convert DDT to DDD almost completely (90%) and to small amounts of DDE. Metabolism does not occur under aerobic conditions (246). Also, DDD is the principal metabolite obtained in equal amounts from DDT-resistant and susceptible strains of stable flies following topical application with radioactive DDT. Common commercial yeast has been shown to convert DDT to a single product DDD (247). Microorganisms that exist in mammals also effect this transformation. *E. coli* and *Aerobacter aerogenes* found in the gastrointestinal tract of rats were able to effect DDD formation (248). DDT to DDD conversion by bovine rumen fluid also has been demonstrated (249). Further, the presence of DDD as the major metabolite in the feces and the liver of rats given DDT by stomach tube has been demonstrated. No DDD was found when DDT was given intraperitoneally, indicating that DDD formation is effected by intestinal flora. Others (250) have reported the presence of DDD in rat liver and assumed that transformation occurred in the liver. More recently, the in vitro transformation of DDT to DDD has been demonstrated with homogenates and slices of pigeon and rat liver(251). As is reported above for microorganisms, aerobic conditions inhibited DDD formation by pigeon liver but activity was very high under anaerobic conditions. Because conversion to DDD still occurred after a homogenate had been maintained under nitrogen at 75°C for 10 min it was suggested that the reaction was nonenzymatic, perhaps analogous to the reaction between reduced porphyrins and DDT to give DDD(249). DDD apparently has been found in high concentrations in fish liver oils(250). The presence of substantial amounts of DDD residues in different tissues of cowbirds given oral dosages of DDT has been reported also(252).

B. Benzenehexachloride (BHC)

Technical benzenehexachloride, obtained by the chlorination of benzene, consists of a number of geometrical isomers. For unexplained reasons the γ-isomer or lindane is insecticidally active, the others have little or no activity. As with DDT; insects have also developed resistance to lindane. Although a single early report(253) indicates the absence of lindane degradation in lindane-resistant houseflies, subsequent reports clearly show that resistance is in large part attributable to increased metabolism to nontoxic compounds(254–257). Rapid metabolic degradation of the α-, β- and δ-isomers of BHC-resistant flies compared to non-resistant flies also has been well demonstrated.

One of the initial products formed by lindane metabolism in houseflies appears to be the mono-dehydrochlorinated substance, pentachlorocyclohexene(258). By an indirect and perhaps inconclusive method, this substance was shown to be the major product isolated from lindane-resistant flies at the 0.5 μg/fly dosage, even after as short a time as 1 hr after treatment. Under the same conditions, no pentachlorocyclohexene could be found in susceptible flies. This compound was considered to be an intermediate metabolite since the total amount of lindane and pentachlorocyclohexene recovered from the fly decreased with time while the amount of the latter remained fairly constant. Although DDT dehydrochlorinase was believed not to be involved in lindane dehydrochlorination to the cyclohexene, no direct evidence supporting this belief was given. The high values of pentachlorocyclohexene isolated from flies reported above by Sternburg and Kearns are in some disagreement with that found by others(259, 260). Bradbury and Standen(261), using ^{14}C-labeled lindane were never able to show more than 3% of the applied dose converted to pentachlorocyclohexene. The large discrepancy between the

English and American workers probably is attributable to the different techniques used, although there may be intrinsic differences in the strains of houseflies used. Bridges(259) has suggested that the high pentachloro-cyclohexene values obtained by Sternburg and Kearns may be attri-butable to the presence of other metabolites, including water-soluble metabolites, which appear as pentachlorocyclohexene in the analytical procedure. The absence of large amounts of the cyclohexene by the English workers has led them to believe that monodehydrochlorination is not the first step in any major pathway for the metabolism of lindane in house flies. The transitory nature of pentachlorocyclohexene, particularly under alkaline conditions(262–264), however, cannot be ignored and it is possible that this compound may be formed initially from lindane, which in turn is rapidly converted to other metabolites, thus keeping its concentration to a minimal level.

The bulk of lindane or other BHC isomers absorbed and metabolized by houseflies is converted into water-soluble metabolites. By using both ^{14}C- and ^{36}Cl-labeled lindane and α-isomer, at least 11 different metabo-lites have been detected by two-dimensional paper chromatography (265); six or seven of these appeared to be common to both BHC isomers. In the course of metabolism, four or more equivalents of chloride ions are formed per molecule of BHC metabolized. Alkaline hydrolysis of the water-soluble metabolites from flies treated with α-isomer and lindane gave a mixture of the various isomeric dichlorothiophenols, which constituted 50–60% of the total water-soluble metabolites. It is probable that the other water-soluble products are the isomeric tri-chlorophenols. Of the six isomeric dichlorothiophenols detected, the 2,4-, 2,5-, and the 3,4-isomers were found in largest abundance for α-BHC and the 2,4- and 2,5-isomers for γ-BHC. Mechanistically it is believed that BHC reacts with a sulfhydryl compound by displacement of chlorine with subsequent dehydrochlorination as shown below. How-ever, the possibility exists that conjugation with sulfhydryl takes place with the pentachlorocyclohexene as is discussed later.

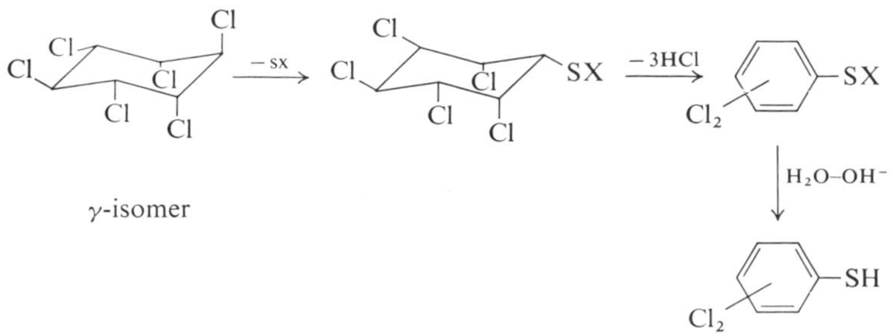

The nature of SX is unknown but it may in some way be related to gluthathione since the presence of this substance was found necessary in effecting in vitro conversion of BHC to dichlorothiophenol using housefly macerates. More recently, the major metabolic product isolated from cattle ticks (*Boophilus decoloratus*), locusts, and houseflies treated with lindane or from glutathione-fortified enzyme preparations from these insects has been demonstrated to be S-(2,4-dichlorophenyl) glutathione (266). Although none of the other dichlorophenyl isomers were detected, the isolation of a glutathione derivative adds support to earlier conjecture concerning the nature of the SX moiety.

An enzyme or enzymes capable of metabolizing certain BHC isomers to water-soluble metabolites has been isolated recently from the housefly and characterized(267, 268). The enzyme preparation initially isolated and partially purified by salt fractionation and gel filtration had a molecular weight of about 54,000 and showed a pH optimum of 7.8 for lindane. This preparation was shown later to consist of at least three different enzymes, all of which metabolize both α-BHC and lindane but at different rates. The original partially purified enzyme metabolized the BHC isomers at rates according to the following order, $\alpha > \gamma > \delta$. The β-isomer was resistant to metabolism under the in vitro conditions used. Curiously enough, this enzyme metabolized the γ- and δ-1,3,4,5,6-pentachloro-1-cyclohexene isomers at rates much faster than that of the corresponding BHC isomers. This finding is in agreement with that of Bradbury and Standen(260) and Bridges(259) who could detect only small amounts of pentachlorocyclohexene in flies after BHC treatment (about 1% of applied dose) and is in disagreement with Sternburg and Kearns(258) who reported large amounts of pentachlorocyclohexene. On the basis of this evidence, pentachlorocyclohexene, if it is formed initially from BHC, should never attain a high concentration since it is rapidly metabolized further to water-soluble materials. The enzyme specifically requires reduced gluthathione for reaction, and although the water-soluble metabolites were not identified, it seems likely that dichlorothiophenols are the ultimate metabolites, as was found by Bradbury(265). Further, the BHC metabolizing enzyme was shown to be active also in catalyzing the dehydrochlorination of DDT. Enzyme preparations from several other insects, including *Drosophila melanogaster*; *Stomoxys calcitrans*; German, Madeira, and American cockroaches; and honey bee, and from rat and rabbit liver and rabbit kidney were inactive or of low activity in metabolizing lindane under the same test conditions. Thus, this enzyme appears to be highly specific to the housefly.

Relatively little has been reported on BHC metabolism in mammalian systems. As mentioned above, enzyme preparations obtained from rat or rabbit livers and rabbit kidney were virtually inactive in metabolizing α-BHC or lindane(267). On the other hand, all three preparations are moderately effective in metabolizing δ-pentachlorocyclohexene but less effective in metabolizing the γ-isomer. The δ- and γ-isomers of BHC have been reported to be metabolized to several water-soluble compounds and eliminated through the kidney, one of which was believed to be 2,4,6-trichlorophenol(269, 270). Both 2,3,5- and 2,4,5-trichlorophenol have been detected in the urine of rats treated intraperitoneally with γ-BHC or γ-2,3,4,5,6-pentachlorocyclohex-1-ene(271). The sulfate and glucuronide conjugates of these phenols also were detected. In addition, 2,4-dichlorophenylmercapturic acid was isolated as a metabolite of γ-pentachlorocyclohex-1-ene and a compound with similar chromatographic properties was isolated from the urine of γ-BHC-treated rats. The phenolic metabolites of γ-BHC and γ-2,3,4,5,6-pentachlorocyclohex-1-ene isolated from urine were similar to those of 1,2,4-trichlorobenzene (272) indicating that the latter two compounds are intermediates in γ-BHC metabolism in rats. The metabolic pathways of γ-BHC in the rat are depicted as follows.

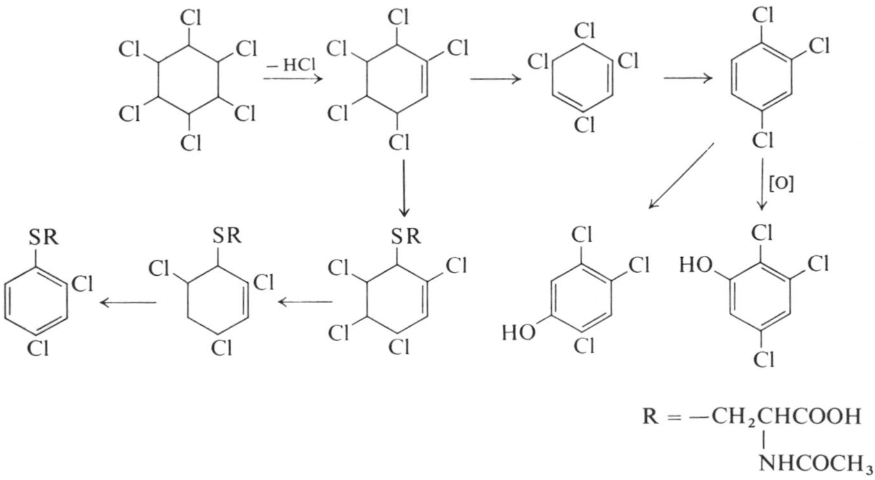

$$R = -CH_2CHCOOH$$
$$\quad\quad\quad | $$
$$\quad\quad NHCOCH_3$$

The in vitro enzymatic conjugation of γ-2,3,4,5,6-pentachlorocyclohex-1-ene with glutathione has been demonstrated with a rat-liver enzyme(273). Conjugation with glutathione, however, does not occur with γ-BHC with this enzyme preparation.

C. Cyclodienes

1. Aldrin and Dieldrin. Unlike DDT, BHC, and numerous other insecticides, the development of insect resistance to certain of the cyclodiene insecticides such as aldrin and dieldrin cannot be explained on the basis of increased metabolism of the insecticide. In the various susceptible and resistant insects examined for aldrin metabolism, with few notable exceptions, the bulk of the evidence has shown that aldrin is metabolized to dieldrin and dieldrin in turn is resistant to metabolism (274–278).

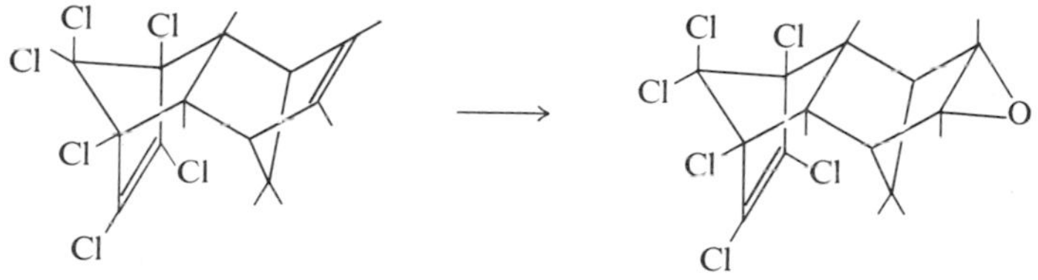

aldrin   dieldrin

In view of the toxic nature of dieldrin an explanation for resistance must be found elsewhere than in metabolism. The fact that susceptible houseflies are capable of converting aldrin to dieldrin has led some investigators to suggest that the toxicity of aldrin is in part attributable to dieldrin formed within the animal(276, 278). This hypothesis is supported by the observation that houseflies treated with aldrin exhibit delayed symptoms of intoxication.

The epoxidation of aldrin to dieldrin probably occurs by an enzymic reaction in living systems. Support for this is found in the fact that neither aldrin or isodrin is converted to the epoxide in the tissues of heat-killed houseflies(275). Stronger support for an enzymic process is found in more recent work in which the epoxidation of aldrin, isodrin, and heptachlor by microsomal fractions from rat and rabbit liver, fat body homogenates of the American cockroach, and homogenates of whole houseflies or fly abdomens was demonstrated(279). Rabbit and rat liver microsomes required NADPH as a cofactor and no reaction occurred when heated microsomes were used; both findings indicate that the epoxidation reaction is enzymic. Further, the epoxidation of heptachlor and aldrin was shown to be inhibited by SKF 525A and piperonyl butoxide, each known to be inhibitors of microsomal oxidases. Neither heptachlor epoxide nor dieldrin was further metabolized by rabbit or rat liver microsomes.

The epoxidation of aldrin and also heptachlor has been shown to occur in soil(280–283). Aldrin epoxidation is known to occur more readily in moist soil, particularly in soil rich in microorganisms(283). In contrast, autoclaved soil, dry soils, and sand of low organic matter are poor in metabolizing aldrin. Evidently soil microorganisms are in a large part responsible for aldrin or heptachlor epoxidation. In this regard, cultures of certain microorganisms, including *Aspergillus niger*, *Penicillium notatum*, *P. chrysogenum*, and *P. vermiculatum* have been shown to metabolize aldrin to dieldrin(284). In addition to dieldrin, however, these microorganisms also produced four other metabolites from aldrin but, strangely enough, were not able to metabolize dieldrin. On the basis of this evidence it is probable that other metabolites of aldrin are present in soil.

Aldrin to dieldrin metabolism also has been demonstrated in plants, including alfalfa, soybeans, and corn(285). As much as 3.24 ppm of dieldrin has been found in alfalfa treated with aldrin at 1 lb per acre one day after treatment. In comparison the residue level of aldrin was 5.9 ppm. The rapid accumulation of dieldrin in the plant precludes the possibility that the aldrin is first metabolized in the soil and then taken up by the plant.

There are relatively few examples in which metabolism of aldrin other than to dieldrin or further metabolism of dieldrin to other substances has been demonstrated in insects. The conversion of aldrin into dieldrin and at least four other metabolites of unknown structure has been reported for the mosquito larvae *Aedes aegypti*(284). Dieldrin also was metabolized apparently to the same four products, two of which were demonstrated to be more polar than either aldrin or dieldrin by paper chromatography. Surprisingly, the total amount of all four metabolites constituted a substantial portion of the amount applied. The metabolism of ^{14}C-dieldrin by dieldrin-resistant strains of mosquito adults, *Culex pipiens quinquefasciatus*, into a more polar compound with similar chromatographic properties as 6,7-dihydroxydihydro-aldrin also has been demonstrated(286).

The metabolism of aldrin and dieldrin is in general more complex in mammals than in insects. As in insects, the metabolism of aldrin to dieldrin has been well demonstrated in a variety of mammals, and dieldrin has been found in the tissues and products of several different mammals treated orally or subcutaneously with aldrin(287). In addition to aldrin-to-dieldrin conversion, other metabolic conversions are known to occur also. In rats, dieldrin is metabolized into a number of different products with varying polar properties and excreted(288). Very little unchanged dieldrin is eliminated and 70% of the dose after oral and intravenous administration is present in the feces, indicating that bile is the most important route of elimination. About 65% of the metabolites present in the feces was a single compound, somewhat more polar than dieldrin but extractable from water by hydrocarbon solvents, 6% was present as a more polar but also neutral compound, 12% as an acidic compound, 14% as highly polar substances unextractable from water by ether, and 3% as dieldrin. Seventy per cent of the metabolites in the bile was a single compound and appeared to be the glucuronide of the major metabolite found in feces. From solvent partition properties, the structure of this major metabolite was suggested as the 6,7-dihydroxy-dihydro-aldrin derivative. Aldrin, like dieldrin, is also metabolized in rats into a number of polar products, 80–90% of which is found in the feces(289). Seventy-five per cent of the metabolites in feces and 95% of the metabolites in urine consisted of products more polar than the parent compound.

More recently, the identification of one of the six dieldrin metabolites obtained from the urine of rabbits treated orally with ^{14}C-labeled dieldrin has been reported(290). This major metabolite, which consisted of 86% of the total metabolites, was isolated in crystalline form and characterized as one of the two enantiomorphic isomers of 6,7-trans-dihydroxydihydro-aldrin with a specific rotation $[\alpha]_a^{20}$ of $+13.7°$. The structure is given below, and one would surmise that it is formed by opening of the epoxide ring.

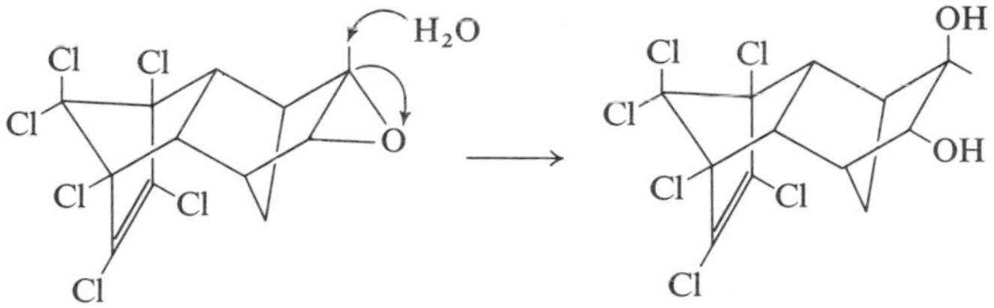

(or its mirror image)

It seems highly probable that this metabolite is identical to the principal metabolite present in rat feces as reported earlier. The other five remaining metabolites, representing 14% of the total radioactivity, were found in approximately equal but in small amounts.

The urine obtained from human males occupationally exposed to dieldrin has been shown to contain at least two neutral, polar metabolites of unknown structure(291). Gas chromatography also gave evidence of the presence of a conjugate that appeared to be unstable in the chromatographic column. One of the neutral metabolites is probably 6,7-trans-dihydroxydihydro-aldrin.

2. Isodrin and Endrin. Considerably less work has been carried out on the metabolism of isodrin and endrin, the respective geometrical isomers of aldrin and dieldrin. The conversion of isodrin to endrin, analogous to the conversion of aldrin to dieldrin, has been demonstrated in both susceptible and resistant strains of houseflies(275). In addition to isodrin epoxidation, susceptible and resistant flies topically treated

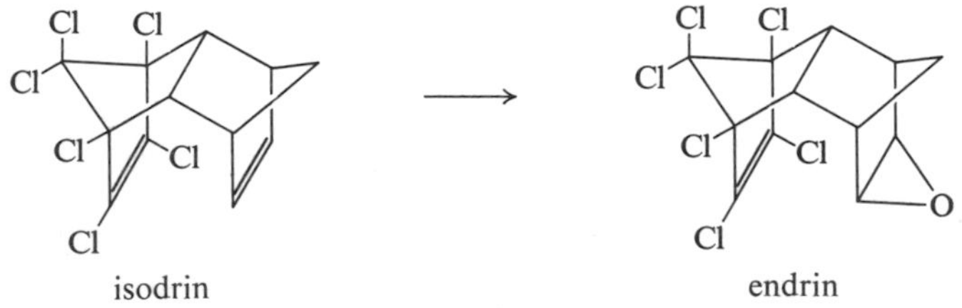

isodrin   endrin

with isodrin and endrin contained small but similar amounts (3–4% of applied dose) of a water-insoluble substance that appeared to be identical to the semi-caged keto derivative (structure given below) as judged from its paper chromatographic behavior. This compound is a known rearrangement product of endrin.

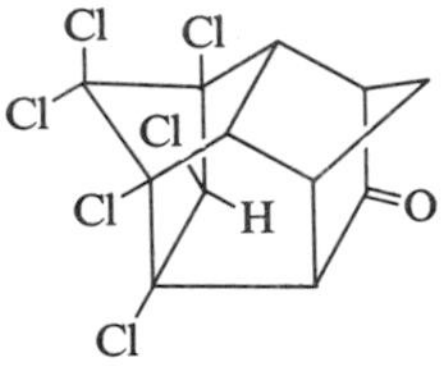

As mentioned earlier, isodrin, like aldrin and heptachlor, is converted to endrin by rabbit liver microsomes. Apparently no further metabolism of the product endrin occurs with this enzyme system(279). In view of the similar structural relationship between isodrin and aldrin it is likely that the enzymatic epoxidation of isodrin takes place in a variety of animals as demonstrated with aldrin.

Isodrin, because of its structural configuration, is not a stable molecule and readily undergoes isomerization to a caged compound of structure given below(292). This reaction is induced by the action of light or acid,

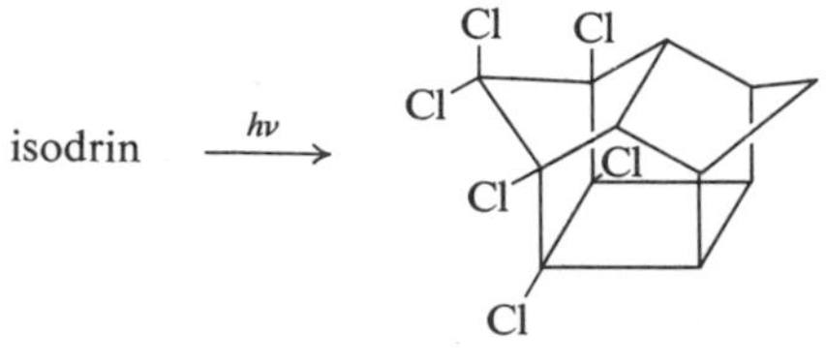

isodrin $\xrightarrow{h\nu}$

hence one might expect it to occur under field conditions, e.g., on leaf or other surfaces where exposure to sunlight prevails.

3. Chlordane and Heptachlor. Technical chlordane, prepared by the chlorination of chlordene (4,5,6,7,8,8-hexachloro-3a,4,7,7a-tetrahydro-4,7,7a-methanoindene), is a mixture of at least five compounds including heptachlor, α- and β-chlordane (empirical formula $C_{10}H_6Cl_8$), and a $C_{10}H_5Cl_9$ compound. Because of the complexity of technical chlordane, very little metabolism work of any significance has been carried out with this material. By means of bioassay methods the metabolism of chlordane to presumably nontoxic substances has been demonstrated in resistant houseflies(293).

Little information is available regarding the metabolism of the individual constituents of chlordane. The epoxidation of chlordene, a constituent of low insecticidal activity, has been demonstrated in house flies(294). α-Chlordane administered to rats intravenously is metabolized to "hydrophilic" products of unknown structure(295). Analysis of the animal body showed large amounts of metabolites in different tissues, particularly in the alimentary tract. Of the total dose 80% was excreted, most of it in the form of metabolites.

Heptachlor, like aldrin and isodrin, contains an olefinic bond and is epoxidized in living systems to heptachlor epoxide. The metabolism of heptachlor to the epoxide has been demonstrated in insects(296, 297), mammals(298–300), plants(301), soil(283), and by microsomal enzyme systems(279).

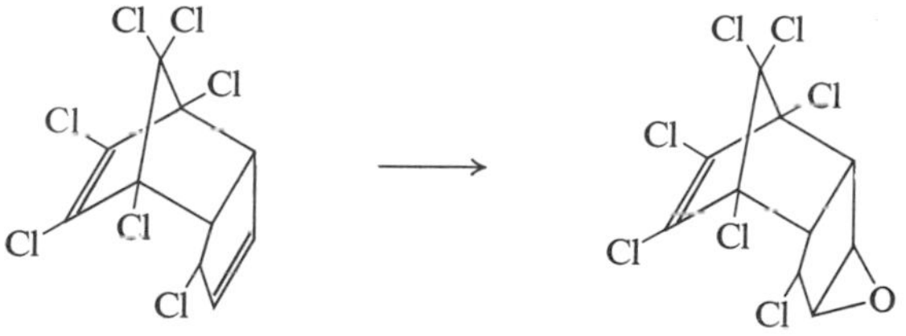

heptachlor   heptachlor epoxide

It is noteworthy that in mammals heptachlor epoxide is stored principally in fat(300). The epoxide is found also in milk, particularly in the butterfat, obtained from cows fed heptachlor(299).

4. Isobenzan or TELODRIN. The metabolism of isobenzan, a cyclodiene insecticide of structure shown below, has been investigated in mosquito larvae, *Aedes aegypti*, and the microorganisms *Aspergillus niger*, *A. flavus*, *Penicillium chrysogenum*, and *P. notatum*(302). At low concentrations

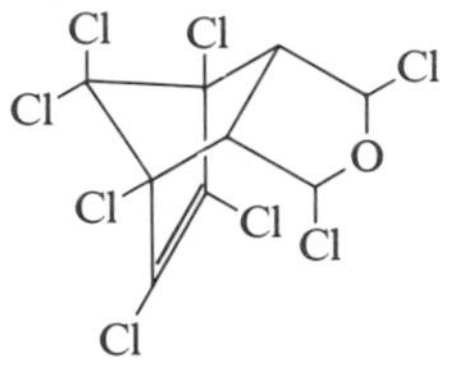

isobenzan

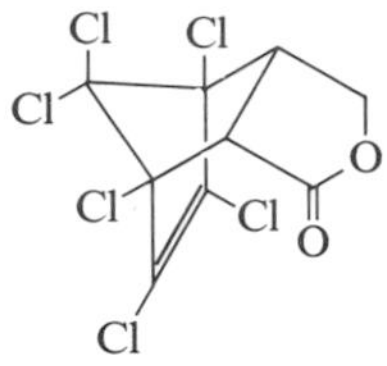

lactone

of isobenzan in medium containing the microorganism conversion
to a polar metabolite A (presumably a conjugate) occurred up to 99%.
A, upon hydrolysis with hydrochloric acid, gave a less polar product of
unknown structure. Mosquito larvae converted isobenzan to three highly
polar products. Hydrolysis of two of these with hydrochloric acid did
not change their chromatographic behavior; the third, however, gave a
product that in all respects was identical to the lactone of structure given
above.

5. Endosulfan. Endosulfan or THIODAN consists of a mixture of two
geometric isomers I and II whose structures have recently been eluci-
dated(*303*). The lower melting endosulfan I has been found to dissipate
more rapidly as plant residues than the higher melting endosulfan II.

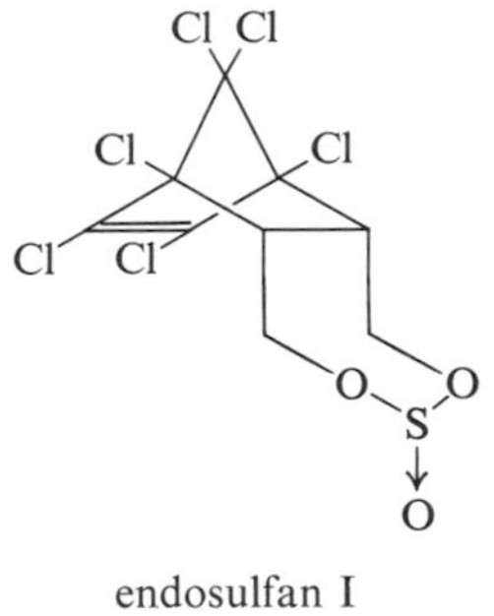

endosulfan I

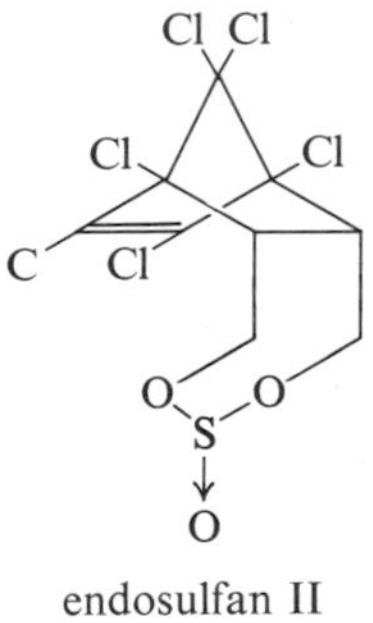

endosulfan II

Plants treated with either I or II gave a single metabolite whose structure
has been confirmed as the corresponding sulfate ester by TLC, GLC,

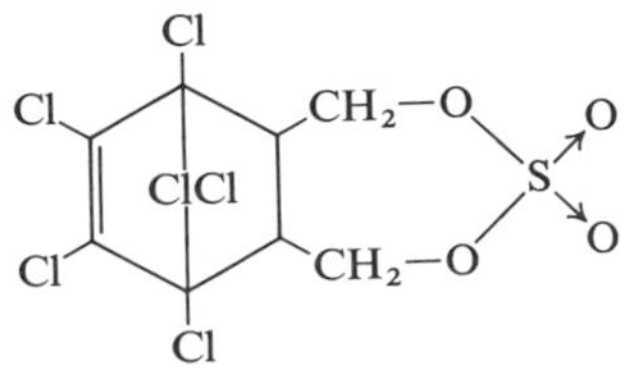

endosulfan sulfate

and infrared spectra(*304*). Apparently no hydrolysis product of the sulfite or sulfate ester (diol) was detected. Endosulfan sulfate appears to have the same toxicity as technical endosulfan to rats and insects.

Endosulfan metabolism has been investigated also in insects. In both susceptible and cyclodiene resistant houseflies, endosulfan I is metabolized more rapidly than the high melting isomer II(*305*). As far as could be determined by radioisotopic methods, no metabolites other than the sulfate could be detected in the hexane extract of treated flies. Further, no metabolite, not even the sulfate, was found in the excreta of treated flies. In comparison, endosulfan ether (two-dimensional structure given below), a known impurity in technical endosulfan, is metabolized and excreted from houseflies as two polar metabolites of unknown structure, one soluble in acetone and the other soluble in water.

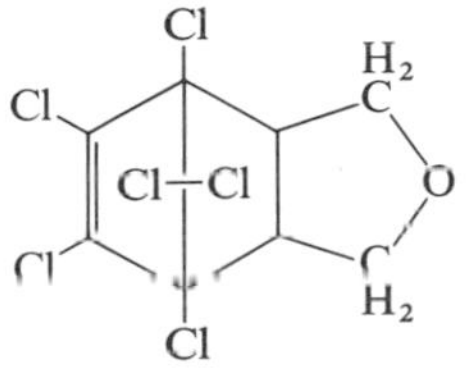

endosulfan ether

The water-soluble metabolite is believed to be a conjugate of some sort. Perhaps some light is shed on the identity of these unknown metabolites by more recent work on the metabolism of endosulfan in the imago of the locust, *Pachytilus migratorius migratorioides*(*306*). By gas chromatographic methods using three types of columns, endosulfan sulfate, endosulfan ether, and two other metabolites (structures given below) were separated from locust excreta and identified with reasonable certainty.

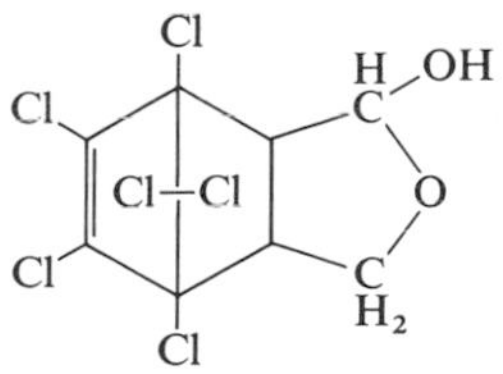

hydroxy metabolite

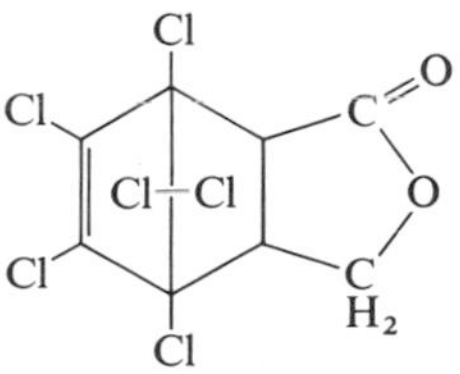

lactone metabolite

The establishment of endosulfan ether and its oxidation products as metabolites of endosulfan suggests that the sulfite or the sulfate ester ring is opened, then closed by displacement of the sulfite or sulfate group to form the ether, which is subsequently oxidized to the lactone through the hydroxylated intermediate. Thus, one may conjecture on not too uncertain grounds that the water-soluble metabolite found in houseflies is a conjugate of the hydroxy compound and the acetone-soluble metabolite is the lactone. It is curious that the lactone metabolite isolated from endosulfan is identical to the metabolite found for TELODRIN.

2-5　BOTANICAL INSECTICIDES AND ANALOGS

A. Pyrethroids

The natural pyrethrins are esters of substituted cyclopropane carboxylic acid (chrysanthemum carboxylic acid) and substituted keto-cyclopentenols (pyretholones) of general structure shown below.

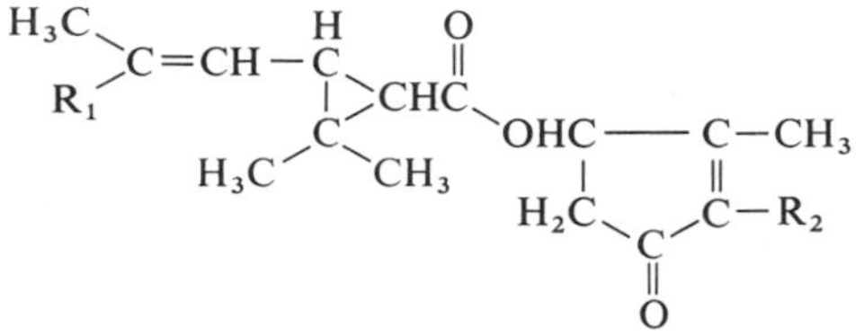

Pyrethrin I
$R_1 = CH_3$
$R_2 = CH_2CH=CHCH=CH_2$
Pyrethrin II
$R_1 = COOCH_3$
$R_2 = CH_2CH=CHCH=CH_2$
Cinerin I
$R_1 = CH_3$
$R_2 = CH_2CH=CHCH_3$
Cinerin II
$R_1 = COOCH_3$
$R_2 = CH_2CH=CHCH_3$

The insecticidal activity of the pyrethrins decreases with storage and exposure to the environment probably because of the various labile groups present in the molecule that may serve as avenues for degradation (*307–309*).

The pyrethrins are highly active as a contact poison against many insects but appear to be poor stomach poisons because of their rapid degradation upon ingestion. The degradative metabolism of crude pyrethrum flower extract was demonstrated early in the imported cabbage worm, *Pieris rapae*(*310*), and in the southern army worm, *Prodenia eridania*(*311*). Inactivation of pyrethrum was shown by bio-assay of excreta against mosquito larvae. Incubation of pyrethrum extract with different tissues of the southern army worm also resulted in partial degradation of the extract. Lipase extracts of the American cockroach and housefly also degrades pyrethrin to nontoxic substances(*312*), presumably to hydrolytic products.

The metabolism of biosynthetically prepared ^{14}C-labeled pyrethrins has been studied in the American cockroach. By paper chromatography, evidence was obtained that indicated that the ^{14}C-labeled pyrethrin mixture administered to the roach is converted by hydrolysis to the respective keto alcohols, chrysanthemum carboxylic acid, and several unidentified metabolites(313). However, because of the questionable purity of the radiolabeled mixtures used and the poor chromatographic resolution, amplified by severe interference from cockroach extract, the validity of these results is in doubt. Considerable quantities of unchanged pyrethrin esters (58%) was recovered unchanged, and up to 12% of the radioactive material was reported to be excreted as carbon dioxide. However, other investigators(314) using the same biosynthetic pyrethrin preparation found no evidence for the presence of chrysanthemumic acid in houseflies treated topically or by injection but did find evidence for the keto alcohol. Also, these investigators were unable to detect any evolution of $^{14}CO_2$ from treated flies. It was demonstrated that piperonyl cyclonene, a pyrethrin synergist, inhibits in vivo metabolism of the labeled pyrethrin and also allethrin, suggesting that the principal route of metabolism is oxidative and not hydrolytic. There is ample evidence that supports the view that the methylenedioxy synergists such as piperonyl cyclonene are inhibitors of oxidative metabolism (2, 193a, 315).

More recent investigations using purified samples of ^{14}C-labeled pyrethrin I and cinerin I in houseflies have shown that each of these compounds is converted into five different metabolites plus a small amount of chrysanthemumic acid(316). The amount of free chrysanthemumic acid never exceeded 2.6% of the applied dose and the majority of the absorbed dose was in the form of one of the metabolites. More than 96% of the absorbed dose of pyrethrin I or cinerin I was metabolized within 4 hr after topical treatment. These results add support to the belief that the principal route of metabolism is not hydrolytic. Of the five metabolites, three were intact esters with no structural changes in the chrysanthemumic acid moiety as demonstrated by positive Deniger color reaction. Therefore, metabolic alteration of the molecule must have occurred in the pyretholone or cinerolone moiety. The major metabolite, consisting of 18 to 37% of the applied dose for pyrethrin I and 12 to 28% for cinerin I, was substantially more polar than the others and gave a negative test for the chrysanthemumate moiety, indicating alteration in this part of the molecule. The use of sesamex with either pyrethrin I or cinerin I strongly depressed detoxication rates in houseflies, the amount of intact pyrethrin or cinerin being 9 and 12 times greater, respectively, in sesamex treated flies than in untreated flies.

The metabolism of allethrin in houseflies is similar to that found for

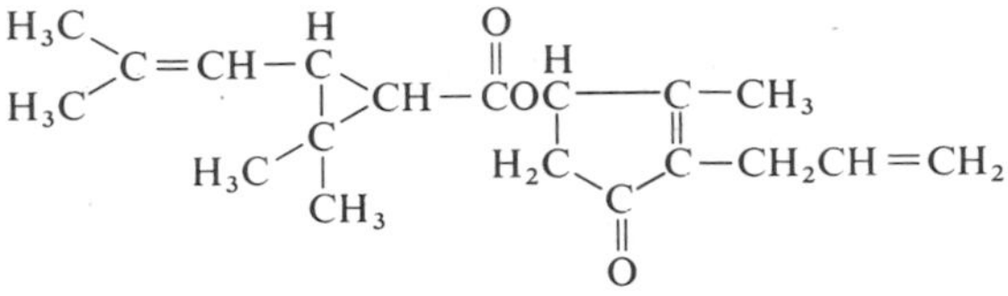

allethrin

pyrethrins(*314, 317, 318*). Using allethrin labeled in the ketocyclopen-
tenyl portion of the molecule, a metabolite that behaved as the ketocyclopentenol was isolated by paper chromatography(*314*). Other investigators using allethrin labeled in the chrysanthemumic acid portion of the
molecule were able to detect only traces of the acid in housefly homogenates or excreta(*317*). Also, only traces of unchanged allethrin were
recoverable and the bulk of the recovered material was a substance that
behaved like the ketocyclopentenol as found by paper chromatography.
However, since the allethrin was labeled in the acid moiety, this metabolite cannot be the alcohol and must be a derivative of the intact ester
or of the acid.

The metabolism of barthrin and dimethrin, two synthetic pyrethrin
analogs, has been studied in rabbits.

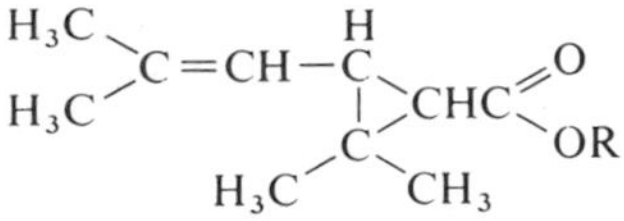

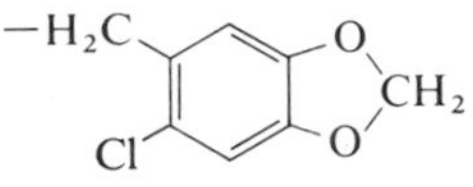

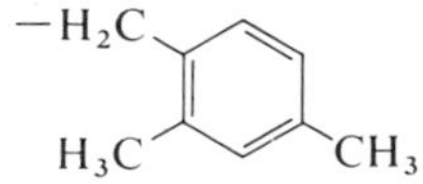

In very preliminary studies, ether and ethyl acetate extracts of urine
from rabbits administered barthrin gave positive color tests for free
chrysanthemumic acid and for methylenedioxy in the piperonyl moiety
(*319*). Similar studies with dimethrin also produced evidence for chrysanthemumic acid in the urine(*320*). More detailed investigations(*321*) by
actual isolation of the metabolic products and unequivocal identification
have proven the presence of chrysanthemumic acid, 6-chloropiperonyl
alcohol, and 6-chloropiperonylic acid in urine from barthrin treated
rabbits and chrysanthemumic acid and 3,5-dimethylbenzoic acid from
dimethrin. These results suggest that at least in rabbits, barthrin and
dimethrin are first hydrolyzed to chrysanthemumic acid and substituted
benzyl alcohol, followed by oxidation of the alcohol to the corresponding
acid.

B. Nicotine

Very little work has been done on nicotine metabolism in insects. In the American cockroach, nicotine is metabolized to cotinine as the

nicotine → cotinine

principal metabolite (322). The principal metabolite in the German cockroach is a substance with similar paper chromatographic properties as cotinine but with slightly different ultraviolet spectral characteristics. Two other metabolites present in minor proportion and of unknown structure were present in both insects. In contrast to roaches, the southern armyworm, *Prodenia eridania*, metabolizes nicotine into at least nine metabolites, none of which has been identified. It is probable that some of these metabolites are identical to those subsequently isolated from mammals.

Nicotine metabolism in mammals has been widely investigated and appears to be quite complex. Studies in the rat (323) and dog (324) with biosynthetic ^{14}C-nicotine have shown that practically all of the injected radioactivity is eliminated in the urine. Cotinine has been isolated from dog urine and is believed to occur by spontaneous lactamization of γ-(3-pyridyl)-γ-methylaminobutyric acid, another metabolite that also has been isolated from dog urine (325). Cotinine is further metabolized in dogs and rats to a number of products. The metabolic pathway for nicotine in mammals is depicted by the following equations. The establishment of the structure of the various metabolites is attributable mainly

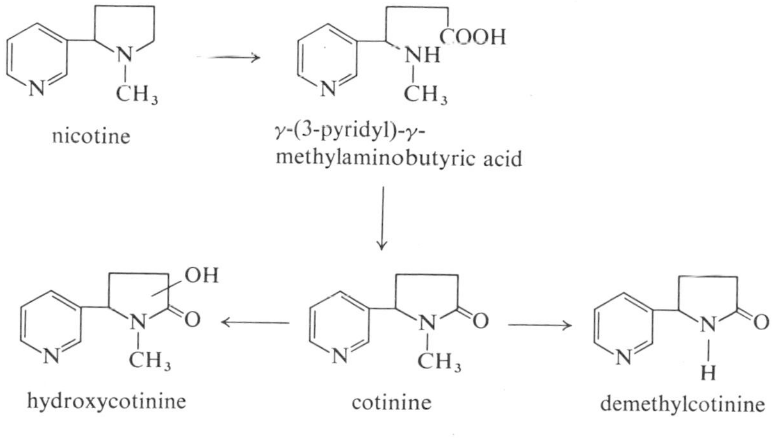

nicotine → γ-(3-pyridyl)-γ-methylaminobutyric acid

hydroxycotinine ← cotinine → demethylcotinine

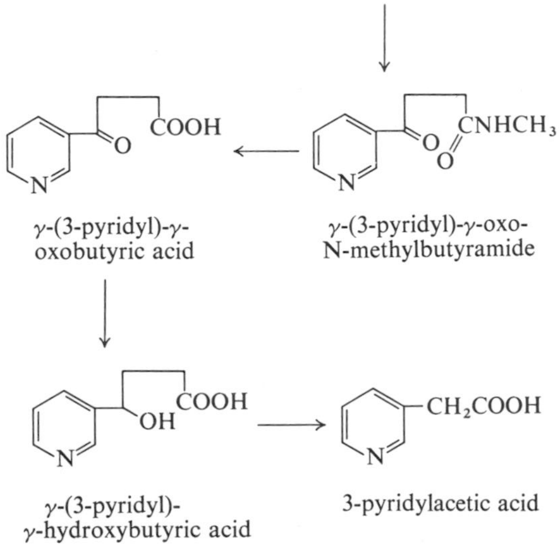

γ-(3-pyridyl)-γ-
oxobutyric acid

γ-(3-pyridyl)-γ-oxo-
N-methylbutyramide

γ-(3-pyridyl)-
γ-hydroxybutyric acid

3-pyridylacetic acid

to the efforts of McKennis and co-workers (*326–330*). The metabolic products in the above scheme have all been isolated and identified after administration of nicotine or one of the intermediate products. 3-Pyridylacetic acid appears to be the ultimate metabolic product.

2.6 PHENOL INSECTICIDES

Dinitrophenols

4,6-Dinitro-*o*-cresol (DNOC) has been used effectively in the past for insect control, particularly against locusts. 4,6-Dinitro-*o*-cresol is metabolized in locusts, *Schistocerca gregaria* and *Locusta migratoria*, into 6-acetamido-4-nitro-*o*-cresol, presumably via reduction to the 6-amino intermediate and subsequent acylation (*331*). Treatment of the locusts with 6-amino-4-nitro-*o*-cresol and analysis of the intestinal tract showed that this compound is largely converted to the corresponding 6-acetamido-4-nitro-*o*-cresol and in lesser amounts to 6-amino-4-nitro-*o*-cresyl sulfate and glucoside. Thus, the metabolic pathway for dinitro-*o*-cresol in locust may be given as follows.

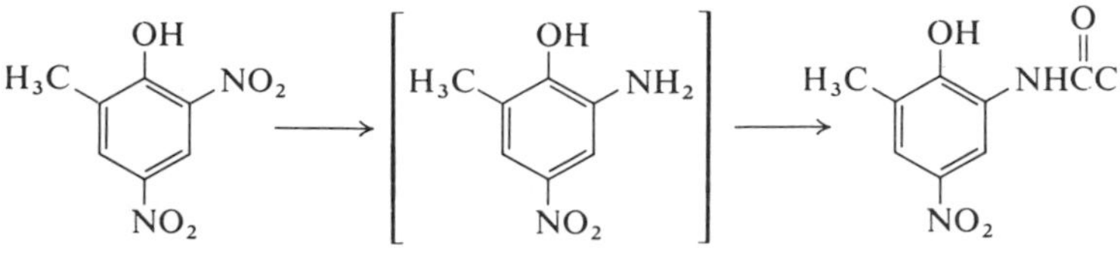

It is possible, however, that the 6-acetamido-derivative is formed directly from DNOC since neither the amino compound nor the conjugates could be detected after treatment with DNOC.

DNOC metabolism in mammals is more complex. The analysis of urine from rabbits treated orally with DNOC has been shown to contain unchanged DNOC (5%), conjugated DNOC (1%), 6-amino-4-nitro-o-cresol (11–12%), 6-acetamido-4-nitro-o-cresol (1–1.5%), 6-acetamido-4-nitro-o-cresol O-conjugates (10%), and trace amounts of 3-amino-5-nitro-salicylic acid and 4-amino-6-nitro-o-cresol(*332*). Parenthetical values give the percentage of total dosage found for each metabolite. Further, the urine of rabbits treated with 6-amino-4-nitro-o-cresol contained the N-acetylated derivative, the unchanged aminonitrocresol and its glucuronide. Treatment of the rabbit with 6-acetamido-4-nitro-o-cresol, however, gave only the unchanged substance and its glucuronide, suggesting that 6-amino-4-nitro-o-cresol is formed first from DNOC and is then acylated to the 6-acetamido derivative.

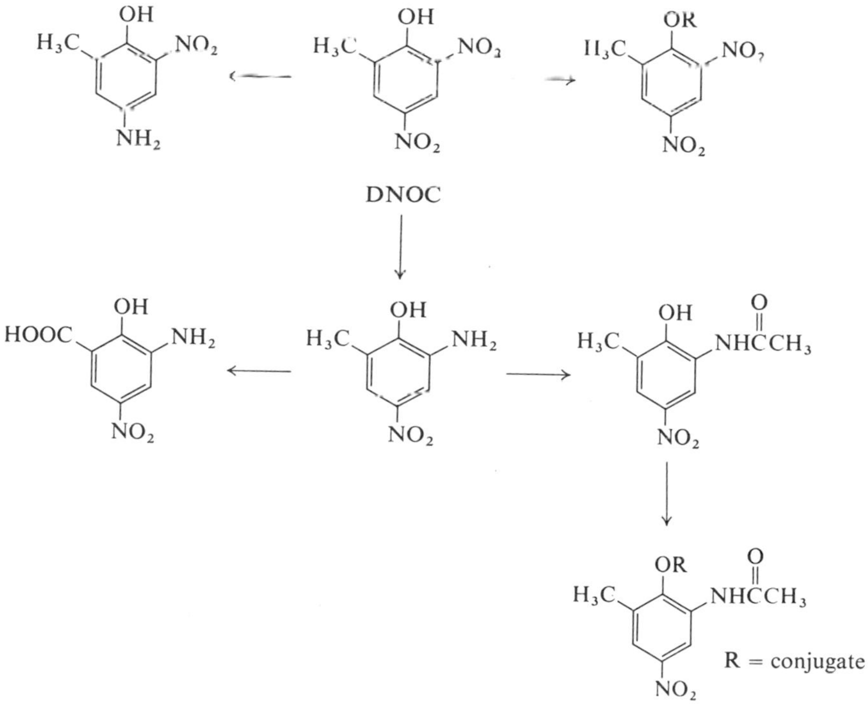

DNOC is rapidly decomposed by a soil microorganism related to *Corynebacterium simplex* to a colorless substance and nitrite ion(*333*). Determination of nitrite ion showed that both nitro groups on the ring were being removed.

2-7 FUNGICIDES

There seems to be a definite paucity of meaningful data concerned with the metabolism of fungicides. In general, investigators working with fungicides are content merely with measuring the rate of disappearance of fungicides from plants or soils by bioassay methods, and the impression is obtained from their work that the compounds simply vanish. It should be very clear just from common sense alone that fungicides do not disappear! There are several available fates for these chemicals: (a) some may be inert and resist change (b) some may be utilized by an organism as a food source and thus be degraded to their elements and reassembled as new compounds or (c) some may undergo one or more simple chemical changes that detoxify or possibly activate them. Thus typical studies using a bioassay to detect *fungicidal* metabolic products answer only part of the question. What happens to the compound that is no longer fungitoxic? Why is the compound no longer fungitoxic? Are the non-fungitoxic compounds produced toxic to other forms of life? The answers to these and related questions are generally not known for fungicides.

Fortunately several groups of investigators have taken up studies to answer these questions but the field is by no means crowded. Only cases in which the actual metabolism of fungicides has been demonstrated are discussed here. The disappearance type of report and others where identification of metabolites have not been made may be found by consulting the recent reviews of Sijpesteijn and Van der Kerk(7, 8). If anything, this brief review serves to illustrate the need for more activity in this field.

A. 6-Azauricil

In studying the mode of action of 6-azauricil (**I**) against powdery mildew of cucumber and other plants, it was found(*334*) that a new compound appeared in the leaves of cucumber plants whose roots had been placed in a solution of **I**. Paper chromatography of the leaf juice of these plants revealed spots having the same R_f value as **I** and its riboside 6-azauridine (**II**). Both compounds were active as fungicides.

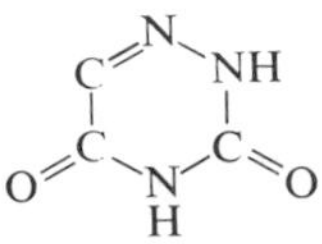

(**I**) 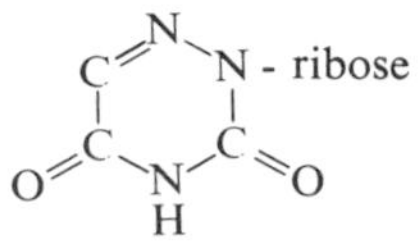   (**II**)

B. Sulfanilamide

The treatment of broad beans and wheat with sulfanilamide I revealed (*335, 336*) that the compound could be acetylated on both nitrogen atoms (**II**) and (**III**). The acetylation reactions were found to take place in the roots. Deacetylation was also shown to take place, occurring mainly in the stems and leaves.

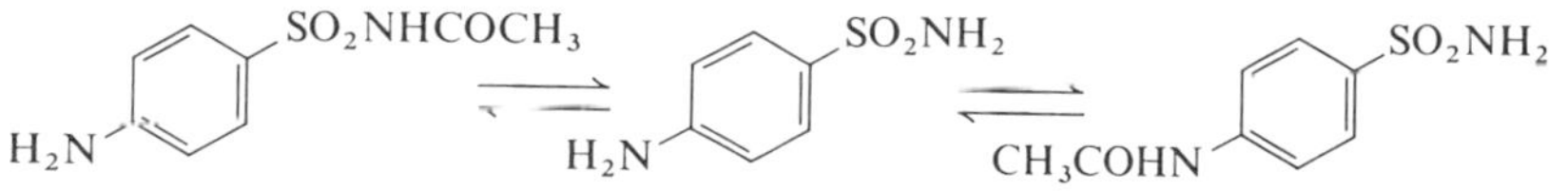

(**II**)   (**I**)   (**III**)

Treatment of the plant with sulfanilamide (**I**) tended to give an equilibrium between compounds (**I**), (**II**), and (**III**). Acetylation of (**III**) took place in the plant to give N,N-diacetyl sulfonamide (**IV**). But no **IV** was reported to be formed by feeding of either **I** or **II**.

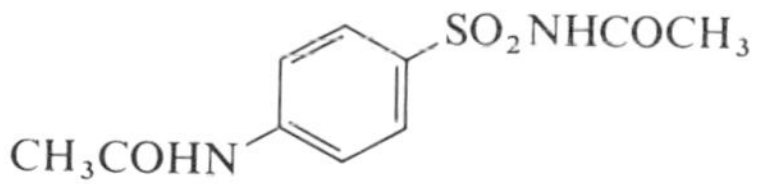

(**IV**)

C. Captan

No actual metabolic work has been carried out with captan but probable metabolic breakdown may be inferred from studies on the reactions of captan with various mercaptans (*337, 338*). These reactions produced, tetrahydrophthalimide, chloride ion, and thiophosgene. The mercaptan used was converted to a disulfide in addition to reacting with the thiophosgene to produce a trithiocarbonate. In the special case of the amino acid cysteine, the expected cystine was formed plus a second product 2-thiazolidinethione-4-carboxylic acid. The latter compound was shown to result from the treatment of cysteine with thiophosgene but also could have resulted from direct reaction of the amino acid with captan. Since the thiophosgene formed is a relatively reactive compound, it may react with any available nucleophiles, of which there are many in normal biological situations. It also condenses with itself to form a dimer, dithiophosgene, which slowly decomposes in water solution to form carbon disulfide and hydrogen sulfide. Carbon disulfide was evolved when yeast cells were treated with captan. Captan also slowly decomposes in water to yield as primary products, tetrahydrophthalimide, thiophosgene, and HCl. These reactions are summarized below.

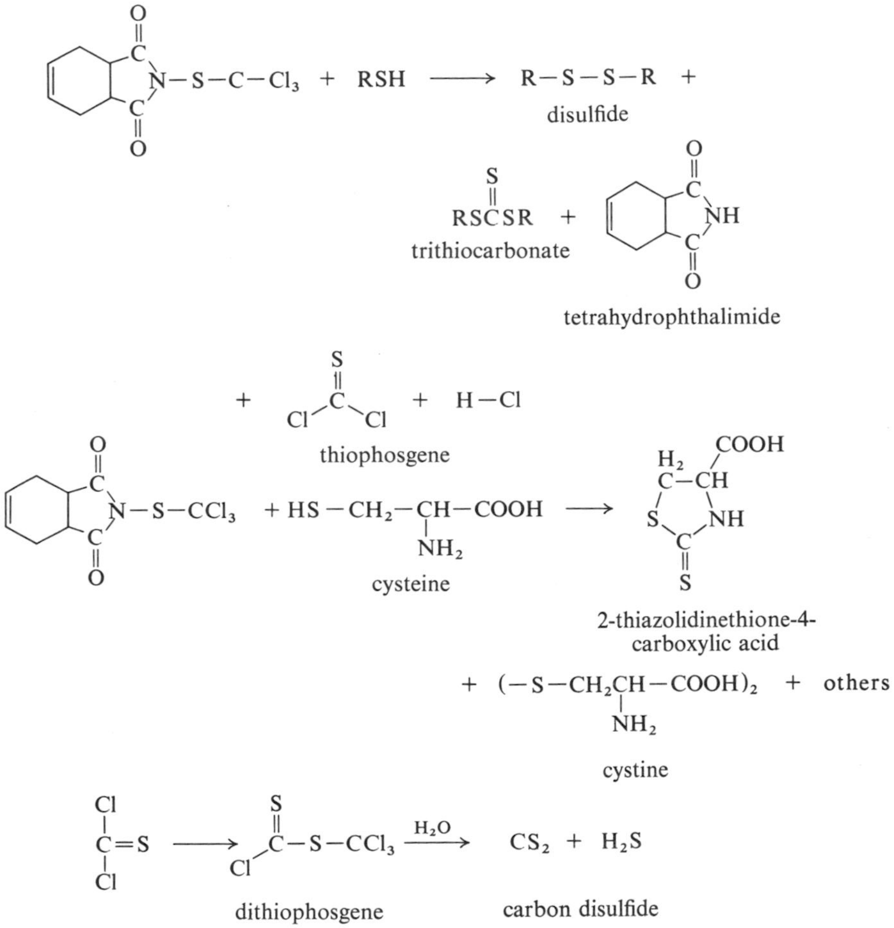

R—S—S—R +
disulfide

$$RS\overset{\text{S}}{\underset{\|}{C}}SR \quad +$$
trithiocarbonate

tetrahydrophthalimide

+ thiophosgene + H—Cl

+ HS—CH$_2$—CH—COOH ⟶
                    |
                   NH$_2$
cysteine

2-thiazolidinethione-4-carboxylic acid

$$+ \quad (-S-CH_2CH-COOH)_2 \quad + \quad \text{others}$$
                        |
                       NH$_2$

cystine

dithiophosgene   carbon disulfide

$$CS_2 + H_2S$$

D. Thiocarbamic Acid Derivatives

Metham (VAPAM), sodium N-methylthiocarbamate, is degraded in soil to methyl isothiocyanate and H_2S (339). One also expects that simple hydrolysis should form methyl amine and carbon disulfide.

Dimethyldithiocarbomates can not give isothiocyanates but should hydrolyze to give dimethylamine and carbon disulfide. The metabolism of sodium dimethyldithiocarbamate has been shown to give several conjugates (7, 8). In plants it was converted to 1-(dimethyldithio carbamoyl)-β-glucoside, β-(dimethylthiocarbamoylthio)-L-alanine, and a third yet unknown compound. All of these compounds were fungitoxic. A fourth, nonfungitoxic compound formed nonenzymatically was thiazolidine-2-thione-4-carboxylic acid. When sodium dimethyldithiocarbamate was incubated with yeast cells, two different metabolites, γ-(dimethylthiocarbamoylthio)-α-aminobutyric acid and γ-(dimethylthiocarbamoylthio)-α-ketobutyric acid were isolated.

The closely related ω-(N,N-dimethylthiocarbamoylthio)alkane carboxylic acids undergo two different reactions.

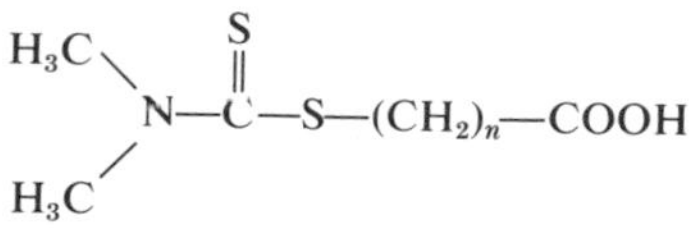

$$\underset{H_3C}{\overset{H_3C}{\diagdown}}N-\overset{\overset{S}{\|}}{C}-S-(CH_2)_n-COOH$$

N,N-dimethylthiocarbamoylthioacetic acid ($n = 1$) is converted by cucumbers and broad beans to the nonfungitoxic compound N-methylrhodamine shown below, in which the original fungicide has been demethylated and cyclized(*7, 8*).

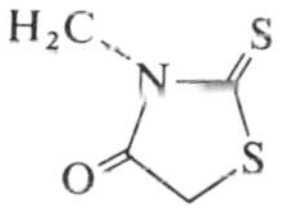

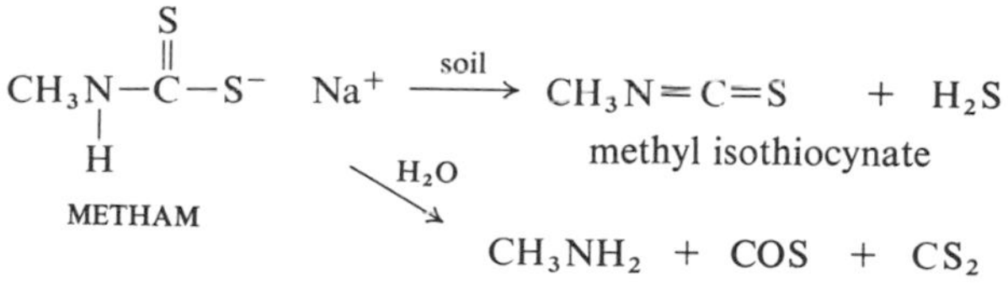

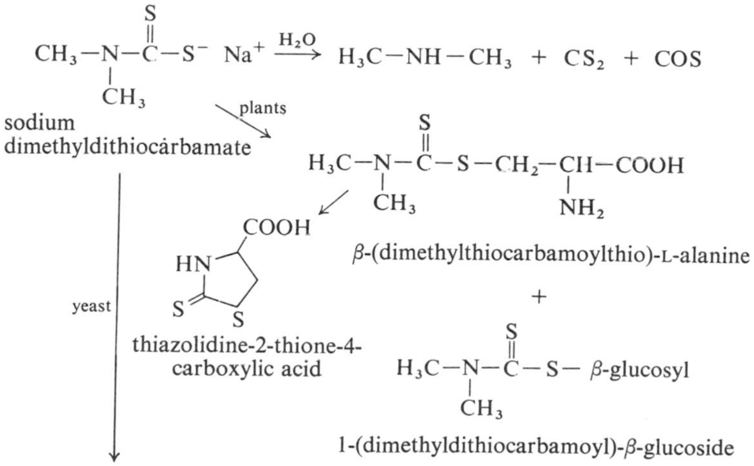

$$\underset{O}{\overset{H_2C}{\diagdown}}\overset{S}{\underset{S}{N-\|}}$$

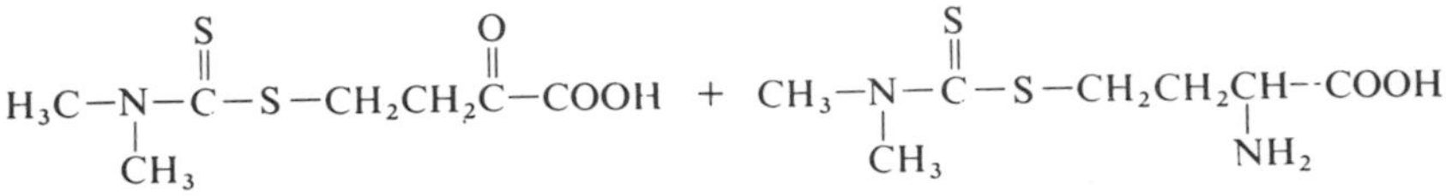

$$
\underset{\substack{|\\ CH_3}}{\overset{\overset{\displaystyle S}{\|}}{H_3C-N-C}}-S-CH_2CH_2\overset{\overset{\displaystyle O}{\|}}{C}-COOH \quad + \quad \underset{\substack{|\\ CH_3}}{\overset{\overset{\displaystyle S}{\|}}{CH_3-N-C}}-S-CH_2CH_2\underset{\substack{|\\ NH_2}}{CH}-COOH
$$

γ-(dimethylthiocarbamoylthio)-   γ-(dimethylthiocarbamoylthio)-
α-ketobutyric acid   α-aminobutyric acid

The reaction is specifically a demethylation since the diethyl compound was not de-ethylated and the ethyl methyl compound was only demethylated.

The second reaction that these compounds undergo is β-oxidation. That is the butyric acid derivative ($n = 3$) is converted to the corresponding acetic acid derivative ($n = 1$) (340).

Another group of thiocarbamic acid derivatives are the ethylene bis-dithiocarbamates (nabam = disodium salt, maneb = manganese salt, zineb = zinc salt). These compounds have a complex breakdown pattern that is qualitatively the same for each different salt. Nabam for instance undergoes hydrolysis in water to the dithiocarbamic acid which then may decompose by several pathways as given below (341–343).

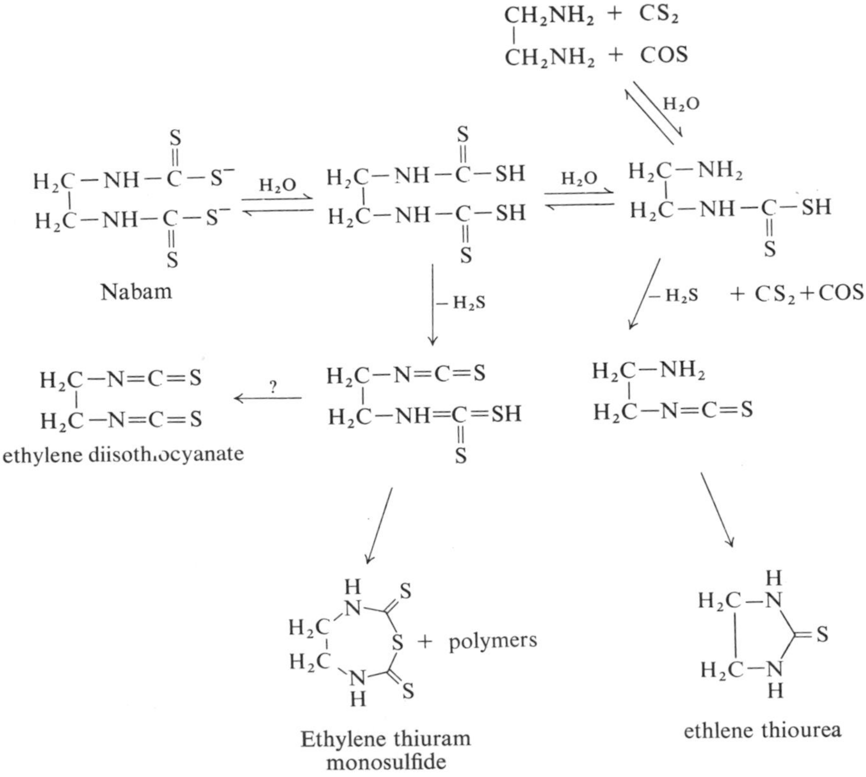

ethylene diisothiocyanate

Ethylene thiuram
monosulfide

ethlene thiourea

The metabolism of tetra-alkylthiuram disulfides is essentially similar to that of the corresponding dialkyldithiocarbamates. The disulfides react either by a free radical process or by nucleophilic substitution on sulfur, with mercaptans to form a new mixed disulfide plus a dialkyl-dithiocarbamate (8, 340).

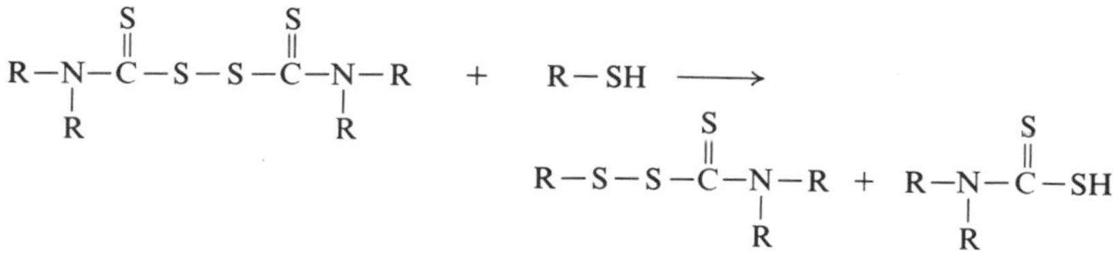

Dazomet (MYLONE), 3,5-dimethyltetrahydro-1,3,5,2H-thiadiazine-2-thione, depends upon hydrolytic breakdown for its fungitoxic action (344). The primary hydrolytic processes release a precursor of the proposed fungitoxic agent, methylisothiocynate.

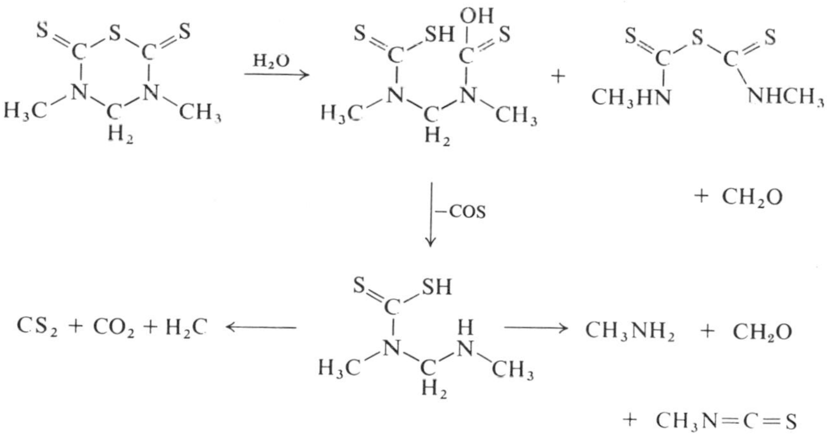

REFERENCES

1. C. M. Menzie, *Metabolism of Pesticides*, Special Scientific Report—Wildlife No. 96, U.S. Dept. Interior, May 1966.
2. F. P. W. Winteringham, *Studies in Comparative Biochemistry* (K. A. Munday, ed.), Pergamon, Oxford, England, 1965, pp. 107–152.
3. A. S. Perry, *J. Agr. Food Chem.*, **8**, 266 (1960).
4. R. L. Metcalf, *Scientific Aspects of Pest Control*, Publ. No. **1402**, National Academy of Sciences, National Research Council, 1966.
5. R. D. O'Brien, *Toxic Phosphorus Esters*, Academic, New York, 1960.
6. D. F. Heath, *Organophosphorus Poisons*, Pergamon, London, 1961.
7. A. K. Sijpesteijn and J. Kaslander, *Outlook on Agriculture*, **4**, 119 (1964).

8. A. K. Sijpesteijn and G. J. M. Van der Kerk, *Ann. Rev. Phytopathol.*, **3**, 127 (1965).
9. T. R. Fukuto, *Advan. Pest Control Res.*, Vol. I, Wiley (Interscience), New York, 1957, pp. 147–192.
10. W. Diggle and J. C. Gage, *Biochem. J.*, **49**, 491 (1951).
11. W. M. Diggle and J. C. Gage, *Nature*, **168**, 998 (1951).
12. R. L. Metcalf and R. B. March, *J. Econ. Entomol.*, **46**, 288 (1953).
13. R. M. Hollingworth, T. R. Fukuto, and R. L. Metcalf, *J. Agr. Food Chem.*, **15**, 235 (1967).
14. W. Chamberlain and W. M. Hoskins, *J. Econ. Entomol.*, **44**, 177 (1951).
15. J. C. Gage, *Biochem. J.*, **54**, 430 (1953).
16. J. Kubistova, *Arch. Intern. Pharmacodyn.*, **118**, 308 (1959).
17. W. A. Brindley and P. A. Dahm, *J. Econ. Entomol.*, **57**, 47 (1964).
18. P. A. Dahm, B. E. Kopecky, and C. B. Walker, *Toxicol. Appl. Pharmacol.*, **4**, 683 (1962).
19. T. Nakatsugawa and P. A. Dahm, *J. Econ. Entomol.*, **55**, 594 (1962).
20. T. Nakatsugawa and P. A. Dahm, *J. Econ. Entomol.*, **58**, 500 (1965).
21. T. L. Hopkins, *J. Econ. Entomol.*, **55**, 334 (1962).
22. S. D. Murphy and K. P. DuBois, *J. Pharmacol.*, **119**, 572 (1957).
23. A. N. Davison, *Biochem. J.*, **61**, 203 (1955).
24. B. E. Hietbrink and K. P. DuBois, *Radiation Res.*, **22**, 598 (1964).
25. R. A. Neal and K. P. DuBois, *J. Pharm. Exptl. Therap.*, **148**, 185 (1965).
26. R. D. O'Brien, *Biochem. J.*, **79**, 229 (1961).
27. R. D. O'Brien, *J. Econ. Entomol.*, **59**, 159 (1957).
28. J. Fukami and T. Shishido, *Botyu Kagaku*, **28**, 63 (1963).
29. G. Schmidt and M. Laskowski, Sr., *The Enzymes*, Vol. 5, 2nd ed. (P. D. Boyer, H. Lardy and K. Myrbäck, eds.), Academic, New York, 1961.
30. M. Cohn, *J. Cellular Comp. Physiol.*, **54**, Suppl. 1, 17 (1959).
31. B. B. Brodie, J. R. Gillette, and B. N. LaDu, *Ann. Rev. Biochem.*, **27**, 427 (1958).
32. W. N. Aldridge, *Biochem. J.*, **53**, 117 (1953).
33. A. R. Main, *Biochem. J.*, **74**, 10 (1960).
34. A. R. Main, *Biochem. J.*, **75**, 188 (1960).
35. E. G. Erdos and L. E. Boggs, *Nature*, **190**, 716 (1961).
36. L. A. Mounter, *The Enzymes*, Vol. **4**, 2nd ed., (P. D. Boyer, H. Lardy, and K. Myrbäck), Academic, New York, 1960.
37. J. E. Casida, *J. Agr. Food Chem.*, **4**, 772 (1956).
38. R. L. Metcalf, M. Maxon, T. R. Fukuto, and R. B. March, *Ann. Entomol. Soc. Am.*, **49**, 274 (1956).
39. H. R. Krueger and J. E. Casida, *J. Econ. Entomol.*, **54**, 239 (1961).
40. H. R. Krueger and R. D. O'Brien, *J. Econ. Entomol.*, **52**, 1063 (1959).
41. F. Matsumura and C. J. Hogendijk, *J. Agr. Food Chem.*, **12**, 447 (1964).
42. K. van Asperen and F. J. Oppenoorth, *Entomol. Exptl. Appl.*, **2**, 48 (1959).
43. K. van Asperen and F. J. Oppenoorth, *Entomol. Exptl. Appl.*, **3**, 68 (1960).
44. F. J. Oppenoorth and K. van Asperen, *Entomol. Exptl. Appl.*, **4**, 311 (1961).
45. F. W. Plapp and J. E. Casida, *J. Econ. Entomol.*, **51**, 800 (1958).
46. T. Shishido and J. Fukami, *Botyu Kagaku*, **28**, 69 (1963).
47. J. Fukami and T. Shishido, *Botyu Kagaku*, **28**, 77 (1963).
48. E. Hodgson and J. E. Casida, *J. Agr. Food Chem.*, **10**, 208 (1962).
49. J. W. Cook, *J. Agr. Food Chem.*, **5**, 859 (1957).
50. J. E. Pankaskie, F. C. Fountaine, and P. A. Dahm, *J. Econ. Entomol.*, **45**, 51 (1952).
51. M. K. Ahmed, J. E. Casida, and R. E. Nichols, *J. Agr. Food Chem.*, **6**, 740 (1958).

52. E. P. Lichtenstein and K. R. Schulz, *J. Econ. Entomol.*, **57**, 618 (1964).
53. D. D. Questel and R. V. Connin, *J. Econ. Entomol.*, **40**, 914 (1947).
54. M. M. Grange and R. W. Leiby, *Agr. Chem.*, **4**, (2), 34–35, 79–81, 83, 85 (1949).
55. O. Starnes, *J. Econ. Entomol.*, **43**, 338 (1950).
56. W. A. L. David and W. N. Aldridge, *Ann. Appl. Biol.*, **45**, 332 (1957).
57. J. B. Knaak, M. A. Stahmann, and J. E. Casida, *J. Agr. Food Chem.*, **10**, 154 (1962).
58. R. M. Hollingworth, R. L. Metcalf, and T. R. Fukuto, *J. Agr. Food Chem.*, **15**, 242 (1967).
59. R. M. Hollingworth, R. L. Metcalf, and T. R. Fukuto, *J. Agr. Food Chem.*, **15**, 250 (1967).
60. A. Vardanis and L. G. Crawford, *J. Econ. Entomol.*, **57**, 136 (1964).
61. G. Schrader, *Die Entwicklung neuer Insektizide auf Grundlage organischer Fluor – und Phosphorverbindungen*, Monographien 62, Parts 1 and 2, Weinheim/Bergstr., Germany, Verlag Chemie., 1952.
62. K. Gardner and D. F. Heath, *Anal. Chem.*, **25**, 1849 (1953).
63. T. R. Fukuto and R. L. Metcalf, *J. Am. Chem. Soc.*, **76**, 5103 (1954).
64. D. F. Heath, P. O. Park, L. A. Lickerish, and R. F. Edson, Pest Control Ltd. (Mimeo report) (1953).
65. R. L. Metcalf, R. B. March, T. R. Fukuto, and M. G. Maxon, *J. Econ. Entomol.*, **47**, 1045 (1954).
66. T. R. Fukuto, R. L. Metcalf, R. B. March, and M. G. Maxon, *J. Econ. Entomol.*, **48**, 347 (1955).
67. R. L. Metcalf, R. B. March, T. R. Fukuto, and M. G. Maxon, *J. Econ. Entomol.*, **48**, 364 (1955).
68. T. R. Fukuto, J. P. Wolf, III, R. L. Metcalf, and R. B. March, *J. Econ. Entomol.*, **49**, 147 (1956).
69. T. R. Fukuto, J. P. Wolf, III, R. L. Metcalf, and R. B. March, *J. Econ. Entomol.*, **50**, 399 (1957).
70. R. B. March, R. L. Metcalf, T. R. Fukuto, and M. G. Maxon, *J. Econ. Entomol.*, **48**, 355 (1955).
71. R. Mühlmann and H. Tietz, *Höfchen-Briefe*, **2**, 1 (1956).
72. D. F. Heath and M. Vandekar, *Biochem. J.*, **67**, 187 (1957).
73. T. R. Fukuto and R. L. Metcalf, *J. Am. Chem. Soc.*, **76**, 5103 (1954).
74. A. Henglein and G. Schrader, *Z. Naturforsch.*, **10B**, 12 (1955).
75. D. F. Heath, *J. Chem. Soc.*, **1958**, 1643.
76. R. L. Metcalf, T. R. Fukuto, and R. B. March, *J. Econ. Entomol.*, **50**, 338 (1957).
77. R. L. Metcalf, H. T. Reynolds, M. Y. Winton, and T. R. Fukuto, *J. Econ. Entomol.*, **52**, 435 (1959).
78. D. L. Bull, *J. Econ. Entomol.*, **58**, 249 (1965).
79. J. S. Bowman and J. E. Casida, *J. Agr. Food Chem.*, **5**, 192 (1957).
80. J. S. Bowman and J. E. Casida, *J. Econ. Entomol.*, **51**, 838 (1958).
81. M. K. Ahmed and J. E. Casida, *J. Econ. Entomol.*, **51**, 59 (1958).
82. J. J. Menn. *J. Econ. Entomol.*, **55**, 90 (1962).
83. D. E. Coffin, *J. Am. Assoc. Agr. Chem.*, **47**, 662 (1964).
84. J. E. Francis and J. M. Barnes, *Bull. World Health. Organ.*, **29**, 205 (1963).
85. H. Niessen, H. Tietz, and H. Frehse, *Pflanzenschutz-Nachrichten*, Bayer, **15**, 125 (1962).
86. R. L. Metcalf, T. R. Fukuto, and M. Y. Winton, *Bull. World. Health. Organ.*, **29**, 219 (1963).
87. C. Tomizawa, *Japan. J. Appl. Entomol. Zool.*, **6**, 237 (1962).

88. H. Fikudo, T. Masuda, and Y. Miyahara, *Japan. J. Appl. Entomol. Zool.*, **6**, 230 (1962).

89. T. R. Fukuto, E. O. Hornig, and R. L. Metcalf, *J. Agr. Food Chem.*, **12**, 169 (1964).

90. U. E. Brady and B. W. Arthur, *J. Econ. Entomol.*, **54**, 1232 (1961).

91. E. Benjamini, R. L. Metcalf, and T. R. Fukuto, *J. Econ. Entomol.*, **52**, 94 (1959).

92. E. Benjamini, R. L. Metcalf, and T. R. Fukuto, *J. Econ. Entomol.*, **52**, 99 (1959).

93. R. B. March, T. R. Fukuto, R. L. Metcalf, and M. G. Maxon, *J. Econ. Entomol.*, **49**, 185 (1956).

94. R. D. O'Brien, *J. Econ. Entomol.*, **50**, 159 (1957).

95. F. W. Seume and R. D. O'Brien, *J. Agr. Food Chem.*, **8**, 36 (1960).

96. J. P. Frawley, H. N. Fuyat, E. C. Hagan, J. R. Blake, and O. C. Fitzhugh, *J. Pharmacol. Exptl. Therap.*, **121**, 96 (1957).

97. J. B. Knaak and R. D. O'Brien, *J. Agr. Food Chem.*, **8**, 198 (1960).

98. R. D. O'Brien, W. C. Dauterman, and R. P. Niedermeier, *J. Agr. Food Chem.*, **9**, 39 (1961).

99. H. R. Krueger, R. D. O'Brien, and W. C. Dauterman, *J. Econ. Entomol.*, **52**, 25 (1960).

100. T. Uchida, W. C. Dauterman, and R. D. O'Brien, *J. Agr. Food Chem.*, **12**, 48 (1964).

101. D. L. Bull, D. A. Lindquist, and J. Hocskaylo, *J. Econ. Entomol.*, **56**, 129 (1963).

102. U. E. Brady and B. W. Arthur, *J. Econ. Entomol.*, **58**, 477 (1963).

103. T. Uchida, H. S. Rahmati, and R. D. O'Brien, *J. Econ. Entomol.*, **58**, 831 (1965).

104. D. M. Sanderson and E. F. Edson, *Brit. J. Ind. Med.*, **21**, 52 (1964).

105. J. N. Kaplanis, W. E. Robbins, D. I. Darson, D. E. Hopkins, R. E. Monroe, and G. Treiber, *J. Econ. Entomol.*, **52**, 1190 (1959).

106. W. C. Dauterman, J. E. Casida, J. B. Knaak, and T. Kowalczyk, *J. Agr. Food Chem.*, **1**, 188 (1959).

107. W. F. Chamberlain, P. E. Gatterdam, and D. E. Hopkins, *J. Econ. Entomol.*, **61**, 733 (1961).

107a. J. Hocskaylo and D. L. Bull, *J. Agr. Food Chem.*, **11**, 464, (1963).

108. W. C. Dauterman, G. B. Viado, J. E. Casida, and R. D. O'Brien, *J. Agr. Food Chem.*, **8**, 115 (1960).

109. R. Santi and R. Giacomelli, *J. Agr. Food Chem.*, **10**, 257 (1962).

110. H. Gysin and A. Margot, *J. Agr. Food Chem.*, **6**, 900 (1958).

111. A. Margot and H. Gysin, *Helv. Chem. Acta.*, **40**, 1562 (1957).

112. J. P. Vigne, R. L. Tabeau, J. Chouteau, and J. Fondarai, *Radioisotopes Sci. Research, Proc. Intern. Conf., Paris, 1957*, Vol. III, pp. 45–61.

113. W. E. Robbins, T. L. Hopkins, and G. W. Eddy, *J. Agr. Food Chem.*, **7**, 509 (1957).

114. J. W. Ralls, D. R. Gilmore, and A. Cortes, *J. Agr. Food Chem.*, **14**, 387 (1966).

114a. D. A. Lindquist, E. C. Burns, C. P. Pant, and P. A. Dahm, *J. Econ. Entomol.*, **51**, 204 (1958).

115. H. R. Krueger, J. E. Casida, and R. P. Niedermeier, *J. Agr. Food Chem.*, **7**, 182 (1959).

116. D. S. Vickery and B. W. Arthur, *J. Econ. Entomol.*, **53**, 1037 (1960).

117. J. N. Kaplanis, D. E. Hopkins, and G. H. Treiber, *J. Agr. Food Chem.*, **7**, 483 (1959).

118. W. E. Robbins, T. L. Hopkins, D. I. Darrow, and G. W. Eddy, *J. Econ. Entomol.*, **52**, 214 (1959).

119. R. D. O'Brien and L. S. Wolfe, *J. Econ. Entomol.*, **52**, 692 (1959).

120. R. D. Radeleff and H. V. Claborn, *J. Agr. Food Chem.*, **8**, 437 (1960).

121. H. W. Dorough, U. E. Brady, Jr., J. A. Timmerman, Jr., and B. W. Arthur, *J. Econ. Entomol.*, **54**, 25 (1961).

122. H. W. Dorough, U. E. Brady, Jr., J. A. Timmerman, Jr., and B. W. Arthur, *J. Econ. Entomol.*, **54**, 97 (1961).
123. S. D. Murphy and K. P. DuBois, *J. Pharm. Exptl. Therap.*, **119**, 572 (1957).
124. K. P. DuBois, D. R. Thursh, and S. D. Murphy, *J. Pharm. Exptl. Therap.*, **119**, 208 (1957).
125. T. Nakatsugawa and P. A. Dahm, *J. Econ. Entomol.*, **55**, 594 (1962).
126. L. J. Everett, C. A. Anderson, and D. MacDougall, *J. Agr. Food Chem.*, **14**, 47 (1966).
127. F. W. Plapp and J. E. Casida, *J. Agr. Food Chem.*, **6**, 662 (1958).
128. J. J. Menn and J. B. McBain, *J. Agr. Food Chem.*, **12**, 162 (1964).
129. J. M. Ford, J. J. Menn, and G. D. Meyding, *J. Agr. Food Chem.*, **14**, 83 (1966).
130. W. F. Chamberlain, *J. Econ. Entomol.*, **58**, 51 (1965).
131. J. J. Menn, J. B. McBain, B. J. Adelson, and G. G. Patchett, *J. Econ. Entomol.*, **58**, 875 (1965).
132. W. F. Barthel, B. H. Alexander, P. A. Giang, and S. A. Hall, *J. Am. Chem. Soc.*, **77**, 2424 (1955).
133. A. M. Mattson, J. T. Spillane, and G. W. Pearce, *J. Agr. Food Chem.*, **3**, 319 (1955).
134. W. Lorenz, A. Henglein, and G. Schrader, *J. Am. Chem. Soc.*, **77**, 2554 (1955).
135. R. L. Metcalf, T. R. Fukuto, and R. B. March, *J. Econ. Entomol.*, **52**, 44 (1959).
136. W. R. Robbins, T. L. Hopkins, and G. W. Eddy, *J. Econ. Entomol.*, **49**, 801 (1956).
137. B. W. Arthur and J. E. Casida, *J. Agr. Food Chem.*, **5**, 186–192 (1957).
138. J. Miyamoto, *Botyu Kagaku*, **24**, 130 (1959).
139. A. Hassan, S. M. A. D. Zayed, and F. M. Abdel-Hamid, *Biochem. Pharmacol.*, **14**, 1577 (1965).
140. S. M. A. D. Zayed and A. Hassan, *Can. J. Biochem.*, **43**, 1257 (1965).
141. I. Y. Mostafa, A. Hassan, and S. M. A. D. Zayed, *Z. Naturforsch.*, **20b**, 67 (1965).
142. A. Hassan and S. M. A. D. Zayed, *Can. J. Biochem.*, **43**, 1271 (1965).
143. A. Hassan, S. M. A. D. Zayed and S. Hashish, *Biochem. Pharmocol.*, **14**, 1692 (1965).
144. B. W. Arthur and J. E. Casida, *J. Agr. Food Chem.*, **6**, 360 (1958).
145. A. Hassan, S. M. A. D. Zayed, and F. M. Abdel-Hamid, *Can. J. Biochem.*, **43**, 1263 (1965).
146. E. Hodgson and J. E. Casida, *J. Agr. Food Chem.*, **10**, 208 (1962).
147. J. E. Casida, L. McBride, and R. P. Niedermeier, *J. Agr. Food Chem.*, **10**, 370 (1962).
148. G. K. Kohn, D. E. Pack, and J. N. Ospenson, 138th Meeting, American Chemical Society, New York, September, 1960, Abstracts, p. 9A.
149. A. R. Stiles, C. A. Reilly, G. R. Pollard, C. H. Tieman, L. F. Ward, Jr., D. D. Phillips, S. B. Soloway, and R. R. Whetstone, *J. Org. Chem.*, **26**, 3960 (1961).
150. T. R. Fukuto, E. O. Hornig, R. L. Metcalf, and M. Y. Winton, *J. Org. Chem.*, **26**, 4620 (1961).
151. J. E. Casida, P. E. Gatterdam, L. W. Getzin, Jr., and R. K. Chapman, *J. Agr. Food Chem.*, **4**, 236 (1956).
152. E. Y. Spencer and J. R. Robinson, *J. Agr. Food Chem.*, **8**, 293 (1960).
153. R. Anliker, E. Beriger, M. Geiger, and K. Schmid, *Helv. Chim. Acta.*, **44**, 1622 (1961).
154. R. Jacques and H. Beim, *Arch. Toxikol.*, **18**, 316 (1960).
155. R. E. Menzer and L. P. Ditman, *J. Agr. Food Chem.*, **11**, 130 (1963).
156. D. L. Bull and D. A. Lindquist, *J. Agr. Food Chem.*, **12**, 310 (1964).
157. R. E. Menzer and J. E. Casida, *J. Agr. Food Chem.*, **13**, 102 (1965).
158. W. E. Hall and Y. P. Sun, *J. Econ. Entomol.*, **58**, 845 (1965).
159. L. Shuster, *Ann. Rev. Biochem.*, **33**, 571 (1964).
160. D. L. Bull and D. A. Lindquist, *J. Agr. Food Chem.*, **14**, 105 (1966).
161. W. F. Chamberlain, *J. Econ. Entomol.*, **57**, 119 (1964).

161a. W. F. Chamberlain, *J. Econ. Entomol.*, **57** 329 (1964).

162. P. E. Gatterdam, M. W. Bullock, and W. H. Linkenheimer, Entomological Society of America National Meeting, St. Louis, December, 1963.

163. R. D. O'Brien, E. C. Kimmel, and P. R. Sferra, *J. Agr. Food Chem.*, **13**, 366 (1965).

164. B. W. Arthur and J. E. Casida, *J. Econ. Entomol.*, **52**, 20 (1959).

165. W. F. Chamberlain, P. E. Gatterdam, and D. E. Hopkins, *J. Econ. Entomol.*, **53**, 672 (1960).

166. G. J. Marco and E. G. Jaworski, *J. Agr. Food Chem.*, **12** 305 (1964).

167. R. D. O'Brien and E. Y. Spencer, *J. Agr. Food Chem.*, **1**, 946 (1953).

168. B. Kilbey, *Chem. Ind.*, **1953**, 856.

169. J. E. Casida and M. Stahmann, *J. Agr. Food Chem.*, **1**, 883 (1953).

170. R. D. O'Brien and E. Y. Spencer, *J. Agr. Food Chem.*, **3**, 56 (1955).

171. W. Ripper, R. Greenslade, C. Hartley, *Bull. Entomol. Res.*, **40**, 481 (1950).

172. J. E. Casida, R. K. Chapman, M. A. Stahmann, and T. C. Allen, *J. Econ. Entomol.*, **47**, 64 (1954).

173. D. F. Heath, D. W. J. Lane, and P. O. Park, *Phil. Trans.*, *Series B*, *239*, 191 (1951).

174. K. Cheng, *Brit. J. Exptl. Pathol.*, *32*, 444 (1951).

175. K. P. DuBois, J. Doull and J. M. Coon, *J. Pharmacol. Exptl. Therap.*, *99*, 376 (1950).

176. H. Tsuyuki, M. A. Stahmann, and J. E. Casida, *J. Agr. Food Chem.*, **3**, 922 (1955).

177. H. Tsuyuki, M. A. Stahmann, and J. E. Casida, *Biochem. J.*, **59**, IV (1955).

178. E. Y. Spencer, R. D. O'Brien and R. W. White, *J. Agr. Food Chem.*, **5**, 123 (1957).

179. G. S. Hartley, Intern. Congr. Pure and Appl. Chem., 15th New York, September, 1951, Section 13, Pesticides.

180. J. E. Casida, T. C. Allen, and M. A. Stahmann, *J. Am. Chem. Soc.*, **74**, 5548 (1952).

181. G. S. Hartley, *J. Soc. Chem. Ind.*, **1954**, 529.

182. E. Y. Spencer, *Chem. Soc., London, Spec. Publ.*, No. **8**, 171 (1957).

183. J. E. Casida, T. C. Allen, and M. A. Stahmann, *Nature*, **172**, 243 (1953).

184. W. R. Baïriedel and M. G. Swank, *J. Agr. Food Chem.*, **10**, 150 (1962).

185. J. R. Buttram and B. W. Arthur, *J. Econ. Entomol.*, **54**, 456 (1961).

186. U. E. Brady and B. W. Arthur, *J. Econ. Entomol.*, **55**, 833 (1962).

187. J. E. Casida and K.-B. Augustinsson, *Biochem. Biophys. Acta*, **36**, 411 (1959).

188. J. E. Casida and K.-B. Augustinsson, *J. Econ. Entomol.*, **53**, 205 (1960).

189. K.-B. Augustinsson and J. E. Casida, *Biochem. Pharmacol.*, **3**, 60 (1959).

190. E. Hodgson and J. E. Casida, *Biochem. Pharmacol.*, **8**, 179 (1961).

191. E. Hodgson and J. E. Casida, *Biochem. Biophys. Acta*, **42**, 184 (1960).

192. J. G. Krishna and J. E. Casida, *J. Agr. Food Chem.*, **14**, 98 (1966).

193. R. L. Metcalf, M. F. Osman, and T. R. Fukuto, *J. Econ. Entomol.*, **60**, 445 (1967).

193a. R. L. Metcalf and T. R. Fukuto, *J. Agr. Food Chem.*, **13**, 220 (1965).

194. M. E. Eldefrawi and W. M. Hoskins, *J. Econ. Entomol.*, **54**, 401 (1961).

195. H. W. Dorough and J. E. Casida, *J. Agr. Food Chem.*, **12**, 294 (1964).

196. N. C. Leeling and J. E. Casida, *J. Agr. Food Chem.*, **14**, 281 (1966).

196a. A. M. Abdel-Wahab, R. J. Kuhr, and J. E. Casida, *J. Agr. Food Chem.*, **14**, 290 (1966).

197. J. B. Knaak, M. J. Tallant, W. J. Bartley, and L. J. Sullivan, *J. Agr. Food Chem.*, **13**, 537 (1965).

198. E. A. Williams, R. W. Meikle, and C. T. Redemann, *J. Agr. Food Chem.*, **12**, 453 (1964).

199. E. A. Williams, R. W. Meikle, and C. T. Redemann, *J. Agr. Food Chem.*, **12**, 457 (1964).

200. J. A. Dawson, D. F. Heath, J. R. Rose, E. M. Thain, and J. B. Ward, *Bull. World Health Org.* **30**, 127 (1964).

201. M. Y. Zubairi and J. E. Casida, *J. Econ. Entomol.*, **58**, 403 (1965).
202. L. K. Payne, Jr., H. A. Stansbury, Jr., and M. H. J. Weidin, *J. Agr. Food Chem.*, **14**, 573 (1966).
203. R. L. Metcalf, T. R. Fukuto, C. Collins, K. Borck, J. Burk, H. T. Reynolds, and M. F. Osman, *J. Agr. Food Chem.*, **14**, 579 (1966).
204. J. B. Knaak, M. J. Tallant, and L. J. Sullivan, *J. Agr. Food Chem.*, **14**, 573 (1966).
205. A. W. A. Brown, *Ann. Rev. Entomol.*, **5**, 301 (1960).
206. C. W. Kearns in *Origins of Resistance to Toxic Agents*, Academic, New York, 1955.
207. F. P. W. Winteringham and J. M. Barnes, *Physiol. Revs.* **35**, 701 (1955).
208. R. L. Metcalf, *Organic Insecticides, Their Chemistry and Mode of Action*, Wiley (Interscience) New York, 1955.
209. J. Sternburg, C. W. Kearns, and W. Bruce, *J. Econ. Entomol.*, **43**, 214 (1950).
210. A. S. Perry and W. M. Hoskins, *Science*, **111**, 600 (1950).
211. A. S. Perry and W. M. Hoskins, *J. Econ. Entomol.*, **44**, 839 (1951).
212. A. S. Perry and W. M. Hoskins, *J. Econ. Entomol.*, **44**, 850 (1951).
213. R. B. March, *Bull. Natl. Res. Council, Publ.* **219**, 45 (1952).
214. F. P. W. Winteringham, *Bull. Natl. Res. Council, Publ.* **219**, 61 (1952).
215. T. Fletcher, *Trans. Roy. Soc. Trop. Med. Hyg.*, **46**, 6 (1952).
216. A. S. Tahori and W. M. Hoskins, *J. Econ. Entomol.*, **46**, 829 (1953).
217. W. M. Hoskins and J. M. Witt, Proc. Intern. Congr. Entomology, X^th, **2**, 151 (1958).
218. A. S. Perry, J. A. Jensen, and G. W. Pearce, *J. Agr. Food Chem.*, **3**, 1008 (1955).
219. L. C. Terriere and R. D. Schonbrod, *J. Econ. Entomol.*, **48**, 736 (1955).
220. J. Sternburg, C. W. Kearns, and H. J. Moorefield, *J. Agr. Food Chem.*, **2**, 1125 (1954).
221. J. Sternburg, E. B. Vinson, and C. W. Kearns, *J. Econ. Entomol.*, **46**, 513 (1953).
222. H. Lipke and C. W. Kearns, *J. Biol. Chem.*, **234**, 2123 (1959).
223. H. Lipke and C. W. Kearns, *Bull. Entomol. Soc. Am.*, **4**, 95 (1958).
224. A. W. A. Brown and A. S. Perry, *Nature*, **178**, 368 (1956).
225. T. Kimura and A. W. A. Brown, *J. Econ. Entomol.*, **57**, 710 (1964).
226. T. Kimura, J. R. Duffy and A. W. A. Brown, *Bull. World Health Organ.*, **32** (4), 557 (1965).
227. H. L. Bami, M. I. D. Sharma, and R. L. Kalra, *Bull. Natl. Soc. India Malaria Mosquito Borne Diseases*, **5**, 246 (1957).
228. H. Lipke and J. Chalkley, *Bull. World Health Organ.*, **30** (1), 57 (1964).
229. A. N. Chattoraj and C. W. Kearns, *Bull. Entomol. Soc. Am.*, **4**, 95 (1958).
230. D. A. Lindquist and P. A. Dahm, *J. Econ. Entomol.*, **49**, 579 (1956).
231. A. S. Perry, *Entomol. Soc. Am., Misc. Publ.*, **2**, 119 (1960).
232. M. Tsukamoto, *Botyu Kagaku*, **24** (3), 141 (1959).
233. M. Tsukamoto, *Botyu Kagaku*, **26** (3), 74 (1961).
234. D. B. Menzel, S. M. Smith, R. Miskus, and W. M. Hoskins, *J. Econ. Entomol.*, **54**, 9 (1961).
235. W. E. Robbins and P. A. Dahm, *J. Agr. Food Chem.*, **3**, 500 (1955).
236. M. Agosin, D. Michaeli, R. Miskus, S. Nagasawa, and W. M. Hoskins, *J. Econ. Entomol.*, **54**, 340 (1961).
237. W. J. Hayes, Jr., "Pharmacology and toxicology of DDT," in *Das Insektizid Dichlorophenyl-trichloroathan and Seine Bedentung* (P. Muller, ed.), Verlag Birkhauser, Basel, Switzerland.
238. M. I. Smith and E. F. Stohlman, *U.S. Public Health Repts.*, **59**, 984 (1944).
239. E. P. Lang, *J. Pharmacol. Exptl. Therap.*, **121**, 55 (1957).
240. M. I. Smith and E. F. Stohlman, *U.S. Public Health Repts.*, **60**, 289 (1945).
241. W. C. White and T. R. Sweeney, *U.S. Public Health Repts.*, **60**, 66 (1946).

242. P. A. Neal, T. R. Sweeney, S. S. Spicer, and W. F. von Oettingen, *U.S. Public Health Repts.*, **61**, 403 (1946).
243. E. C. Burns, P. A. Dahm, and D. A. Lindquist, *J. Pharmacol. Exptl. Therap.*, **121**, 55 (1957).
244. J. A. Jensen, C. Cueto, W. E. Dale, C. F. Rothe, G. W. Pearce, and A. M. Mattson, *J. Agr. Food Chem.*, **5**, 919 (1957).
245. J. E. Peterson and W. H. Robison, *Toxicol. Appl. Pharmacol.*, **6**, 321 (1964).
246. J. H. V. Stenersen, *Nature*, **207**, 660 (1965).
247. B. I. Kallman and A. K. Andrews, *Science*, **141**, 1050 (1963).
248. J. L. Mendel and M. S. Walton, *Science*, **151**, 1527 (1966).
249. R. P. Miskus, D. P. Blair, and J. E. Casida, *J. Agr. Food Chem.*, **13**, 481 (1965).
250. P. R. Datta, E. P. Lang, and A. P. Klein, *Science*, **145**, 1052 (1964).
251. P. J. Bunyan, J. M. J. Page, and A. Taylor, *Nature*, **210** 1048 (1966).
252. L. F. Stickel, W. H. Stickel, and R. Christensen, *Science*, **151**, 1549 (1966).
253. F. R. Bradbury, P. Nield, and J. F. Newman, *Nature*, **172**, 1052 (1953).
254. K. van Asperen and F. J. Oppenoorth, *Nature*, **173**, 1000 (1954).
255. F. R. Bradbury and H. Standen, *J. Sci. Food Agr.*, **7**, 389 (1956).
256. F. J. Oppenoorth, *Nature*, **175**, 124 (1955).
257. J. R. Busvine and M. G. Townsend, *Bull. Entomol. Res.*, **53**, 763 (1963).
258. J. Sternburg and C. W. Kearns, *J. Econ. Entomol.*, **49**, 548 (1956).
259. R. G. Bridges, *Nature*, **184**, 1337 (1959).
260. F. R. Bradbury and H. Standen, *J. Sci. Food Agr.*, **9**, 203 (1958).
261. F. R. Bradbury and H. Standen, *Nature*, **183**, 983 (1959).
262. S. J. Cristol, *J. Am. Chem. Soc.*, **69**, 338 (1947).
263. K. C. Kauer, R. D. DuVall, and F. N. Alquist, *Ind. Eng. Chem.*, *Intern. Ed.* **39**, 1335 (1947).
264. E. D. Hughes, C. K. Ingold, and R. Pasternak, *J. Chem. Soc.*, **1953**, 3832.
265. F. R. Bradbury, *J. Sci. Agr. Food*, **8**, 90 (1957).
266. A. G. Clark, M. Hitchcock, and J. N. Smith, *Nature*, **209**, 103 (1966).
267. M. Ishida and P. A. Dahm, *J. Econ. Entomol.*, **58**, 383 (1965).
268. M. Ishida and P. A. Dahm, *J. Econ. Entomol.*, **58**, 602 (1965).
269. W. Koransky, J. Portig, and G. Munch, *Arch. Exptl. Pathol. Pharmakol.*, **244**, 564 (1963).
270. W. Koransky, J. Portig, H. W. Vohland, and I. Klempau, *Arch. Exptl. Pathol. Pharmakol.* **247**, 49 (1964).
271. P. L. Grover and P. Sims, *Biochem. J.*, **96**, 521 (1965).
272. W. R. Jondorf, D. V. Parke, and R. T. Williams, *Biochem. J.*, **61**, 512 (1955).
273. P. Sims and P. L. Grover, *Biochem. J.*, **95**, 156 (1965).
274. O. Gianotti, R. L. Metcalf, and R. B. March, *Ann. Entomol. Soc. Am.*, **49**, 588 (1956).
275. G. T. Brooks, *Nature*, **186**, 96 (1960).
276. G. T. Brooks and A. Harrison, *Nature*, **198**, 1169 (1963).
277. N. W. Earle, *J. Agr. Food Chem.*, **11**, 281 (1963).
278. A. S. Perry, G. W. Pearce, and A. J. Buckner, *J. Econ. Entomol.*, **57**, 867 (1964).
279. T. Nakatsugawa, M. Ishida, and P. A. Dahm, *Biochem. Pharmacol.*, **14**, 1853 (1965).
280. C. A. Edwards, S. D. Beck, and E. P. Lichtenstein, *J. Econ. Entomol.* **50**, 622 (1957).
281. N. Gannon and J. H. Bigger, *J. Econ. Entomol.*, **51**, 1 (1958).
282. W. B. Bollen, J. E. Roberts, and H. E. Morrison, *J. Econ. Entomol.*, **51**, 214 (1958).
283. E. P. Lichtenstein and K. R. Schulz, *J. Econ. Entomol.*, **53**, 192 (1960).
284. F. Korte, G. Ludwig, and J. Vogel, *Ann. Chem.*, **656**, 135 (1962).
285. N. Gannon and G. C. Decker, *J. Econ. Entomol.*, **51**, 8 (1958).

286. E. S. Oonnithan and R. Miskus, *J. Econ. Entomol.*, **57**, 425 (1964).
287. J. M. Bann, T. J. Decino, N. W. Earle, and Y. P. Sun, *J. Agr. Food Chem.*, **4**, 937 (1956).
288. D. F. Heath and M. Vandekar, *Brit. J. Ind. Med.*, **21**, 269 (1964).
289. G. Ludwig, J. Weiss, and F. Korte, *Life Sci.*, **3**, 123 (1964).
290. F. Korte and H. Arent, *Life Sci.*, **4**, 2017 (1965).
291. C. Cueto, Jr., and W. J. Hayes, Jr., *J. Agr. Food Chem.*, **10**, 366 (1962).
292. R. C. Cookson and E. Crundwell, *Chem. Ind. (London)*, (1958) 1004.
293. R. A. Hoffman and A. W. Lindquist, *J. Econ. Entomol.*, **45**, 233 (1952).
294. G. T. Brooks and A. Harrison, *Biochem. J.*, **87**, 5P (1963).
295. N. H. Poonawalla and F. Korte, *Life Sciences*, **3**, 1497 (1964).
296. A. S. Perry, A. M. Mattson, and A. J. Buckner, *J. Econ. Entomol.*, **51**, 364 (1958).
297. M. C. Bowman, F. Acru, Jr., C. S. Lofgren, and M. Beroza, *Science*, **146**, 1480 (1964).
298. B. Davidow and J. L. Radomski, *J. Pharmacol. Exptl. Therap.*, **107**, 259 (1953).
299. B. Davidow, J. L. Radomski, and R. Ely, *Science*, **118**, 383 (1953).
300. J. Radomski and B. Davidow, *J. Pharmacol. Exptl. Therap.*, **107**, 266 (1953).
301. N. Gannon and G. C. Decker, *J. Econ. Entomol.*, **51**, 3 (1958).
302. F. Korte and M. Stiasni, *Ann. Chem.*, **673**, 146 (1964).
303. S. E. Forman, A. J. Durbetaki, M. V. Cohen and R. A. Olofson, *J. Org. Chem.*, **30**, 169 (1965).
304. C. C. Cassil and P. E. Drummond, *J. Econ. Entomol.*, **58**, 356 (1965).
305. W. W. Barnes and G. Ware, *J. Econ. Entomol.*, **58**, 286 (1965).
306. K. Ballschmiter and G. Tolg, *Angew. Chem., Internat. Ed.* Engl., **5**, 730 (1966).
307. C. B. Gnadinger, *Pyrethrum Flowers*, 2nd ed., McLaughlin, Ghormley and King, Minneapolis, Minn., 1936.
308. F. Tattersfield and J. T. Martin, *J. Agr. Sci.*, **24**, 598 (1934).
309. W. O. Negherborn, *Handbook of Toxicology*, Vol. III, *Insecticides*, W. B. Saunders Co., Philadelphia and London, 1959.
310. M. C. Swingle, *J. Econ. Entomol.*, **27**, 1101 (1934).
311. P. A. Woke, *J. Agr. Res.*, **58**, 289 (1939).
312. R. W. Chamberlain, *Am. J. Hyg.*, **52**, 153 (1952).
313. M. I. Zeid, P. A. Dahm, R. E. Hein, and R. H. McFarland, *J. Econ. Entomol.*, **46**, 324 (1953).
314. F. P. W. Winteringham, A. Harrison, and P. M. Bridges, *Biochem. J.*, **61**, 359 (1955).
315. R. L. Metcalf, T. R. Fukuto, C. F. Wilkinson, M. H. Fahmy, S. A. El-Aziz, and E. R. Metcalf, *J. Agr. Food Chem.*, **14**, 555 (1966).
316. S. C. Chang and C. W. Kearns, *J. Econ. Entomol.*, **57**, 397 (1964).
317. T. L. Hopkins and W. E. Robbins, *J. Econ. Entomol.*, **50**, 684 (1957).
318. P. M. Bridges, *Biochem. J.*, **66**, 316 (1957).
319. A. M. Ambrose, *Toxicol Appl. Pharmacol.*, **5**, 414 (1963).
320. A. M. Ambrose, *Toxicol Appl. Pharmacol.*, **6**, 112 (1964).
321. M. S. Masri, F. T. Jones, R. E. Lundin, G. F. Bailey, and F. DeEds, *Toxicol. Appl. Pharmacol.* **6**, 711 (1964).
322. F. E. Guthrie, R. L. Ringler, and T. G. Bowery, *J. Econ. Entomol.*, **50**, 821 (1957).
323. A. Ganz, F. E. Kelsey, and E. M. Geiling, *J. Pharmacol. Exptl. Therap.*, **103**, 209 (1951).
324. D. R. Bennett, R. E. Tedeschi, and P. S. Larson, *Arch. Intern. Pharmacodyn.*, **98**, 221 (1954).
325. H. McKennis, Jr., L. B. Turnbull, and E. R. Bowman, *J. Am. Chem. Soc.*, **79**, 6342 (1957).

326. H. McKennis, Jr., L. B. Turnbull, E. R. Bowman, and E. Wada, *J. Am. Chem. Soc.*, **81**, 3951 (1959).

327. H. McKennis, Jr., L. B. Turnbull, S. L. Schwartz, E. Tamaki, and E. R. Bowman, *J. Biol. Chem.*, **237**, 541 (1962).

328. H. McKennis, Jr., E. R. Bowman and L. B. Turnbull, *J. Am. Chem. Soc.*, **84**, 4598 (1962).

329. H. McKennis, Jr., S. L. Schwartz, L. B. Turnbull, E. Tamaki, and E. R. Bowman, *J. Biol. Chem.*, **239**, 3981 (1964).

330. H. McKennis, Jr., S. L. Schwartz, and E. R. Bowman, *J. Biol. Chem.*, **239**, 3990 (1964).

331. T. Kikal and J. N. Smith, *Biochem. J.*, **71**, 48 (1959).

332. J. N. Smith, R. H. Smithies, and R. T. Williams, *Biochem. J.*, **54**, 225 (1953).

333. H. L. Jensen and K. Gundersen, *Nature*, **175**, 341 (1955).

334. J. Dekker and A. J. P. Oort, *Phytopathology*, **54**, 815 (1964).

335. D. Rudd Jones and J. Wignall, *Nature*, **175**, 207 (1955).

336. S. H. Crowdy and D. Rudd Jones, *J. Exptl. Botany*, **9**, 220 (1958).

337. R. J. Lukens and H. D. Sisler, *Phytopathology* **48**, 235 (1958).

338. R. G. Owens and G. Black, *Contrib. Boyce Thompson Inst.*, **20**, 475 (1960).

339. D. E. Munneke, K. H. Domsch, and J. W. Eckert, *Phytopathology*, **52**, 1298 (1962).

340. J. L. Garraway, *Chem. Ind.*, **1965**, 1880.

341. W. Moje, D. E. Munneke, and L. T. Richardson, *Nature*, **202**, 831 (1964).

342. R. A. Ludwig, G. D. Thorn, and C. H. Unwin, *Can. J. Botany*, **33**, 42 (1955).

343. A. K. Sijpesteijn and G. J. M. Van der Kerk, *Biochem. Biophys. Acta.*, **13**, 545 (1954).

344. D. E. Munneke and J. P. Martin, *Phytopath.*, **54**, 941 (1964).

Chapter 3

METABOLISM OF HERBICIDES

*J. E. Loeffler and J. van Overbeek**

SHELL DEVELOPMENT COMPANY
BIOLOGICAL SCIENCES RESEARCH CENTER
MODESTO, CALIFORNIA

3-1   INTRODUCTION

The use of herbicides, i.e., chemical compounds that more or less selectively interfere with the growth of uneconomical plant species ("weeds"), is an integral part of modern agricultural practice. Although financial losses attributable to weeds are not so eminently and directly perceptible as those caused by insects, they are just as real. According to USDA estimates, the annual loss through weeds on farm land alone— not including range land—was close to four billion dollars in 1954(*1*).

*Present address: Texas A & M University, Institute of Life Science, College Station, Texas.

A more vivid impression can be gained by another estimate: without weed control, corn yields would be reduced to less than 15% of the yields in weed-free fields(2). Even if this estimate should be somewhat optimistic, the need for development of ever more effective herbicides is obvious.

One of the major difficulties in the development of herbicides for use in agriculture is the problem of species selectivity. Crop plants and weeds are growing side by side and a herbicide must not injure the crop plant, but it should destroy or suppress all unwanted species of weeds even if they are closely related to the economic species.

Selectivity can sometimes be achieved by proper timing of the application and/or placement of the herbicide. A safer approach to selectivity would be to make use of characteristic differences in the metabolism of a crop plant compared to that of the undesirable species. Simazine or atrazine, which can be safely used for weed control in corn, provide an example of one approach to selectivity; only the economic plant transforms a biologically active compound into a nonherbicidal form(3). The other alternative would be a chemical that is inactive as such but that can be transformed into a herbicide only by weeds but not by the crop plant. An example of this is 2,4-DB (2,4-dichlorophenoxybutyric acid), which can be safely used to control dicotyledonous (broadleaf) weeds in some legumes such as certain clovers and alfalfa. Although these legumes are broadleaves also, they have one metabolic peculiarity: they cannot perform the process of β-oxidation sufficiently fast to generate toxic concentrations of the herbicide 2,4-D (2,4-dichlorophenoxyacetic acid) from the corresponding butyric acid. Many of the important weeds in these crops can and do perform this reaction quite rapidly so that they produce, by their own metabolic activity, enough of the toxicant to kill them(4).

This aspect of selective toxicity alone would supply compelling reasons for studying the fate of herbicidal chemicals in a wide variety of both susceptible and resistant plant species. The rational use of herbicides for crop protection involves, however, more than just the inter-relationship between herbicides and plants. It is equally important that the consumer of a treated crop be protected from ingesting possible harmful amounts of toxic compounds. It is thus necessary to perform residue analysis in order to know the amount of a crop-protecting agent present as such in a particular crop at the time of harvest. This is elementary, although the difficulties encountered in the analytical determination of such residues in concentrations in the order of parts per million or even parts per billion may be formidable. It is also necessary to know whether compounds derived from the applied herbicide are present in the crop.

Such derivatives can be formed by a variety of processes. They may result from the metabolic activities of the plant, or from nonenzymatic reactions with plant constituents. The herbicide may be transformed outside the plant via photochemical reactions, by chemical reactions with components of the soil, or by microorganisms, and the transformation products may be taken up by the plant and may be modified there further. The term "metabolism studies," which is frequently used for investigations concerned with the fate of a chemical in organisms, is therefore too narrow if one includes all these factors. Calling such investigations "transformation studies" as has been suggested by some authors would be more descriptive. The term "transformation" is used in this chapter for the description of any type of change of the original molecule, while the term "metabolism" is restricted to those transformations that actually have been proved to be attributable to metabolic activities of the treated plant.

Organic herbicides, like any other organic compound, are thermodynamically unstable in an oxygen-containing environment above the temperature of absolute zero. The actual degree of stability varies with the molecular structure and depends on the physical and chemical environment. Information on the chemical stability or reactivity of a herbicide is important from many different points of view. The manufacturer has an understandable interest in the chemical stability of the compound as such. The shelf life has to be long enough so that the product will reach the user before it decomposes. The manufacturer has to be sure also that the formulation and the material used for application have no detrimental effect on the activity of the compound, and from his knowledge will counsel the formulator and applicator with regard to recommended procedures. The purity of the technical product and the amount and type of possible by-products from the manufacturing process have to be assessed, and information has to be obtained on the possible toxicity of such by-products either to crops or to animals, including man. Similar information has to be collected with regard to formulations, including a check on activating or synergistic effects attributable to ingredients present in the formulated product. Such effects would, of course, have an effect on the recommended dosage.

Compatibility with other agricultural chemicals, fertilizers, or pesticides has to be established long before a new product actually comes on the market. Similarly, the effects of heat, moisture, sunlight, soil, microorganisms, and plants are investigated at early stages in the development of a new herbicide. The amount and the nature of transformation products, if any are formed, are determined and their behavior toward plants and animals is established. Answers have to be obtained for questions like:

"Are transformation products, if formed outside the plant, taken up by the crop? Are they further transformed within the plant? What is the toxicity of stable transformation products to various life forms?" If the crop or part of it is used as feeds, there is the additional question of metabolism by the animal of transformation products present in the crop.

This line of investigation has, by necessity, to cease before answers for all possible queries can be obtained, not because the research chemists, analysts, biochemists, physiologists, pathologists, etc., do not wish to pursue it any further, but because of the physical limitations encountered. The number of transformation products formed becomes larger and larger with each additional step and, even more important, the amount of each of these products becomes too small to be detected by even the most sophisticated modern methods of analysis. Aside from the logical problem of when to decide to stop investigating transformation products of transformation products, there is a final barrier of sensitivity of available methods and techniques beyond which no study is possible. This limit it not absolute; it depends on the state of the art, both with regard to the actual determination of the pure compound itself and to the degree to which it is possible to separate the compound from a tremendous excess of all the other organic compounds that constitute the organisms of interest. At present, the limit of detection lies—for most practical purposes—between one and one hundred parts per billion. Figures like these, although they are used routinely, are so far outside the range of everyday experience that one needs some comparison to get an approximate idea of what they mean. One such comparison: You, the reader, and the two authors of this chapter represent about one part per billion of the total human population of the earth.

3-2 PRINCIPLES AND PROBLEMS OF TRANSFORMATION STUDIES

At this point it may be well to stop for a moment and examine the premises on which transformation and residue studies are based. Knowledge of the metabolic differences that do or may exist in different plant groups, families, species, and even varieties is of importance if we pursue the goal of making herbicides available that are selectively destroyed by the crop plant. This desirable goal has been expressed by, among others, the chairman of the 1964 Corvallis symposium on metabolism of herbicides, who stated in his introduction: "Unquestionably, as our knowledge of this field matures, we shall be seeking molecules that may be selectively metabolized by the crop we wish to protect, thus virtually eliminating residue problems" (5).

Unfortunately, so far, most of the work in this field has been done with only a very limited number of plant species, so that no systematic comparison is possible between the metabolic capacities and capabilities of different species. Besides, many of the studies have not been extended beyond determination of rate of disappearance or inactivation of a herbicide. They were mainly conducted for practical purposes, to make sure that not enough of a particular herbicide is left in the field to interfere with the growth of next year's crop and to ascertain that any organic or inorganic compound applied to a crop, for a particular reason, does not reach the consumer in amounts large enough to affect his health.

There are two major problems that confront both regulatory agencies and the pesticide industry. One is how to decide on a safety factor that can be applied to either chronic or acute toxicity data in order to recommend maximum tolerable amounts of the pesticide itself and of its transformation products. The other problem, intimately connected with the first one, concerns the question of how much information should be available before such a decision is made. Ideally, this is not difficult: Get all the information that can be obtained at the present level of scientific knowledge. In practice, one has to deal realistically with the fact that only a finite amount of manpower is available. Can it be justified to spend 10, 20, or more man-years of industrial or government research workers on the elucidation of structure and biological activity of minor metabolites if the parent compound itself has only a negligible toxicity, particularly if, based on existing knowledge, one cannot expect that any of the known transformation processes would cause an increase in toxicity? Is it justifiable to draw this intellectual potential away from more meaningful investigations with far greater potentialities for the benefit of society? There is, of course, no formula to answer these questions. As is the case with most problems in our world, a compromise must be sought. The solution must be based upon facts interpreted to the best knowledge of competent workers in the areas involved, and it must be free from emotions and subjectivity.

Obviously, the current standard of living bears upon the standards adopted for the quality of food and the amounts of residues tolerated. As Dr. Henry Hurtig pointed out in a scholarly address before the Fifth International Pesticide Congress, "In those countries that no longer have to be concerned themselves with the meaning of food shortages and hunger of their populations, there will be a more sophisticated demand for knowledge of the dynamics of pesticide behavior in defining how pesticides can be used in the best interests of the community as a whole" (6). The United States fits the qualifications mentioned by Dr. Hurtig, and both industry and regulatory agencies are spending increasing amounts of effort to gain increased knowledge in the field of pesticide transformations. It should, however, be made very clear that it is practically impossible to prove that there is no effect—beneficial or detrimental—on the metabolism of any individual person exerted by any minute amount of a compound taken into his system. It is a well known fact that many constituents in our average diet, when ingested in large amounts, will cause illness or death. Ethyl alcohol, Vitamin D, raw egg white, or plain table salt are common examples. The alkaloids in lettuce, or the cyanide-producing glycosides of many plants (bitter almonds, sorghum) are other examples of compounds which, if ingested with food or feeds in amounts exceeding a certain threshold value, may produce detrimental effects. The point is that the animal organism is able to excrete, or modify into an excretable form, small amounts of virtually any chemical compound. The source where the compound came from, whether it was produced by a living cell, in a lysing bacterium, or in a laboratory, is immaterial. The organism cannot, however, cope fast enough with large quantities of chemical compounds, not even of those that are indispensable—in small amounts—for sustaining life. The amount that can be handled by the body metabolism is not the same for all compounds. It is not the same for each individual; it depends on the state of health, the diet, physical activities, exposure to light, or practically any hereditary or environmental influence.

Since it would not only be impossible but also meaningless to test every compound on every individual, some forms of limited tests were devised to find the concentration that, on the average, can be tolerated—or that which can not. The LD_{50} value, for example, denotes that amount of a chemical that will cause 50% of an experimental group of animals to die within a given time. By determining similarly LD_{25}, LD_{10}, and LD_5 values, the upper limit of tolerated dose with a given probability of survival can be estimated by extrapolation for a given species. Another set of data can be obtained from residue analysis, which provides information on the amount of herbicide present in the crop after harvest. Multiplication of this amount with the average daily intake of the particular food derived from the treated crop gives the amount of herbicide that might actually be ingested. A large multiple of this calculated amount is then used in studies to detect chronic toxicity in a number of laboratory animals. More and more emphasis is placed on the determination of the "no-biological effect" level by feeding increasing concentrations of the compound under investigation. All biological effects are considered, e.g., growth, reproduction, pathological changes in organs, and the possibility of effects carried over to future generations. A thorough discussion of these aspects of metabolism studies is presented elsewhere in this treatise.

Results from these and similar studies will enable regulatory agencies to come to decisions on recommended use on certain crops, tolerated residue limits, and so on. There is no doubt that information about the possible transformations of a compound is one of the prerequisites for such decisions. It might therefore be worthwhile to consider some of the problems that confront the chemist engaged in transformation studies, the tools at his disposal, and the limitations of his methods, in order to find out what questions can be asked to get meaningful answers.

3-3 METHODS AND TECHNIQUES

A. Isotope Labeling

In order to follow the path and the fate of a compound in the tremendously complex mixture with other, often rather similar compounds present in a living organism, the compound of interest has to be tagged with a marker by which it, or an essential part of it, can be recognized. The marker should interfere as little as possible with the behavior of the compound, i.e., the compound carrying a marker should undergo qualitatively and, at least within a tolerable limit, also quantitatively, the same reactions as the same compound without the marker. The best, in fact the only, possibility of achieving this consists in substitution of one or more atoms in the molecule by isotopic atoms, that is, atoms which are chemically identical—having the same distribution of electrons surrounding the nucleus—but having a nucleus of different mass. There are two types of isotopes available. Stable isotopes, which can be distinguished from the average atom population only by their different weight, and unstable, radioactive isotopes which, because their nucleus is inherently unstable, will spontaneously change into, or "decay" to a more stable configuration. The isotopes most commonly used in the study of biochemical pathways are β-emitters, which emit negatively charged electrons during their decay. Their half-lives, i.e., the time in which half of the radioactive atoms present in a sample will decay into a more stable form, vary from isotope to isotope. Of isotopes frequently used in metabolism studies, chlorine-36 has a half-life of 308,000 years, carbon-14: 5600 years, hydrogen-3 (tritium): 12.46 years, sulfur-35: 86.7 days, and phosphorus-32 only 14.3 days. Corrections for changes in concentration attributable to decay have to be applied in biological work only in those experiments involving phosphorus-32, and in some experiments of longer duration with sulfur-35. Stable isotopes like nitrogen-15 and oxygen-18, two elements of importance in life processes for which no radioactive isotopes with sufficiently long half-lives are available, are hardly used in metabolism studies. The reason for this is that stable isotopes can be detected and determined only with the aid of a mass spectrograph. This is a rather expensive instrument which, unfortunately, is not suited for the detection of unknown compounds in the complex mixtures one obtains by extraction of biological material. Radioactive isotopes on the other hand can be detected relatively easily by nondestructive means with rather simple instruments. Radioactivity is measured in units of curies or, more frequently, as used in biological experiments, in millicuries, abbreviated as mCi. One millicurie denotes that amount of radioactive material such that 2.22×10^9 atoms will disintegrate every minute. The specific activity, expressed in millicuries per millimole, of a compound labeled with carbon-14 is usually in the order of magnitude of 0.1 to 10 mCi/mmole. A conventional GM (Geiger Mueller) tube, which is used in inexpensive apparatus for detection of radioactivity, has a detector efficiency of a few per cent. A gasflow GM counter with an ultra-thin window can pick up about 10 to 20% of all disintegrating atoms and up to about 80% of disintegrating atoms can be detected by a liquid scintillation detector, (in which the energy of the emitted electrons is converted into light, the amount of which is then measured after amplification by a photomultiplier tube). Since 10 disintegrations per minute above the natural background radiation can be measured with a reasonably small error, provided the necessary corrections are made for self-absorption, quenching, and so on, it follows that as little as 1 nanogram (10^{-9} g) of a compound with a molecular weight of 200 and a specific activity of 1 mCi/mmole can be determined without too much difficulty. A comparison might be helpful again: The smallest speck of dust one can see with the unaided eye weighs approximately $1 \mu g$, that is 1000 ng.

A further increase in sensitivity could be achieved in two ways. One would be to increase the specific radioactivity of the herbicide under investigation. For all practical purposes, one cannot achieve more than about a tenfold gain in sensitivity by this approach. It might be mentioned here that radioactive compounds, especially those with high specific activities, are not as stable as their "cold," i.e., unlabeled counterparts. The energy liberated during the disintegration of the radioactive atoms can result in the rupture of covalent bonds with the result that transformation products are produced right in the bottle that contains the "pure" compound. It is therefore imperative that every radioactive compound be checked for radiochemical purity every time it is used for experiments in transformation studies.

Another way to increase sensitivity and to find still smaller amounts of transformation products would be to increase the amount of plant material to be treated with the herbicide. Although there is no principal limit to this approach, the difficulties of extracting and fractionating kilogram amounts of plant tissue in the search for compounds with unknown characteristics are considerable if quantitative data are desired. Most investigators work, therefore, with relatively small samples, from a few grams to a few hundred grams fresh weight, for general metabolism studies.

B. Application of Labeled Herbicides to Plants

The form of application of a compound to a plant is an important factor in transformation studies. The pattern of transformation products obtained may depend on the way the compound has been applied. There are many possibilities: injection into the stem, uptake into an excised leaf via the cut end of the petiole, foliar application on the intact plant by putting droplets of a herbicide solution on the leaf blade or on the petiole (the solvent in which the herbicide is dissolved may influence the rate of penetration and the time the compound is in contact with the epidermis and its enzymes), uptake through the roots of plants grown in nutrient solution to which the herbicide is added at a given time, pretreatment of soil in which seeds are planted and allowed to grow, etc. To this list can be added all the methods in which excised parts of the plants (root tips, hypocotyls, epicotyls, cotyledons, leaf punches, etc.) are floated on, immersed in, or are vacuum infiltrated with a herbicide solution. The choice of the method of application depends primarily on the type of information one wishes to obtain, but it is imperative to recognize the limitations and possible pitfalls of each method. One of the easiest ways to get information rapidly on the type and number of transformation products expected in a particular plant is to allow the herbicide to be taken up into a leaf through the cut end of the petiole. Large numbers of leaves can be treated in this way, and the total amount of radioactive material absorbed can be determined accurately. This method is especially advantageous in ascertaining how and to what extent different plant species transform the same herbicide. Results obtained from this type of application may be different, however, from those obtained, for example, after surface application. To obtain quantitative information on the number and the amount of the various transformation products that are present in the crop at harvest or when it reaches the consumer, the method of treatment of the plant should parallel as closely as possible the actual field application. Legal restrictions necessary to ensure the safe use of radioactive isotopes may necessitate some small changes. Plants will have to be grown in containers, for example, in the case of soil treatment with pre-emergent herbicides labeled with carbon-14. This means that care must attend the watering regime. Surface watering or watering from below alters the concentration of the compound in the soil profile and, therefore, may change also, the amount taken up by shallow-rooted or deep-rooted plants.

C. Isolation of Transformation Products

In order to obtain a correct picture of the fate of a herbicide after it has been applied to the plant, it is necessary to avoid possible changes of the compound itself and of its transformation products during extraction and fractionation(7). The safest procedure for minimizing the formation of artifacts is to work in the cold, at temperatures where enzymatic and chemical reactions proceed only very slowly, and also in the dark, to prevent photochemical reactions that may proceed rapidly even at very low temperatures. This is quite impractical and it is necessary, therefore, to compromise. Enzymatic reactions can be kept to a minimum by rapidly inactivating the enzymes. Since all enzymes are sensitive to heat, although in various degrees, application of hot solvents is the most universal way to achieve denaturation of all enzymes. There is, however, one problem involved. The hot solvent does not penetrate instantaneously into the interior of the tissue. A temperature gradient will establish itself initially and the overall effect will be that the enzymes, most of which increase their activity at slightly elevated temperatures, will first be stimulated to maximum activity before they become inactivated at still higher temperatures. This can be overcome if the tissue is finely divided before denaturation by hot solvents is attempted, e.g., by homogenizing the treated plant material immediately after harvest in liquid nitrogen with a suitable high-speed homogenizer. The resulting fine powder can then be treated with a solvent heated to at least 80°C using, for example, boiling isopropanol to inactivate the enzymes. The hot solvent may, however, increase the rate of nonenzymatic reactions. Only when it has been established that different methods of extraction produce identical results and extreme care has been taken to prevent the formation of artifacts can such results be considered reliable and to give a true picture of the fate of a herbicide in a plant. Unfortunately, such precautions are not always applied, which often makes it difficult to judge the reliability of published data and is probably the major cause of controversy among authors. The extraction procedure and the solvents used for extraction depend, to some extent, upon the type of compound originally applied, as well as upon the preference of the individual investigator. Since it cannot be generally predicted what transformation products will be found, however, the use of a universal extraction scheme using a series of solvents which extract both lipophilic and hydrophilic compounds is recommended. If alcohols are used and if there is a possibility of esterification or transesterification, then a parallel sample should be extracted with solvents not containing such esterifiable groups in order to establish whether extraction with alcohols is permissible. Acidic and basic conditions must be avoided until sufficient information is obtained to ensure that their use does not produce artifacts. The use of ion exchange resins also falls in this category. Conjugates of herbicides or their hydroxylation products with glucose are sometimes formed in plants. They can, therefore, be wholly or partially split by extremes in pH and thus escape detection. Subsequent separation procedures will be determined mainly by individual preference and previous experience. Separations can be achieved by taking advantage of differences in solubility, size, ionic charges, adsorption, and volatility. Solvent partition, direct or reversed phase liquid–liquid chromatography, adsorption chromatography, electrophoresis, ion exchange procedures, paper, thin-layer and gas–liquid chromatography are some of the techniques that can be used to fractionate the extract(8) enough so that only one radioactive transformation product is present in any one fraction. This does not mean that this particular compound is present in pure form; it just means that it is the only *radioactive* compound present.

There still may remain a very large amount of plant extractives in such a fraction. The presence of these other compounds prevents the use of physical measurements (spectral absorption in the visible, ultraviolet, or infrared region of the spectrum, NMR, melting point, molecular weight determination, etc.). Identification is therefore somewhat difficult. The compound can be characterized by repeated chromatography with different solvent systems and on different support materials (paper, silica gel, polyamide powder, etc.). If one of a large number of suspected transformation products that have been synthesized in the laboratory shows identical behavior in all the different chromatographic systems used, then there is a high degree of probability that the two compounds are identical. This suspected compound can now be added to the radioactive unknown and the mixture recrystallized. If the specific activity of the mixture stays constant through several recrystallizations, identity of the two compounds can be assumed. This is the textbook procedure. The fact that only very few of all the reported metabolites have been identified and that the majority of them are still called Unknown X, Y, or Z, etc., suggests that a complication exists. This is probably attributable to the small amount of such transformation products (X, Y, or Z) present. A potent herbicide that is used in amounts of 1 to 2 lb/acre cannot be applied to a crop plant at levels that much exceed this value. On average, from about 10 to 100 μg of a specific herbicide per gram of leaf is the upper limit that can be tolerated by a plant; otherwise, toxic effects will appear, and the metabolism in a sick or dying plant is certainly not the same as in a healthy, growing one. Transformation studies performed with toxic levels of a herbicide are justified only if they are part of a concentration range study.

3-4   METABOLIC REACTIONS

Of the metabolic reactions that are known, or can be expected to participate in transforming herbicides in plants, the types of transformations encountered are really quite limited. Their combination, however, can make for a bewildering number of transformation products. The most important of these reactions are, in decreasing order: Oxidation including hydroxylation, synthesis of conjugates and other synthetic reactions, hydrolysis, and reduction. Most of these reactions are enzymatic ones, but nonenzymatic reactions occur too; for example, replacement of chlorine with a hydroxyl group is likely to proceed via primary reaction with a sulfhydryl group of a protein, or with such reactive plant constituents as the oxazinones.

Interest in the transformation by plants of externally applied chemical compounds, so-called foreign compounds, is relatively recent if compared to that in the detoxification mechanisms in animals. The latter have received powerful stimuli by the development of modern drugs. Since plants do not have a typical excretory system, there was admittedly not very much that could be done in this area before the availability of isotopes, except with inorganic materials; but even now, the gap between our knowledge of the pathways used in transformation of foreign materials by plants, and the information available on those occurring in animals, is still very large. For example, a monograph from the year 1959 on the metabolism and detoxication of drugs and other foreign compounds in animals(9) lists approximately 1100 publications that appeared after 1950, whereas a recent (1965) rather comprehensive publication on the metabolic fate of herbicides in plants(10) lists only 178 references published after 1950. Much of the discussion in the next pages is based therefore, not only upon studies with plants, but also draws heavily on information obtained from research on animals.

Many, if not most, biochemical reactions are qualitatively similar in most organisms. The transition from structural similarity and biochemical diversity, which was characteristic of primordial organisms, to structural diversity and biochemical similarity over the course of evolution has long been accomplished. One would, therefore, expect that a compound externally applied will be metabolized in an essentially similar way by most higher organisms, and this has been found, in general, to be true. Quantitative differences may, however, be quite pronounced. An example is the formation of a glucoside of amiben, which proceeds nearly quantitatively in soybeans, whereas in barley only a small percentage of amiben is transformed into this glucoside (11). Organisms with an effective excretory mechanism will, as a rule, not degrade a compound that does not fit into the normal sequence of metabolic events as completely as do those organisms in which this means of disposing of a compound is absent. A compound with sufficient water solubility to allow excretion with the body fluids requires less energy than do the large number of chemical reactions required to degrade a compound into such small molecular fragments as to be identical with the normal biochemical building blocks, or even to gaseous products. One of the more typical reactions found in mammals is the synthesis of glucuronides from compounds containing a suitable functional group (e.g., carboxyl or hydroxyl) in the molecule, either present originally or introduced metabolically. The glucoronide is frequently more water soluble than the original compound and is, therefore, excreted in the urine. Higher plants usually do not have such an efficient excretion apparatus. Although there may be some loss of material from both the leaves and the roots, this type of excretion is not particularly selective and does not appear to be quantitatively important. Defoliation is, of course, quite efficient but can be disregarded in this context since we are interested mainly in effects during one season. In spite of the absence of an excretory system, one of the more typical reactions in plants is the synthesis of glycosides from compounds containing a suitable functional group in the molecule, in a manner similar to the synthesis of glucuronides in mammals. Microsomal enzymes are in both cases responsible for the oxidation or hydroxylation reaction and probably also for the synthesis of the conjugate molecule. The question where these glycosides (the sugar involved is not necessarily always glucose) are stored, whether they are slowly hydrolyzed, and the aglycon then metabolized in some other way, does not seem to have been investigated. In general, one would be inclined to think that the question of whether a compound will participate at all in enzyme catalyzed reactions, whether then their metabolites will take part in biosynthetic pathways, whether they will be further degraded, or whether they are end-products, will depend primarily on their structure, on their ability to fit closely enough into an active center of an enzyme to mimic a natural substrate. It has become quite clear in recent years that biological pathways frequently consist of a series of reactions in which the starting material is handed on from one enzyme to another, modified at each step until finally the end-product is released from the last enzyme in the chain. This end-product can now regulate its own synthesis by a feedback mechanism: If the amount of the end-product produced exceeds a certain threshold, some of the excess end-product can combine reversible with the first enzyme in the biosynthetic chain in such a way as to make this enzyme incapable of accepting new starting material as a substrate, thereby preventing the further production of end-product in this particular assembly line. Such feedback inhibition mechanisms have been demonstrated for quite a number of biosynthetic pathways, mainly in microorganisms(12). But they do exist also in plants, for example, in the biosynthesis of leucine where it has been demonstrated with maize embryos that leucine, the end-product of a biosynthetic chain, inhibits the first enzyme in the sequence, α-isopropylmalatesynthetase(13) and thereby regulates the rate of its own synthesis.

For many of the enzymes that metabolize herbicides and that seem to be largely identical with the group of enzymes that in mammalian systems have been found to act on drugs, the natural substrates are not yet known. It is thus not possible to say whether, in this group of enzymes, sequences of reactions will proceed in a manner similar to that found in known biosynthetic pathways and whether a similar type of regulation by feedback mechanisms is operative. If we assume this to be so, and if the compound acted upon by the first enzyme in such a chain is a foreign compound, then the probability that it too will be passed on and fit properly, like its natural counterpart, into each of the active centers of the following enzymes is rather slim. There will probably be a break somewhere in the sequence so that inhibition by feedback will not become operative, and a substantial amount of foreign compound may be metabolized by the first enzyme. The rate of such enzymatic reactions is governed by the classical kinetic theory, the frequency of collisions being proportional to the concentration of the reaction partners, the concentration of the substrate, and the concentration of the enzyme. Since a cell is not a homogeneous medium, but a rather complicated structure of compartments, it is evident that the localization of a transforming enzyme within the cell is of importance because the solubility characteristics of the compound, its partition coefficient, will determine its concentration in the environment of the enzyme.

Studies on the oxidative dealkylation of a large number of compounds by liver microsomes in vitro, which established a correlation between lipid solubility of the substrate and the rate of its dealkylation(*14–16*), have led to the suggestion that the enzymes involved were either surrounded by a lipid barrier that could be penetrated only by fat soluble compounds or, but less likely, that only nonpolar compounds could interact with the active sites. Some exceptions to this general concept have been reported(*17*). The compounds which fell out of line, however, were all derivatives of purines, pyrimidines, and closely related heterocyclic compounds. It is well known that many compounds in these groups exhibit rather unusual solubility properties. Whereas in the aliphatic and aromatic series, addition of an amino group increases, as a rule, the water solubility (methylamine is more water soluble than methane, aniline more so than benzene), the amino derivative of purine, adenine, is much less water soluble than purine itself. An analogous situation is found in the triazine-type herbicides where replacement of a hydrogen atom by a methyl group can increase the water solubility. The single attribute of more or less solubility in any one solvent may be expected to be insufficient to establish a correlation with the penetration of a compound through the hypothetical lipid – or lipoprotein – barrier. It is probable that no single characteristic of a given compound is sufficient to predict its behavior against isolated microsomal enzyme preparations, and even less probable to make predictions about the fate of the compound in a cell or tissue. In general, however, increased lipoid character favors oxidation by microsomal enzymes. Microsomal preparations are pieces of endoplasmic reticulum *membrane* with attached ribosomes. The membranes of organelles are 50% fat, 50% protein, so it is evident that partitioning is an important property of any compound that reacts with such a membrane system. Sometimes the process whereby a foreign compound is transformed by an organism is called detoxification. This may be a convenient term to use in the laboratory. It seems, however, hardly necessary to invoke such a teleological concept. The existing laws of physics and chemistry seem perfectly adequate to explain the observed transformations of foreign compounds by living organisms. That transformations that will take place cannot always be predicted is attributable to the limited information available about biological structures and pathways. Even a synthetic organic chemist cannot always predict the outcome of a specific reaction although he works with much simpler and better defined systems.

Before proceeding with a more detailed discussion of the ways in which herbicides are or can be metabolized, it is propitious to take a closer look at those cell structures where most of these activities take place(*18*).

The Main Site of Metabolic Reactions

Among the numerous subcellular particles, like nucleus, nucleolus, chloroplasts, starch grains, mitochondria, ribosomes, Golgi apparatus, endoplasmic reticulum, etc., the largest one is the nucleus, the place where the genetic information is stored in the form of deoxyribonucleic acid (DNA) molecules. Electron microscope pictures show that the nucleus is surrounded by a double membrane. It is interesting to note that the outer one of these two membranes (which is about 7 mμ thick) is continuous with membranes that make up the smallest known subcellular structures, the endoplasmic reticulum(19). This endoplasmic reticulum (ER) seems to represent the individuality of the cell, just as the nucleus represents the individuality of the species. It provides a complex, finely divided vacuolar system with many subdivisions, the most constant of which is the nuclear envelope. The form displayed by the rest of the component elements of the ER is highly variable. The endoplasmic reticulum of any particular cell type, however, has a characteristic structural pattern(20). This pattern is, in any one cell, a rather precise mark of differentiation. These patterns even seem to be repeated, at least to some degree, in cells from different animals species if the cells are of the same type, for example, the cells of the retina. There appears to be here, like in so many other instances, some relationship between biological structure and chemical function. The large surfaces of the endoplasmic reticulum can provide for an orderly spatial distribution of a package of sequential enzymes. This is similar in a way to the cristae, the tremendously invaginated inner membrane of the mitochondria where the enzyme packages needed for respiration and energy production are stacked, arranged like machines in an assembly line.

With the electron microscope one can easily see two major types of vesicles, one with a rough and one with a smooth surface. The rough form represents a type of the endoplasmic reticulum that is capable of attaching ribosomes to the vesicular membranes, which gives them the rough appearance. It is here that selected protein synthesis occurs as directed by nuclear DNA via messenger ribonucleic acid (m-RNA) and amino-acid-specific transfer ribonucleic acids. Not much is known as yet on the function of the smooth form of the ER. It seems that most of the oxygen and NADPH-dependent oxidation reactions are catalyzed by enzymes present in this fraction. The term microsomes has been given to these particles that could be obtained by centrifugation at greater than $100,000 \times g$ of the supernatant after all the heavier particles such as nuclei and mitochondria had been removed by centrifugation at lower speeds. It should be remembered that the microsomal fraction obtained this way is a rather poorly defined mixture, the composition of which depends not only on the cell species and cell type but also on the way the cells are broken up and in what medium the separation is carried out (21).

This microsomal fraction is actually nothing but the preparative result of attempts to isolate the endoplasmic reticulum and is made up of its heavier particles. To confuse the situation, the term microsomal fraction is also sometimes given to the whole supernatant liquid obtained after centrifugation of cell homogenates at 40,000 and sometimes even at 10,000 × g. A fraction obtained this way is obviously a very heterogeneous mixture that contains mitochondria and soluble cytoplasmic enzymes besides the membranes of the ER. By careful and repeated centrifugation in density gradients, one can, however, separate not only the rough surfaced vesicles from the smooth ones, but also show that the smooth membrane particles themselves can be separated again into at least six distinct fractions (22). This supports the hypothesis of different functional activities in different parts of the endoplasmic reticulum.

Looking now at the enzymatic capabilities of the endoplasmic reticulum reveals an impressive variety. The rough classification possible at the present time is rather unsatisfactory. There is one rather heterogeneous group of enzymes, the natural substrates for which are known or at least strongly suspected. There is, however, another group, the natural substrates for which are not as yet known. Since pharmacologists, interested in the problem of increasing the time during which a drug is active, have found that the main obstacle to their efforts is the degradation of drugs by enzymes of the endoplasmic reticulum, of the "microsomal fraction", this variety of enzymatic activities has been assembled together under such names as microsomal enzymes, drug metabolizing enzymes, drug enzymes, or detoxifying enzymes. In this group there occurs the enzymatic apparatus for aromatic hydroxylation, side chain oxidation, dealkylation of N-alkyl and O-alkyl groups, cleavage of ether linkages, oxidation of thioethers, reduction of aromatic nitro groups, reductive splitting of azo linkages, oxidative deamination, and glucuronide synthesis. Since it has been observed that dephosphorylation of uridinediphosphoglucose (UDPG) takes place in the ER too, it is quite likely that this is also the site for the synthesis of glycosides. This concept receives some support from the observation that phosphatides, via their cytidinediphospho-(CDP-) and uridinediphospho-(UDP)- intermediates are synthesized there too. Most of the work on these "drug metabolizing enzymes" has been done with preparations from mammalian tissue, particularly liver; but, as is evident from the list of enzymatic activities, nearly all those reactions, which are of major importance in the metabolism of herbicides by plants, can, and probably do, occur within this complex arrangement, the endoplasmic reticulum.

Among the enzymatic activities of the endoplasmic reticulum for which a natural substrate is known or at least strongly suspected are the following: hydroxylation of proline(23), iodination of amino acids or proteins (24), synthesis of ascorbic acid(25), steroid synthesis(26), steroid transformations such as reduction of double bonds and of carbonyl groups(27, 28), aromatization and hydroxylation(29, 30), synthesis of triglycerides (31), and of phosphatides(32) via cytidinediphosphate- and uridinediphosphate intermediates, incorporation of glycerophosphate into phosphatides(33), reactions involving uridinediphosphoglucuronic acid and uridinediphosphoglucose(34), transmethylation of phosphatides(35), metabolism of plasmalogens(36), and of glycolipids(36). Polysaccharide synthesis(37) has been linked to the Golgi apparatus which is considered by some authors to be a part of or derived from the ER(38).

As has been mentioned before, different types of cells may have their own characteristic type of endoplasmic reticulum and consequently their own particular pattern of enzymatic activity. Aquatic animals (fish, turtles, and amphibia), seem to lack most of the so-called drug-metabolizing enzymes(39) (it has been suggested that they can rid themselves of foreign lipid-soluble compounds by diffusion through the gills); whereas they are present in toads and crustaceae. Less drastic but nonetheless important differences may be found in the enzymatic makeup of the ER of different plant families, species, and perhaps varieties, which would help to explain the varying patterns of transformation products frequently produced by plants of different types.

3-5 METABOLISM AND TRANSFORMATION OF HERBICIDES

During a recent symposium on Metabolism of Herbicides, the statement was made that "... despite the excellent research that has been done, in no case is the complete metabolic picture worked out for even one chemical"(5) (this was in 1964). The situation has not changed very much since. There is more information available now on the metabolism of some herbicides in mammals, e.g., a thorough study of the fate of chlorthiamide (2,6-dichlorothiobenzamide) and dichlobenil (2,6-dichloro-benzonitrile) in the rat and dog(40); this approaches a complete metabolic picture for these two related compounds. With regard to the metabolism of herbicides in plants, however, the word "unidentified" in connection with transformation products seems to be — for the reasons given above — still predominant.

It is, therefore, better not to try to present at this time a comprehensive list of the published results of transformation products that have been found in various tissues and species of plants. Instead, attention is focused on the types of reactions that have been, or can be expected to be, encountered. A few representative examples serve to illustrate reactions listed under headings such as photochemical transformations, non-enzymatic chemical reactions, oxidation, conjugation, and hydrolysis. A section on miscellaneous reactions includes some of those reactions that have not been studied intensively or have not been observed as yet, in plants.

A. Photochemical Transformation

Most of the herbicidal chemicals that can absorb sunlight in either the visible part of the spectrum (all colored compounds) or in the ultraviolet (which includes all aromatic compounds) undergo some transformation when exposed to light of such wavelength. Usually only the disappearance of the starting material has been measured, and, in a few cases, some of the transformation products have been identified. For example, illumination of the sodium salt of pentachlorophenol produces, among others, the following compounds (41, 42):

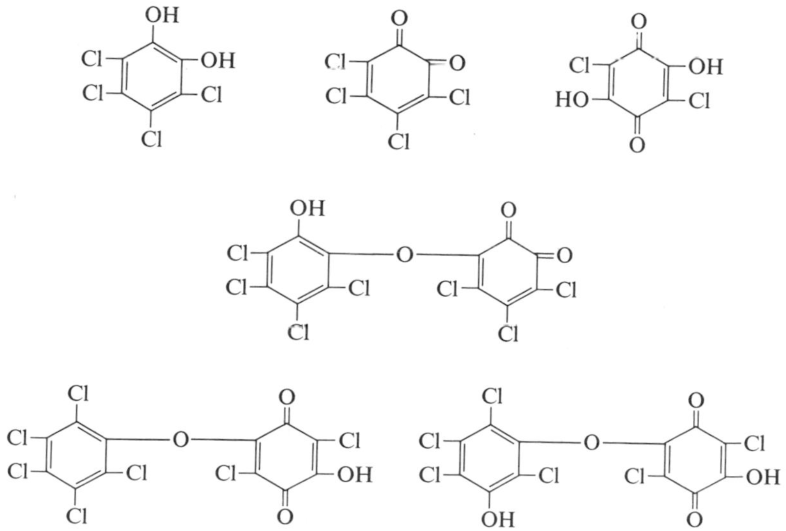

Paraquat, which apparently is not at all metabolized by plants, and which is also excreted unchanged in animals, is broken down in sunlight according to the following scheme (43):

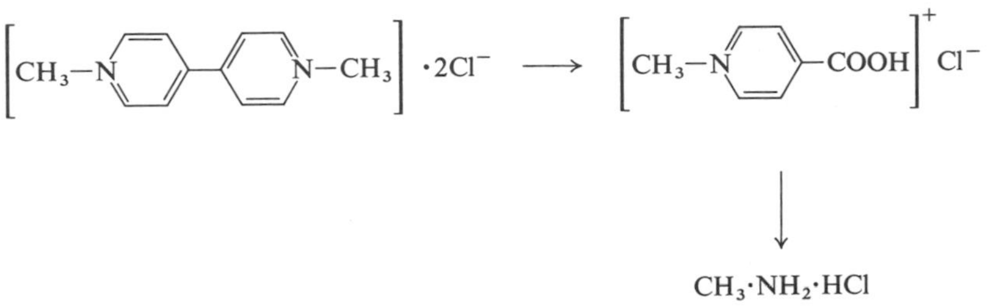

$$\left[CH_3-N \overset{+}{\diagdown}\diagup \diagdown \diagup N-CH_3 \right] \cdot 2Cl^- \longrightarrow \left[CH_3-N \diagdown\diagup-COOH \right]^+ Cl^-$$

$$\downarrow$$

$$CH_3 \cdot NH_2 \cdot HCl$$

The monomethyl analog of trifluralin undergoes an internal oxidation reduction during exposure to ultraviolet light in xylene solution(*44*),

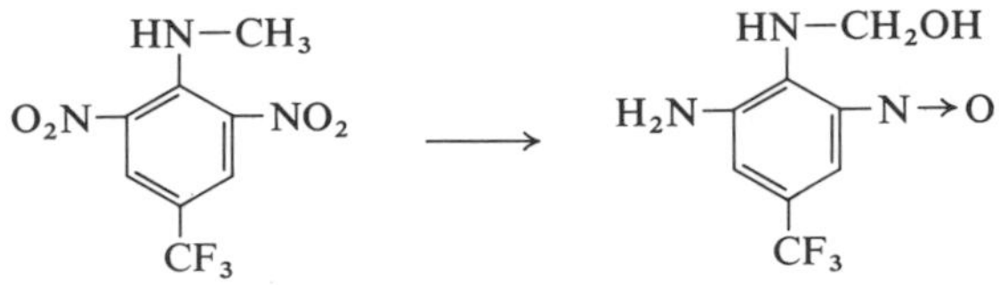

Nearly all herbicides have been reported to undergo inactivation in light in vitro. Among them are amiben(*45*), which produces two major unknown products, trifluralin(*46*), CDEC(*47*), 2,4-D(*48*), chloroxuron (*49*), bromoxynil and ioxynil(*50*), triazines(*51*) and PLANAVIN herbicide (*52*). Flavins (flavinmononucleotide) have been shown to cause photo-inactivation of phenylureas, such as monuron, in vitro(*53–55*). It does not appear, however, that such a process plays a significant role in the inactivation of these herbicides in vivo. Amitrole is also degraded by light in the presence of riboflavin(*56, 57*), and it has been claimed that the toxicity of amitrole to barley seedlings can be reduced by simultaneous application of riboflavin(*58*). This could, however, not be confirmed by another laboratory(*59*).

B. Nonenzymatic Chemical Reactions

The classical example of this group is the transformation of simazine and similar halogenated triazines into inactive hydroxy compounds by replacement of the halogen atom with a hydroxyl group(*3, 60*). Originally this reaction was thought to be an enzymatic one since it proceeded with fresh but not with boiled press juice of maize plants. It was shown later that a nonenzymatic reaction of simazine with a benzoxazinone(*61*) or its glucoside present in maize is responsible for the formation of hydroxysimazine. This benzoxazinone is not stable to heat

(*60*) and is transformed by boiling in water into a benzoxazalone, a stable compound

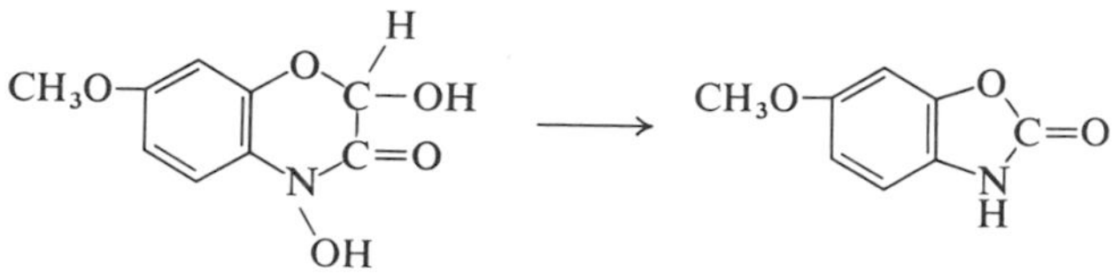

that does not react with simazine. Chloroalkylacetamides, such as CDAA (2-chloro-N,N-diallylacetamide), may be degraded in corn by a similar mechanism(*62*) with subsequent hydrolysis to glycolic acid and diallylamine(*63*).

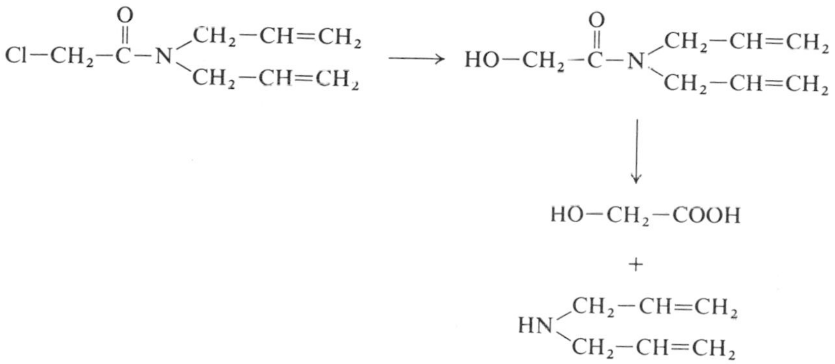

It might be suspected that in this case, chemical reaction with sulfhydryl groups of proteins or with that of reduced glutathione might take place too. This would generate the same final transformation products.

The formation of N-glucuronides (which, in contrast to O-glucuronides, are not hydrolyzed by β-glucuronidase) has been considered to result from a chemical rather than from an enzymatic reaction between amines and glucuronic acid(*64–66*). It proceeds rapidly in urine at a pH between 3 and 4, but is still measurable at neutral pH. The N-glucoside of amitrole is therefore probably formed in plants as well by a nonenzymatic pathway. This is strengthened by reports that glucose itself(*67*) and glucose-1-phosphate(*68*) can act as glucose donors. Enzymatic glucosidation reactions are generally considered to require high-energy glucose in the form of uridine diphosphoglucose(*69*) or adenine diphosphoglucose(*70*). It may, however, be possible that some N-glucosides can also be formed enzymatically. The nearly complete conversion of amiben into the N-glucoside(*11*) in soybean (a crop which is tolerant to amiben) compared to the modest amount formed in barley (which is susceptible) might be taken as an indication for this possibility.

C. Oxidation

Biochemical oxidation reactions of foreign compounds have been studied most frequently in animals and in the microsomal fraction obtained from liver. They have been investigated very little in detail, however, from the point of view of herbicide metabolism in plants.

Before discussing this area more fully, it might be helpful to present a short summary on oxidative enzymes especially with regard to their nomenclature, which is somewhat confusing. This, in turn, is attributable chiefly to the fact that detailed knowledge about some of the subgroups is still rather sparse(71–73).

The best known group of enzymes involved in biological oxidations are the dehydrogenases. They participate mainly in reactions involved in energy production. As their name implies, they remove hydrogen from an activated substrate. If the immediate acceptor for this hydrogen is molecular oxygen, then they are called oxidases. The overall reaction can be expressed by the following equations:

$$R \cdot H_2 + O_2 \longrightarrow R + H_2O_2$$
$$R \cdot H_2 + \tfrac{1}{2}O_2 \longrightarrow R + H_2O$$

In the other major group of oxidative enzymes, it is believed that it is not the organic molecule but molecular oxygen that is activated by the enzymes, perhaps with the aid of transition elements, e.g., via a perferryl ion-protein complex $Fe^{2+}O_2 \cdot$enzyme(71). These enzymes catalyze the incorporation of molecular oxygen into a substrate:

$$R + O_2 \longrightarrow RO_2$$

One of the most thoroughly investigated examples is the ring cleavage of orthodihydroxy compounds. Catechol can thus be split by pyrocatechase to cis, cis-muconic acid(74), or by metapyrocatechase to α-hydroxymuconic acid semialdehyde(71, 75).

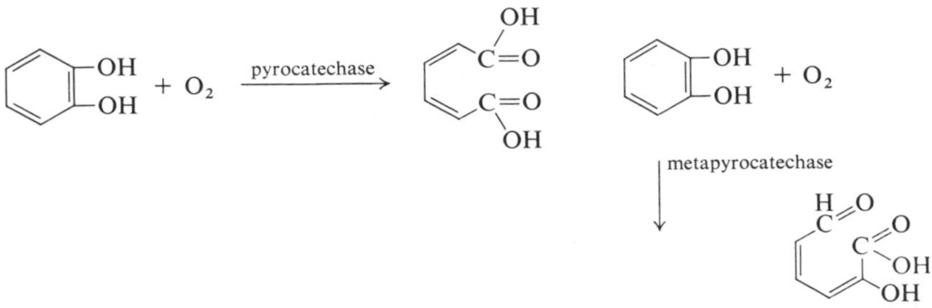

These enzymes have been named true oxygenases, or dioxygenases, to distinguish them from the mono-oxygenases, enzymes that carry out the following type of reaction:

$$R + O_2 + XH_2 \longrightarrow RO + H_2O + X$$

for example, aromatic hydroxylation:

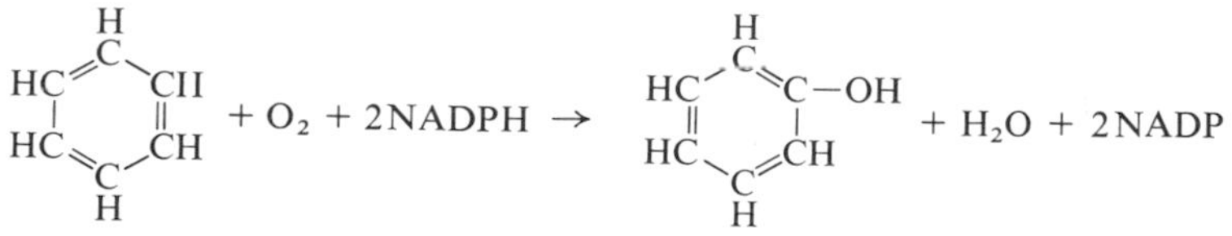

Since these enzymes catalyze not only the incorporation of oxygen into the substrate but also perform the reactions typical of oxidases, they are referred to as "mixed function oxidases." They are also known as hydroxylases, even in such cases where the introduction of a hydroxyl group is a transitory reaction as in the oxidative N- or O-demethylases. There the final product has lost the hydroxyl group.

$$R-\underset{H}{N}-CH_3 \longrightarrow \longrightarrow R-\underset{H}{N}-CH_2OH \longrightarrow \longrightarrow R-NH_2 + HCHO$$

Many of the hydroxylases are found in the endoplasmic reticulum and are therefore included in the term "microsomal enzymes." They are usually difficult to solubilize; consequently they are only poorly characterized. It is known that such mixed function oxidases participate in lipid and steroid metabolism, for example, the omega oxidation of alkyl groups(76, 77), the oxidative desaturation of fatty acids(78), and the hydroxylation of steroids(79). It has been shown recently that plant prolinehydroxylase uses molecular oxygen for the formation of the hydroxyl group of hydroxyproline(23), and therefore also belongs to this group. The list of those enzymes for which a natural substrate becomes known is constantly increasing, but for the present they are still classified into the category of drug-metabolizing enzymes or microsomal enzymes. This classification is being used even though most of them are probably present not so much in the microsomal or rough part of the endoplasmic reticulum, but in the smooth-appearing vesicles of this intricate structure(80, 81).

Oxidative reactions taking place with herbicides in plants are, for example, the demethylation reactions that have been reported for chloroxuron(49) and diphenamide(82). These demethylation reactions proceed probably via the corresponding methylol compounds which, if not stabilized for example by glucosidation, will decompose with formation of a des-methylamine and formaldehyde(82, 83). It might be mentioned here that in contrast to lipophilic N-methyl compounds, hydrophilic compounds containing N-methyl groups are demethylated not by microsomal but by mitochondrial enzymes for which the required cofactor is the reduced form of nicotinamide adenine dinucleotide (NADH) instead of nicotinamide adenine dinucleotide phosphate (NADPH). Ring hydroxylation of aromatic compounds has been investigated again primarily with animal tissues. Apparently it can proceed via two different mechanisms. This has been demonstrated with benzene which can be converted to phenol not only by the direct hydroxylase-mediated hydroxylation mentioned above, but also via a sequence where it is first hydroxymethylated to benzyl alcohol, the benzyl alcohol then undergoes ring hydroxylation, and finally the hydroxyphenyl alcohol is converted to phenol in an unknown sequence of reactions(84). Although this latter pathway seems to be of minor importance, it should be kept in mind that the possibility of finding a metabolite that has been formed by such an unusual reaction does exist. Ring hydroxylation of herbicides in mammals has been demonstrated, for example, with diphenamide(82), chlorthiamide(40), and dichlobenil(40). Both the 3-hydroxy- and 4-hydroxy derivatives of 2,6-dichlorobenzonitrile and the corresponding benzoic acid, as well as their conjugates (glucuronides, ester glucuronides, mercapturic acids) have been identified in the urine excreted by dogs and rats fed with chlorthiamide.

Ring hydroxylation of such aromatic acids as benzoic, cinnamic, or phenoxyacetic acids occurs readily in a variety of plant tissues(62, 85, 86). The herbicidal 2,3,6-trichlorobenzoic acid, however, seems to be quite stable and does not undergo hydroxylation. A very interesting case of ring hydroxylation has been reported to occur with *Aspergillus niger*, a fungus that can convert the herbicide 2,4-dichlorophenoxyacetic acid to 4-hydroxy-2,5-dichlorophenoxyacetic acid(87). Such hydroxylation of 2,4-D, accompanied by a shift of the chlorine atom from position 4 to either position 3 or position 5 of the aromatic ring, has been observed to occur also in the stems of beans but not of barley(88). The two products, 2,3-dichloro-4-hydroxy-phenoxyacetic acid and 2,5-dichloro-4-hydroxy-phenoxyacetic acid were found to be present in the stem tissue as glucosides. In the degradation of amiben, hydroxylation of the ring has been suggested to be the most likely primary reaction, which is then followed by oxidative cleavage(89). But since only benzene soluble metabolites have been checked (and found to be identical with neither 2-chloro-4-nitro aniline, 2,6-dichloro-*p*-phenylenediamine, or *p*-nitroaniline), evidence for this hypothesis is only indirect, namely the presence of a great number of labeled metabolites that are highly polar. Besides direct introduction of a hydroxyl group into an aromatic ring, N-hydroxylation of aromatic amines is known to occur frequently, at least in animals. The resulting substituted hydroxylamine can be either excreted in some conjugate form, or it can rearrange to an ortho-aminophenol, which then can combine with glucuronic acid, or form any of the other possible adducts.

Soil microorganisms can degrade most herbicides to small fragments which, in most cases, have not been identified. Oxidative reactions are certainly involved. For example, in a study(90) of the metabolism of DNOC (3,5-dinitro-o-cresol) in pure cultures of a Pseudomonas and an Arthrobacter, 2,3,5-trihydroxytoluene has been found as an intermediate metabolite in both cultures before ring cleavage occurred. The reaction sequences leading to the hydroxylated product were however different for the two organisms. In Arthrobacter, which lacks a nitroreductase, the following route is indicated:

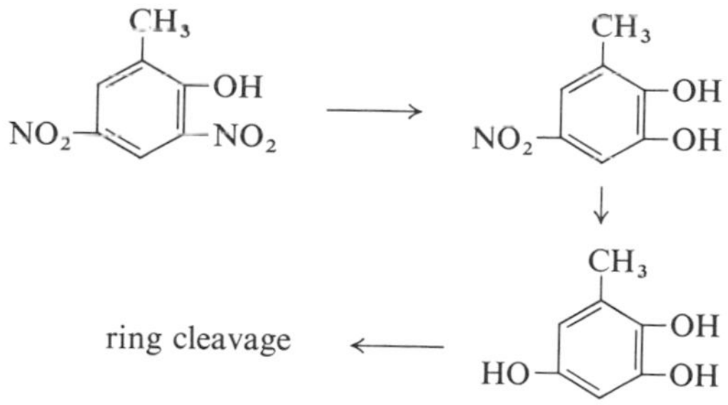

while in the pseudomonad an additional metabolite, 3-methyl-5-amino-catechol was formed by the reduction of the corresponding nitro compound 3-methyl-5-nitrocatechol. The removal of the nitro group is certainly an enzymatic reaction. A soluble enzyme, glutathiokinase, that catalyzes the displacement by reduced glutathione of aromatic halogen or nitro groups, has been isolated from rat liver(91).

β-Oxidation has been investigated rather extensively in plants, mainly in connection with the use of 2,4-dichlorophenoxybutyric acid and other homologs(92, 93) of 2,4-D as selective herbicides. The selectivity of these herbicides depends on the more or less pronounced ability of a plant species to degrade the nonherbicidal butyric acid derivative by β-oxidation to the herbicidal acetic acid derivative, 2,4-D. Similarly, the removal of the sidechain of MCPA (4-chloro-2-methylphenoxyacetic acid) by which this herbicide is rapidly inactivated by some weeds, can be prevented by introduction of an alkyl substituent in the 2-position of the acetic acid sidechain(94), as in CMPP [DL-α-(4-chloro-2-methylphenoxy) propionic acid].

D. Conjugation

The next step, which, at least to some extent, frequently occurs in both animals and plants after a hydroxyl group has been introduced into a foreign molecule, is the formation of conjugates. In plants(95), insects (96), and molluscs(97), the synthesis of glycosides is apparently the most important of these reactions and glucose seems to be the sugar most frequently involved. In mammals, birds, and some fish and amphibia, it is glucuronic acid and not glucose that is combined with hydroxyl- or carboxyl groups to form either a glycosidic or ester-type bond(98). The reaction takes place in the endoplasmic reticulum, where an enzyme, glucuronyltransferase, mediates the transfer of glucuronic acid from uridinediphosphoglucuronic acid to the functional group of the molecule to be conjugated. The glucosyl residue is transferred to a hydroxyl group, predominantly from uridinediphosphoglucose or from adeninedi-phosphoglucose. The other sugar-carrying bases, cytidinediphosphoglu-cose and guaninediphosphoglucose, have only weak activities(70). The pathways employing glucose or glucuronic acid seem to be mutually exclusive.

Herbicides for which the formation of O-glucosides has been proved include maleic hydrazide, which forms the glucoside of the enol form(99), and 2,4-D for which a glucose ester has been reported. Hydroxylated metabolites of 2,4-D, 4-hydroxy-2,3-dichlorophenoxyacetic acid, and 4-hydroxy-2,5-dichlorophenoxy-acetic acid have also been found as glucosides(88). Amiben(11, 100) and amitrole(101) form N-glucosides. The N-glucoside of amitrole has been shown to participate in other reactions. It can be phosphorylated by yeast hexokinase and the phosphorylated compound can be split by yeast aldolase to the glycerol derivative of amitrole(68). The formation of an S-glucoside has been reported in a study of the metabolism of the fungicide dimethyldithio-carbamate(102).

An N-glycoside in which the sugar is not glucose but ribose has been shown to be synthesized from N-6-benzyladenine, a synthetic cytokinin, in cocklebur leaves(103). The formation of glycosides of this cytokinin with sugars other than ribose occurs too, for example, in cruciferae(104). Strangely enough, no glycoside of 2,4-DB seems to have been reported. The reason for this may be that most of the interest in this compound was centered primarily on β-oxidation and not so much on the formation of highly polar metabolites. Neither has there been a report on sugar adducts of simazine and similar triazines nor of the hydroxy compounds that result from their reaction with oxazinones. However, an unidentified, chloroform-soluble compound has been reported as a metabolite of the triazine ipazine(105). This metabolite yielded hydroxy-ipazine after hydrolysis with alcoholic hydrochloric acid. Recently, lipid-like materials have also been found when corn plants were treated with 6-chloropicolinic acid(106), which itself is a metabolite of 2-chloro-6-(trichloromethyl)-pyridine. The picolinic acid could be recovered from these lipid-soluble metabolites after incubation with pancreatic lipase. It will be interesting to see whether the formation of lipid-soluble conjugates is a general reaction of plants that so far has been largely overlooked.

Conjugation with amino acids has been studied especially with amitrole (107), which reacts with serine to form aminotriazolylalanine(108). An aspartic acid derivative of 2,4-D has been identified. Ten other ninhydrin positive metabolites have been reported and said to be amino acid amides of unknown metabolites of 2,4-D(109). Amino acid conjugates of chlorinated herbicidal benzoic acids have not been observed, although benzoic acid itself is known to form benzoylasparagine(85).

Protein conjugates of herbicides in plants have not been characterized well enough to warrant a detailed discussion. A tannin complex with amitrole has also been reported(110). Other reactions leading to the formation of conjugates have been found so far only in animals. Some of these processes are confined largely to particular classes of animals. Glutamine conjugation seems to occur only in man and the chimpanzee, ornithine conjugation in birds and reptiles, and agmatine conjugation has been reported in scorpions(111). More generally, the conjugation of acids with glycine to form hippuric acid, the synthesis of mercapturic acids(112), and the conjugation of sulfate with phenols (for which the term "ethereal sulfate" is still sometimes used) are found. Phosphate conjugation of foreign compounds, however, is seldom encountered in spite of the large number of naturally occurring phosphate derivatives. One case reported is that of β-naphthylamine which is transformed into bis(2-amino-1-naphthyl) hydrogen phosphate and excreted in the urine. Its hydrolysis in the body to 2-amino-1-naphthol is thought to be responsible for the carcinogenic activity of β-naphthylamine(113, 114).

E. Hydrolysis

Acid or base-catalyzed hydrolysis of esters proceeds usually at too slow a rate at physiological pH values to be of significance, compared to enzyme-catalyzed reactions. The esters of 2,4-D, for example, are readily hydrolyzed by esterases located primarily in the epidermis, the cambium, and the phloem(115, 116). Nitriles like the 2,4-dichlorophen-oxyacetonitrile(93, 117) can be hydrolyzed by plants to the corresponding amides and/or acids. Chlorthiamide and dichlobenil are hydrolyzed to some extent to the amide and the acid in mammals. Formation of hydrocyanic acid from these herbicides was not observed(40). N-Substituted alkyl amides of 2,4-D and 2,4,5-T are degraded by plants to the parent acids(118). Ioxynil is hydrolyzed in the soil to the carboxylic acid via the amide. The ability of soil microorganisms to hydrolyze sulfate esters has been used in the design of the nonherbicidal 2,4-dichloro-phenoxyethylsulfate, which, in the soil, is hydrolyzed to the herbicide 2,4-D(119, 120). Cyanamide is probably hydrolyzed to urea. Maleic hydrazide, however, is not hydrolyzed to either maleic or fumaric acid diamide(121). The enzymes involved in the hydrolysis of herbicides in plants have not been studied to any extent. A microsomal esterase has been isolated from rat liver. This enzyme has been shown to be identical with microsomal amidase, amino acid esterase, aliesterase, and procain esterase(122). The low degree of specificity of this esterase reminds one of another group of microsomal enzymes, the oxygen and NADPH-dependent oxidases, some of which exhibit also a wide spectrum of activity.

F. Miscellaneous Reactions

The evolution of carbon dioxide has been studied extensively with herbicides of the phenoxyacetic-type herbicides. Since most of the work involved compounds labeled with carbon-14 in the carboxy group of the acyl side chain, production of radioactive carbon dioxide has sometimes been called decarboxylation. Most authors agree, however, that a true decarboxylation is probably not involved, but that the acetyl group is first removed from the parent molecule(94, 123, 124). The liberated acetic acid can then be oxidized to carbon dioxide via the citric acid cycle. A more detailed study of the mechanism(s) involved in the removal of the side chain would probably be quite rewarding. It might lead to an explanation of apparently contradictory reports on the metabolism of 2,4-D. In some plants evolution of carbon dioxide proceeds twice as fast from carboxyl labeled 2,4-D than from 2,4-D that carries the label in the methylene group(125). [It is known that α-oxidation does occur in leaves(126).] In other plants there is either no difference with regard to the position of the label(124), or there is even a reversal so that the methylene carbon is more efficient in releasing carbon dioxide than the carboxyl group(86).

In general, measurement of evolution of carbon dioxide is used only as an indication of a more or less complete degradation of that part of the molecule that carries the label. This certainly holds true for ring labeled compounds, e.g., simazine(*127*), for which fission of the ring is a prerequisite for carbon dioxide production. Such an extensive degradation is always accompanied by the incorporation of small amounts of radioactivity into a large number of compounds including proteins, polysaccharides, and nucleic acids, an indication that degradation proceeded far enough to produce molecules identical to those used in the synthetic metabolism of the cell. Herbicides that are rapidly degraded, at least in resistant plants, include carbamates, chloroalkylacetamides, and cyanamide.

Deamination has not been reported for herbicides in plants. The presence of ammonia lyases in plants is, however, established(*128*). Tyrase, for example, converts L-tyrosine to *trans-p*-coumaric acid(*129*). Deamination of aromatic amines probably does not play a very important role in the metabolism of foreign compounds in mammals either. There, amines are either transformed to hydroxylated compounds(*130, 131*) that can form conjugates, or they are transformed into acetyl derivatives. Acetylation competes with deacetylation. The amount and/or activity of the enzymes involved varies with the species. This competition determines the ratio of free and acetylated amines in the urine(*132*). In soil, oxidative deamination by microorganisms proceeds readily, for example, in the deamination of 3-methyl-5-amino-catechol, itself a metabolite of DNOC, to trihydroxytoluene(*90*).

Reduction, although so important in the mode of action of the bipyridylium-type herbicides, such as diquat and paraquat, seems not to be of major importance in the breakdown of herbicides. Since it has been shown that isolated, illuminated chloroplasts are capable of reducing 2,4-dinitrophenol to 2-amino-4-nitrophenol(*133*), one would expect that herbicides carrying nitro groups would similarly be reduced. Such reactions do not seem to have been reported. Reduction does play a role in the soil where bacteria possessing the enzyme nitroreductase can reduce aromatic nitro groups(*90*). In contrast to bacterial nitro reductase, which is not inhibited by oxygen, the microsomal nitro reductase of mammals can reduce nitro groups only under anaerobic conditions. The reduction proper is actually nonenzymatic. It is performed by flavin adenine dinucleotide (FAD) after it had been reduced by the reduced form of nicotinamide adenine dinucleotide phosphate (NADPH), a reaction that is catalyzed by the nitro reductase(*134*). Reduction of a disulfide, tetramethylthiuramdisulfide, has been shown to occur in plant tissue(*102*) by reaction with reduced glutathione(*135*).

G. No Transformation

The halogenated aliphatic acids that are used as herbicides, trichloroacetic acid(*136*), 2,2-dichloropropionic acid (dalapon)(*137*), 2,2,3-trichloropropionic acid, 2,2-dichlorobutyric acid, and 2,3-dichloroisobutyric acid are apparently quite stable in plants. They circulate freely within the plant and are to some extent even excreted — or, rather, exuded — by the roots(*138*). In soil, these acids can be degraded by microorganisms (*139, 140*) to inorganic chloride and carbon dioxide. A slower, non-microbial degradation of dalapon in soil to chloride ion and pyruvate has been reported(*141*). The aromatic herbicidal acids 2,3,6-trichlorobenzoic acid and 2,3,5,6-tetrachlorobenzoic acid are also resistant to a large extent to modification by plants(*142*). They too circulate freely in the plant and are exuded by the roots(*143*). It is interesting to compare these results with those obtained with unsubstituted benzoic acid, which is rapidly metabolized to benzoylaspartic acid and benzoylglucoside, to salicylic acid and its glucoside, and to at least six other unidentified compounds(*85*). Although one might expect that steric effects would interfere with hydroxylation of the chlorinated aromatic ring, it is less easily understood why glucosidation or other forms of conjugation should not occur at all with these chlorinated acids. Another important herbicide, paraquat, is metabolized by neither plants nor mammals, which excrete it unchanged. The compound breaks down in sunlight. One reaction that does not seem to have been demonstrated is the methylation of herbicidal phenols and amines by plants. In animals, this is a rather common reaction. The methylation of nor-nicotine to nicotine, or of codeine to morphine is brought about by a nonspecific methyltransferase. It seems strange that in plants that so copiously produce methylated products (lignin, anthocyanins, alkaloids, etc.), methylation apparently cannot compete with any of the other processes for the metabolism of foreign compounds that have been discussed.

3-6   CONCLUSIONS

Herbicides change the metabolism of plants. Plants change the structure of herbicides. Much progress has been made in understanding both these aspects of the intricate interaction between plants and herbicides. The mode of action of many herbicides can be described in physiological terms — they inhibit photosynthesis, mimic hormones, or interfere with the formation of subcellular structures — but complete understanding of these processes on a molecular level has not yet been achieved. Transformation of herbicides was first studied by measuring rates of disappearance of a compound after application to plants or soil. In recent years the general information obtained this way has been supplemented by identification and analysis of the reaction products into which pesticides are transformed. Most of the general reactions involved in the synthesis of these transformations products are already known. The mechanisms involved in these reactions have, however, not yet been studied in great detail. To cope successfully with the many facets of these problems, the analytical chemist must cooperate with the enzymologist, the electron microscopist, the taxonomist, to mention only a few of the scientists involved in a comprehensive analysis of metabolism of pesticides. The field is wide open and the rewards for studying it intensively may be greater than expected.

Note added in proof: Since this chapter was written, more recent work on transformation and metabolism of specific herbicides has been included in reviews by Kearny and Kaufman(*144*), and by Casida and Lykken(*145*). The general topic of enzymic hydroxylation has been reviewed by Hayaishi(*146*).

REFERENCES

1. G. C. Klingman, *Weed Control as a Science*, Wiley, New York, 1961.
2. J. Rosin and M. Eastman, *The Road to Abundance*, McGraw-Hill, New York, 1953.
3. H. Gysin and E. Knuesli, *Advan. Pest Control Res.*, **3**, 289 (1960).
4. R. L. Wain, *The Physiology and Biochemistry of Herbicides* (L. J. Audus, ed.), Academic, New York, 1964.
5. V. H. Freed, *J. Agr. Food Chem.*, **12**, 2 (1964).
6. H. Hurtig, *Intern. Pesticide Congr.*, V, Pesticides Abstracts and News Summary, Feb. 1964, p. 63.
7. Th. Buecher, K. Krejci, W. Ruessmann, H. Schnitger, and W. Wesemann, *Rapid Mixing and Sampling Techniques in Biochemistry* (B. Chance, R. H. Eisenhardt, Q. H. Gibson, and K. K. Lonberg-Holm, eds.), Academic, New York, 1964.
8. C. J. O. R. Morris and P. Morris, *Separation Methods in Biochemistry*, Wiley (Interscience), New York, 1963.
9. R. T. Williams, *Detoxication Mechanisms*, Wiley, New York, 1959.
10. C. R. Swanson, Crops Research ARS 34-66, Agr. Res. Service, USDA, 1965.
11. S. R. Colby, *Science*, **150**, 619 (1965).
12. E. H. Umbarger, *Ann. Rev. Plant Physiol.*, **14**, 19 (1963).
13. M. Strassman and L. N. Ceci, *J. Biol. Chem..*, **238**, 2445 (1963).
14. L. E. Gaudette and B. B. Brodie, *Biochem. Pharmacol.*, **2**, 89 (1959).
15. R. E. McMahon, *J. Med. Pharm. Chem.*, **4**, 67 (1961).
16. R. E. McMahon and N. R. Easton, *J. Med. Pharm. Chem.*, **4**, 437 (1961).
17. P. Mazel and J. F. Henderson, *Biochem. Pharmacol.*, **14**, 92 (1965).
18. K. R. Porter, *The Cell*, Vol. II (J. Bracket and A. E. Mirsky, eds.), Academic, New York, 1961.
19. M. L. Watson, *J. Biophys. Biochem. Cytol..*, **1**, 257 (1955).

20. L. D. Peachey, *Federation Proc.*, **19**, 257 (1960).
21. G. E. Palade, *Microsomal Particles and Protein Synthesis* (R. B. Roberts, ed.), Pergamon, London, 1958.
22. J. Rothschild, *The Structure and Function of the Membranes and Surfaces of Cells*, Biochem. Soc. Symposium No. 22, (D. J. Bell, and J. K. Grant, eds.), Cambridge University Press, London, 1963.
23. D. T. A. Lamport, *J. Biol. Chem.*, **238**, 1438 (1963).
24. R. Ekholm, *J. Ultrastructure Res.*, **5**, 575 (1961).
25. I. B. Chatterjee, G. C. Chatterjee, N. C. Ghosh, J. J. Ghosh, and B. C. Guha, *Biochem. J.*, **76**, 279 (1960).
26. N. L. R. Bucher and K. McGarrahan, *J. Biol. Chem.*, **222**, 1 (1956).
27. E. Forchielli and R. I. Dorfman, *J. Biol. Chem.*, **223**, 443 (1956).
28. H. J. Hubener, D. K. Fukushima and T. F. Gallagher, *J. Biol. Chem.*, **220**, 499 (1956).
29. K. J. Ryan, *J. Biol. Chem.*, **234**, 268 (1959).
30. K. J. Ryan and L. L. Engel, *J. Am. Chem. Soc.*, **78**, 2654 (1956).
31. S. B. Weiss and E. P. Kennedy, *J. Am. Chem. Soc.*, **78**, 3550 (1956).
32. E. P. Kennedy and S. B. Weiss, *J. Biol. Chem.*, **222**, 193 (1956).
33. P. S. Sastry and M. Kates, *Can. J. Biochem.*, **44**, 459 (1966).
34. B. M. Pogell and L. F. Leloir, *J. Biol. Chem.*, **236**, 293 (1961).
35. K. B. Gibson, J. D. Wilson, and S. Udenfriend, *J. Biol. Chem.*, **236**, 673 (1961).
36. J. Y. Kiyasu and E. P. Kennedy, *J. Biol. Chem.*, **235**, 2590 (1960).
37. D. H. Northcote and J. D. Pickett-Heaps, *Biochem. J.*, **98**, 159 (1965).
38. H. H. Mollenhauer and D. J. Morre, *Ann. Rev. Plant Physiol.*, **17**, 27 (1966).
39. R. H. Adamson, R. L. Dixon, F. L. Francis and D. P. Rall, *Proc. Natl. Acad. Sci. U.S.*, **54**, 1386 (1965).
40. M. H. Griffiths, J. A. Moss, J. A. Rose and D. E. Hathway, *Biochem. J.*, **98**, 770 (1966).
41. M. Kuwahara, N. Kato and K. Munakata, *Agr. Biol. Chem.*, **30**, 232 (1966).
42. M. Kuwahara, N. Kato and K. Munakata, *Agr. Biol. Chem.*, **30**, 239 (1966).
43. P. Slade, *Nature*, **207**, 515 (1965).
44. R. E. McMahon, *Tetrahedron Letters*, **21**, 2307 (1966).
45. T. J. Sheets, *Weeds*, **9**, 1 (1961).
46. W. L. Wright and G. F. Warren, *Weeds*, **13**, 329 (1965).
47. R. B. Taylorson, *Weeds*, **14**, 155 (1966).
48. G. R. Bell, *Botany Gaz.*, **118**, 133 (1956).
49. H. Geissbühler, C. Haselbach, H. Aebi and L. Ebner, *Weed Res.*, **3**, 277 (1963).
50. K. Carpenter, H. J. Cottrell, W. H. DeSilva, B. J. Heywood, W. G. Leeds, K. F. Rivett and M. L. Soundy, *Weed Res.*, **4**, 175 (1964).
51. L. S. Jordan, B. E. Day and W. A. Clerx, *Weeds*, **12**, 5 (1964).
52. W. J. Hughes and R. H. Schieferstein, *Proc. South. Weed Control Conf.*, **19**, 170 (1966).
53. L. A. Birk, *Can. J. Agr. Sci.*, **35**, 377 (1955).
54. P. B. Sweetser, *Biochim. Biophys. Acta*, **66**, 78 (1963).
55. W. Weldon and F. L . Timmons, *Weeds*, **9**, 111 (1961).
56. P. Castelfranco, A. Oppenheim and S. Yamaguchi, *Weeds*, **11**, 111 (1963).
57. P. Castelfranco and M. S. Brown, *Weeds*, **11**, 116 (1963).
58. J. L. Hilton, *Plant Physiol.*, **37**, 238 (1962).
59. A. W. Naylor, *J. Agr. Food Chem.*, **12**, 21 (1964).
60. P. Castelfranco, C. L. Foy and D. B. Deutsch, *Weeds*, **9**, 580 (1961).

61. W. Roth and E. Knuesli, *Experientia*, **17**, 312 (1961).

62. L. H. Hannah, *Proc. Northeast. Weed Control Conf.*, **1955**, 15.

63. E. G. Jaworski, *J. Agr. Food Chem.*, **12**, 33 (1964).

64. J. W. Bridges and R. T. Williams, *Biochem. J.*, **83**, 27 P (1962).

65. E. Boyland, D. Manson and S. F. D. Orr, *Biochem. J.*, **65**, 417 (1957).

66. I. M. Arias, *Biochem. Biophys. Res. Commun.*, **6**, 81 (1961).

67. R. A. Herrett and W. P. Bagley, *J. Agr. Food Chem.*, **12**, 17 (1964).

68. J. F. Fredrick, *Physiol. Plantarum*, **15**, 186 (1962).

69. J. B. Pridham, *Phytochemistry*, **3**, 493 (1964).

70. J. C. Trivelloni, E. Recondo and C. E. Cardini, *Nature*, 195, 1202 (1962).

71. O. Hayaishi, *Congr. Biochem.*, Proceedings of the Plenary Sessions, International Union of Biology, 1964, Vol. 33, p. 31.

72. H. S. Mason, *Ann. Rev. Biochem.*, **34**, 595 (1965).

73. J. A. Donkersloot, *Chem. Weekblad*, **61**, 604 (1965).

74. O. Hayaishi and K. Hashimoto, *J. Biochem.*, **37**, 371 (1950).

75. S. Dagley and D. A. Stopher, *Biochem. J.*, **73**, 16P (1959).

76. R. K. Gholson, J. N. Baptist and M. J. Coon, *Biochemistry*, **2**, 1155 (1963).

77. M. Kusunose, E. Kusunose and M. J. Coon, *J. Biol. Chem.*, **239**, 1374 (1964).

78. W. Stoffel, *Biochem. Biophys. Res. Commun.*, **6**, 270 (1961).

79. J. Chiriboga, *Biochem. Biophys. Res. Commun.*, **11**, 277 (1963).

80. H. Remmer, *Arch. Exptl. Pathol. Pharmakol.*, **238**, 36 (1960).

81. J. R. Fouts, *Biochem. Biophys. Res. Commun.*, **6**, 373 (1961).

82. R. E. McMahon and H. R. Sullivan, *Biochem. Pharmacol.*, **14**, 1085 (1965).

83. S. Orrenius, *J. Cell. Biol.*, **26**, 713 (1965).

84. N. H. Sloane, *Biochim. Biophys. Acta*, **107**, 599 (1965).

85. H. D. Klämbt, *Nature*, **196**, 491 (1962).

86. P. W. Morgan and W. C. Hall, *Weeds*, **11**, 130 (1963).

87. J. K. Faulkner and D. Woodcock, *Nature*, **203**, 865 (1964).

88. E. W. Thomas, B. C. Loughman and R. G. Powell, *Nature*, **204**, 884 (1964).

89. A. J. Lemin, *J. Agr. Food Chem.*, **13**, 557 (1965).

90. M. S. Tewfink and W. C. Evans, *Biochem. J.*, **99**, 31 P (1966).

91. J. Booth, E. Boyland and P. Sims, *Biochem. J.*, **79**, 516 (1961).

92. D. L. Linscott, *J. Agr. Food Chem.*, **12**, 7 (1964).

93. C. H. Fawcett, R. M. Pascal, M. B. Pybus, H. F. Taylor, R. L. Wain and F. Wightman, *Proc. Roy. Soc. (London)*, **B150**, 95 (1959).

94. E. L. Leafe, *Nature*, **193**, 485 (1962).

95. J. B. Pridham, *Ann. Rev. Plant Physiol.*, **16**, 13 (1965).

96. I. Ishiguro and B. Linzen, *J. Insect Physiol.*, **12**, 267 (1966).

97. G. J. Dutton, *Biochem. J.*, **96**, 36P (1965).

98. K. J. Isselbacher, M. F. Chrabas and R. C. Quinn, *J. Biol. Chem.*, **237**, 3033 (1962).

99. G. H. N. Towers, A. Hutchinson and W. A. Andreae, *Nature*, **181**, 1535 (1958).

100. F. M. Ashton, *Weeds*, **14**, 55 (1966).

101. J. F. Fredrick and A. C. Gentile, *Phytochemistry*, **4**, 851 (1965).

102. J. Kaslander, A. Kaars Sijpesteijn and G. J. M. Van der Kerk, *Biochim. Biophys. Acta*, **52**, 396 (1961).

103. D. R. McCalla, D. J. Morre and D. J. Osborne, *Biochim. Biophys. Acta*, **55**, 522 (1962).

104. J. E. Loeffler, unpublished.

105. R. H. Hamilton and D. E. Moreland, *Weeds*, **11**, 213 (1963).

106. R. W. Meikle and C. T. Redemann, *J. Agr. Food Chem.*, **14**, 159 (1966).

107. P. Massini, *Biochim. Biophys. Acta*, **36**, 548 (1959).

108. M. C. Carter, *Physiol. Plantarum*, **18**, 1054 (1965).

109. M. K. Bach and J. Fellig, *Nature*, **189**, 763 (1961).

110. E. Kroller, *Residue Rev.*, **12**, 162 (1966).

111. M. Hitchcock and J. N. Smith, *Biochem. J.*, **98**, 736 (1966).

112. E. Boyland, *Metabolic Factors Controlling Duration of Drug Action* (B. B. Brodie, and E . G. Erdos, eds.), Proc. Intern. Pharmacol. Meeting, 1st, Vol. 6, Pergamon, London and Macmillan, New York, 1962, p. 65.

113. E. Boyland, C. H. Kinder and D. Manson, *Biochem. J.*, **78**, 175 (1961).

114. L. Shuster, *Ann. Rev. Biochem.*, **33**, 571 (1964).

115. A. S. Crafts, *Weeds*, **8**, 19 (1960).

116. J.v.d.W. Jooste and D. E. Moreland, *Phytochemistry*, **2**, 263 (1963).

117. K. V. Thimann and S. Mahadevan, *Arch. Biochem. Biophys.*, **105**, 133 (1964).

118. J. Sudi, G. Josepovits and G. Matolcsy, *J. Exptl. Botany*, **12**, 390 (1961).

119. A. J. Vlitos, *Contrib. Boyce Thompson Inst.*, **17**, 127 (1953).

120. L. J. Audus, *Nature*, **171**, 523 (1953).

121. E. V. Parups, I. Hoffman and H. V. Morley, *Can. J. Biochem. Physiol.*, **40**, 1159 (1962).

122. K. Krisch, *Biochem. Z.*, **337**, 546 (1963).

123. R. L. Weintraub, J. W. Brown, M. Fields and J. Rohan, *Am. J. Botany*, **37**, 682 (1950).

124. M. J. Canny and K. Markus, *Australian J. Biol. Sci.*, **13**, 486 (1960).

125. L. C. Luckwill and C. P. Lloyd-Jones, *Ann. Appl. Biol.*, **48**, 626 (1960).

126. C. Hitchcock and A. T. James, *Biochim. Biophys. Acta*, **116**, 413 (1966).

127. M. T. H. Ragab and J. P. McCollum, *Weeds*, **9**, 72 (1961).

128. N. J. Macleod and J. B. Pridham, *Biochem. J.*, **88**, 45P (1963).

129. A. C. Neish, *Phytochemistry*, **1**, 1 (1961).

130. E. Boyland, *The Biochemistry of Bladder Cancer*, Thomas, Springfield, Ill., 1963.

131. W. Troll and N. Nelson, *Federation Proc.*, **20**, 41 (1961).

132. H. A. Krebs, W. O. Sykes and W. C. Bartley, *Biochem. J.*, **41**, 622 (1947).

133. J. S. C. Wessels, *Biochim. Biophys. Acta*, **38**, 195 (1960).

134. J. J. Kamm and J. R. Gillette, *Life Sci.*, **1963**, 254.

135. P. E. B. Lindahl, *Lantbrukshogskolans Annaler*, **30**, 375 (1964).

136. F. A. Blanchard, *Weeds*, **3**, 274 (1954).

137. G. N. Smith and D. L. Dyer, *J. Agr. Food Chem.*, **9**, 155 (1961).

138. C. L. Foy, *Plant Physiol.*, **36**, 688 (1961).

139. H. L. Jensen, *Tidsskr. Planteavl.*, **63**, 470 (1959).

140. A. N. Macgregor, *J. Gen. Microbiol.*, **30**, 497 (1963).

141. A. H. Kutschinski, *Down Earth*, **10**, 14 (1954).

142. P. G. Balayannis, M. S. Smith and R. L. Wain, *Ann. Appl. Biol.*, **55**, 149 (1965).

143. P. J. Linder, J. W. Mitchell and G. D. Freeman, *J. Agr. Food Chem.*, **12**, 437 (1964).

144. P. C. Kearny and D. D. Kaufman (eds.), *Degradation of Herbicides*, Dekker, New York, 1969.

145. J. E. Casida and L. Lykken, *Ann. Rev. Plant Physiol.*, **20**, 607 (1969).

146. O. Hayaishi, *Ann. Rev. Biochem.*, **38**, 21 (1969).